D1085102

Fundamental
Statistical
Concepts

color.

Fundamental Statistical Concepts

Frederic E. Fischer

State University of New York
College at Oswego

Canfield Press   ⌀   San Francisco
A Department of Harper & Row, Publishers, Inc.
New York   •   Evanston   •   London

Fundamental Statistical Concepts

Copyright © 1973 by Frederic E. Fischer

International Standard Book Number: 0-06-382662-3

Library of Congress Catalogue Card Number: 72-3309

Preface

This text is designed to serve as a general introduction to modern statistics for students in all academic disciplines. Although the examples and exercises suggest applications in a wide variety of fields, this text attempts to present to the college student the basic concepts of statistics which are common to all areas of application, including engineering, education, psychology, business, biology, and so on.

At present the market is replete with a variety of *introductory* statistics texts designed for specific academic disciplines. For the following reasons I believe that this approach is both inefficient and academically unsound: (a) there is a basic body of knowledge which is common to all introductory statistics courses, regardless of academic discipline—easily a full semester's worth, and (b) invariably such texts, and the courses which are taught from them, are designed to cover too much material in one semester—the result being that the basic concepts are covered rather quickly and incompletely, and the rest of the material becomes a collection of cookbook methods and procedures for testing various types of statistical hypotheses deemed particularly important to that discipline. The concept of probability, easily the most important aspect of any serious study of modern statistics, often receives only token attention in such discipline-oriented texts. I contend that a student cannot effectively understand and apply statistical methods without first acquiring a thorough knowledge of basic concepts, which includes a strong emphasis on probability.

It is my opinion that an ideal situation at a college or university would be for the mathematics or statistics department to offer a one-semester course in introductory statistics designed to service students in all disciplines that require a basic foundation in statistics. This course should then be followed by a second course—offered by each of the appropriate major departments. In this second course the instructor could cover rather thoroughly a study of the various tests of significance and research designs used most widely in his own discipline.

I consider this text to be a mathematically sound treatment of the basic statistical concepts which should be covered in a first course of the type described above. No college mathematics prerequisite is assumed; a good background in high school algebra is sufficient. The text is divided into three logically distinct parts: 1. Descriptive Statistics, 2. Probability, and 3. Statistical Inference. The first part deals exclusively with the task of describing in various ways the characteristics of sample data. Part 2 is a thorough treatment of the basic concepts of probability and probability functions, both discrete and continuous. In Part 3, the concepts learned earlier are combined to form a basis for the theory of statistical inference. I consider the first ten chapters, and perhaps selected parts of Chapters 11 and 12, a necessary and sufficient body of material for an adequate semester course in introductory statistics.

The format in this text is one of concise language which hopefully realizes a satisfactory compromise between strict mathematical rigor and intuitive discussion. Definitions and theorems form the basic structure of the text. Some theorems are proved by simple, yet rigorous, mathematical deduction, and others are "proved" by intuitive justification. The reader is continually invited to test his understanding of theorems by intuitive reasoning. The understanding of concepts is rated higher in importance than the practice of collecting cookbook formulas. On the other hand, the numerous exercises at the end of each section (with answers in the back) afford the student ample opportunity to test his knowledge in applied contexts. The exercises should be considered an integral part of the book. Summaries at the end of each chapter and a bibliography at the end of the text are also included.

Frederic E. Fischer

Oswego, New York
September 1972

Contents

Part 3. Statistical Inference

Fundamental
Statistical
Concepts

1

Introduction

1.1 What Is (Are) Statistics?

The field of modern statistics is a rapidly growing branch of applied mathematics. Students in almost every academic discipline are discovering that a basic understanding of statistics is essential for successful advanced work. More and more colleges require or strongly recommend at least one undergraduate course in statistics for those who plan to do graduate study.

The term "statistics," as it is used above, is singular and refers to an area of science. On the other hand, the word "statistics" (plural) is commonly used to denote numerical data, such as batting averages, stock quotations, and IQ scores. The gathering of numerical data (statistics) is an important, but rather small, part of the field of statistics. Generally speaking, it can be thought of as a combination of two subfields called *descriptive statistics* and *statistical inference*.

1

1.2   Descriptive Statistics vs. Statistical Inference

The subfield of *descriptive statistics* deals with the collection, tabulation, and presentation of data and the calculation of measures which describe the data in various ways. For example, during and after a national election reams of data are compiled on the distribution of the vote among the candidates, examined with respect to many variables, such as geographical location, age, sex, religion, family income, and so forth. Averages, correlations, and other descriptive measures are calculated. In short, the entire history of the event is presented and described in a concise and informative manner. In a similar way the National Safety Council keeps the public informed by providing data on traffic accidents and results of tests on the latest accident-prevention techniques. As in the first example, these data serve to describe observed happenings or phenomena.

If the data are simply presented and described in various informative ways, then the subfield of descriptive statistics is being utilized. If, on the other hand, the data and the descriptions are used as a basis for making predictions or decisions relating to unobserved events (such as future events), then the subfield of *statistical inference* is involved. The major political parties use data on past voting patterns to help plan future campaign strategy. On election night the television networks feed voting results of a few preselected precincts into giant computers which compare these data with past voting patterns in those same precincts. The computers can often predict winners of various races with a high degree of success when only one or two percent of the total vote has been tabulated. Descriptive data on traffic fatalities are used to predict anticipated death tolls on holiday weekends. Insurance companies use similar statistics in the calculation of car insurance premiums; the premiums must be high enough to provide a fund to cover claims resulting from future highway accidents.

The distinction between descriptive statistics and statistical inference is perhaps better explained by a numerical example. Suppose that an educator wishes to compare two methods of teaching history. He takes two fifth grade classes from a large elementary school and for four weeks teaches the same material to both classes, using teaching Method A in one class and teaching Method B in the other. At the end of the four weeks both classes take the same history achievement test. By averaging the achievement scores for each of the two classes, the educator is able to report the following results:

Teaching Method	Average Score
A	73
B	76

This report simply describes the results of the experiment and thus it falls into the category of descriptive statistics. If, on the other hand, the educator wishes to make a statement based on these observations such as, "Method B is superior to Method A for the teaching of history," then he finds himself involved in the area of statistical inference.

Generally speaking, the area of descriptive statistics involves rather simple and straightforward methods, while statistical inference requires a more subtle and complex approach. It is a simple task to calculate two averages, 73 and 76, and report them. It is much more difficult to justify that Method B is superior to Method A. Is the observed difference between averages (three points) due to the fact that one class is more intelligent than the other? Is it due to the fact that one class knew more history to begin with than the other? If the two classes were virtually equal in ability and experience and the two teaching methods were equally effective, could chance factors (such as individuals feeling ill, misunderstanding directions, or guessing on multiple choice questions) cause a difference as large as the one observed? How large a difference between averages would be necessary before the educator could be reasonably confident that one teaching method is in fact superior to the other?

The statistician can report descriptive statistics with complete certainty (assuming no arithmetical errors), but a decision based on statistical inference can never be made with 100 percent confidence. There is always some uncertainty involved with a statistical inference. The statistician does not base a decision on a statistical inference unless he can reduce the degree of uncertainty to a tolerable minimum. For example, if it could be shown that it is highly likely that a difference between averages as large as three points would occur by chance, assuming that the two teaching methods are of equal effectiveness, then the educator would probably conclude that he has no real evidence that either method is superior. On the other hand, if he could show that a difference between averages as large as three points is very unlikely to occur by chance, assuming equally effective teaching methods, then his results (an actual three-point difference) would be considered good evidence for arguing the superiority of Method B. The educator's decision, therefore, depends upon the *likelihood* that a difference as large as the observed difference could occur, given that the two teaching methods are equally good. Problems involving "likelihood," "uncertainty," "chance," and other such factors require an application of the theory of *probability*, which is a branch of mathematics and easily the most important set of concepts used in the field of statistics. Without a thorough understanding of the basic notions of probability theory, the student of statistics cannot fully understand or appreciate the methods and procedures used in statistical inference. This text, therefore, is divided into three major areas: Part I—Descriptive Statistics, Part II—Probability, and Part III—Statistical Inference.

The remainder of this introductory chapter is devoted to notation, specifically (a) *set notation*, which is the basic language used in the development of probability theory, and (b) *summation notation*, a shortcut symbolism used throughout the text. Since many students are introduced to these notations in secondary school, the coverage here is limited – with a concentration on those concepts and definitions which are directly applicable to the study of statistics.

1.3   Set Notation

A *set* is a collection of objects; the objects are called *elements*. It is customary to use capital letters, such as A, B, C, etc., to denote sets and small letters, numbers, and other symbols to denote elements. The elements of a set are enclosed in braces and are separated by commas. For example, if A is the set of elements a, b, and c, we denote this fact by writing $A = \{a, b, c\}$. If B represents the set of even positive integers smaller than 12, then $B = \{2, 4, 6, 8, 10\}$. The order in which the elements are listed within the braces is irrelevant. For example, A could just as well have been written as $\{b, c, a\}$, or B as $\{4, 10, 2, 8, 6\}$. When elements are used which are thought of as having a commonly accepted order, such as letters of the alphabet or positive integers, it is habit or convenience rather than necessity which causes us to list them in their familiar order. For large sets a short cut notation is occasionally used to indicate the inclusion of elements which are not specifically listed. For example, the set M of all positive integers less than 100 could be expressed as $M = \{1, 2, 3, 4, \ldots, 99\}$, or the set T of all positive integers which are multiples of 3 can be represented by $T = \{3, 6, 9, 12, \ldots\}$. The use of three dots in the notation means essentially "and so forth." It is necessary to list enough elements (usually in a familiar sequence) so that it is clearly understood which elements of the set have not been listed. Some sets, such as A, B, and M above, have a limited number of elements. They are called *finite* sets. Some sets, such as T above, have an unlimited number of elements. T is an example of an *infinite* set.

The notation "$n(D)$" denotes "the number of elements in set D." For example, we can say in the preceding paragraph that $n(A) = 3$, $n(B) = 5$, and $n(M) = 99$. If a set has no elements, then we say that the set is the *null* or *empty set*. We denote the null set by the symbol ϕ. For example, if set E is the set of all female past presidents of the United States, then E is the null set, and we denote this fact by writing $E = \phi$, or equivalently, $n(E) = 0$.

In any specific discussion involving sets, it is customary to define a fixed set of elements to which we limit the discussion. This "largest" set is called the *universal* set, usually symbolized by U. The set of all integers might be considered the universal set in one discussion, while the set of all Harvard freshmen might be the universal set in another. Thus the membership of set U varies with the situation, and once the universal set U is defined for a particular discussion, all other sets mentioned in the same discussion must have their members selected from U.

For our purposes it is necessary for us to understand certain *relations* between sets and certain *operations* that can be performed on sets.

1.3.1 Relations Between Sets

Definition 1.1: Sets A and B are said to be *equal* if they contain exactly the same elements, and we write $A = B$. If the sets A and B do not contain exactly the same elements, then $A \neq B$.

Definition 1.2: Sets A and B are said to be *mutually exclusive* (or *disjoint*) if they contain no common elements.

Definition 1.3: Set A is said to be a *subset* of set B if every element in A is also in B; and we write $A \subseteq B$.*

For example, if $H = \{2, 4, 5\}$ and $J = \{2, 4\}$, then $J \subseteq H$ (and also $J \neq H$). According to Definition 1.2 the following sets are all subsets of H: $\{2\}$, $\{4\}$, $\{5\}$, $\{2, 4\}$, $\{2, 5\}$, $\{4, 5\}$, $\{2, 4, 5\}$, and ϕ. Note that set H is included among the eight subsets of H. *Every set is a subset of itself.* Note also that the null set, ϕ, is listed among the eight subsets of H. *The null set is a subset of every set.*

1.3.2 Operations on Sets

Definition 1.4: The *complement* of set A, denoted by A', is the set of all elements in the universal set U *except* the elements in A.

Definition 1.5: The *intersection* of sets A and B, denoted by $A \cap B$, is the set of all elements that belong to both A and B.

Definition 1.6: The *union* of sets A and B, denoted by $A \cup B$, is the set of all elements that belong to at least one of the two sets (i.e., all elements except those which are in neither of the sets).

Each of the three rectangles in Figure 1.1 represents a universal set, the circles represent sets, and the shaded areas represent A', $A \cap B$, and $A \cup B$, respectively.

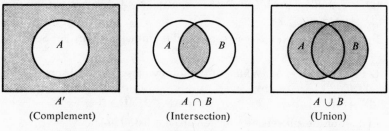

A'	$A \cap B$	$A \cup B$
(Complement)	(Intersection)	(Union)

Figure 1.1.

*An alternate definition of "subset" is: A is a subset of B if A contains no elements which are not in B.

For example, let $U = \{1, 2, 3, 4, 5, 6\}$, $Q = \{1, 2, 3, 4\}$, and $P = \{3, 4, 5\}$. Therefore, $Q' = \{5, 6\}$, $P' = \{1, 2, 6\}$, $Q \cap P = \{3, 4\}$, and $Q \cup P = \{1, 2, 3, 4, 5\}$.

Definition 1.7: The *cartesian product* of sets A and B, denoted by $A \times B$, is the set of all ordered pairs of elements of the form (x, y), such that x is an element of A, and y is an element of B.

For example, if $R = \{2, 3\}$ and $S = \{3, 5, 7\}$, then

$$R \times S = \{(2, 3), (2, 5), (2, 7), (3, 3), (3, 5), (3, 7)\}$$

Note that each element of $R \times S$ is an ordered pair of elements; $R \times S$ has six elements. The order in which these six elements are listed is irrelevant. However, the order in which the numbers are listed in each pair is important. Whereas the set $\{2, 3\}$ equals the set $\{3, 2\}$, the ordered pair $(2, 3)$ does *not* equal the ordered pair $(3, 2)$. The use of parentheses points up the difference in meaning between ordinary sets (in which order is irrelevant) and ordered pairs (in which order is important). Thus, the cartesian product $A \times B$ is not, in general, equal to $B \times A$. In our previous example:

$$S \times R = \{(3, 2), (3, 3), (5, 2), (5, 3), (7, 2), (7, 3)\}$$

Since the only element in common between $R \times S$ and $S \times R$ is the ordered pair $(3, 3)$, $R \times S \neq S \times R$. If, on the other hand, two sets A and B are equal, then sets $A \times B$ and $B \times A$ are also equal.

Exercises

1. Let $U = \{1, 2, 3, \ldots , 9\}$ (Universal Set)
    $A = \{2, 3, 4, 5\}$
    $B = \{2, 4, 6\}$
    $C = \{5, 8\}$
    and find:
    (a) C' (b) $A \cup C$ (c) $A \cap B$
    (d) $A' \cap C$ (e) $B' \cup A$ (f) $A \cap (B \cup C)$
    (g) $C' \cap B'$ (h) $B \cap C$ (i) $A \times C$
    (j) $C \times A$

2. In the previous exercise are sets A and B mutually exclusive? Sets A and C? Sets B and C?

3. Let $N = \{m, 2, *, \Delta\}$. Which of the following sets are subsets of N?
    (a) $\{2, \Delta\}$ (b) $\{m\}$ (c) $\{*, m, \Delta, 2\}$
    (d) $\{*, 5, \Delta\}$ (e) $\{2, *, \Delta, x, m\}$ (f) ϕ

4. List all of the subsets of N which are not listed above.

5. When $R \cap S = \phi$, we say that sets R and S are _____.

6. Which of the following pairs of sets are mutually exclusive? In each case, except (b), draw a diagram similar to Figure 1.2, and shade in each set in the pair given.

(a) M and M' (b) M and ϕ
(c) $M \cap T$ and $M \cup T$ (d) $M \cap T'$ and $M' \cap T$
(e) $(M \cup T)'$ and $M \cap T$

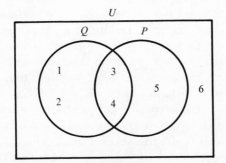

Figure 1.2.

7. Given the universal set U and a set E, which is a subset of U, which of the
    following is (are) always true?
    (a) $E \cup E' = U$ (b) $E \cap E' = \phi$ (c) $U \cup E = U$
    (d) $U \cap E = E$ (e) $U' = \phi$ (f) $E' \subseteq U$
    (g) $\phi \subseteq E$

8. If the universal set U is the set of all integers, i.e., $U = \{. . . , -2, -1, 0, 1, 2,$
    $. . .\}$, and if $F = \{. . . , -2, -1, 0, 1\}$ and $G = \{0, 1, 2, 3, 4, . . .\}$, then find:
    (a) $F \cap G$ (b) $F \cup G$ (c) $F' \cap G$
    (d) $F' \cup G$ (e) $F' \cap G'$

9. Note that a set containing three elements (set H, page 5) has eight, or 2^3,
    different subsets. Note also that set N in Exercise 3 has 16, or 2^4, different
    subsets. How many different subsets does a set containing five elements
    have? How many for a set containing n elements?

10. An alternate definition for set equality (Definition 1.1) could be stated as
    follows: If $A \subseteq B$ and $B \subseteq A$, then $A = B$. State in your own words why this
    definition makes sense.

11. For the sets defined in Exercise 1, write in the elements $1, 2, . . . , 9$ in
    the appropriate regions on the Venn diagram in Figure 1.3.

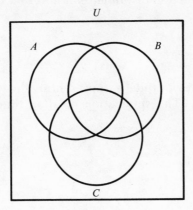

Figure 1.3.

1.4 Summation Notation

In statistical work one deals with many types of variables, such as height, weight, income, IQ, and so forth. Typically, the statistician must collect data by measuring one or more of these variables on a group of people or objects. For example, he might wish to measure the height in inches and weight in pounds of every child in a second grade classroom. The statistician needs a symbolic language which will allow him to refer to these measurements and what he plans to do with them. It is customary for different variables to be assigned different letters, such as x, y, and z. If several people (or objects) are to be measured on the same variable, the several measurements are represented by the same letter with different subscripts. In the second grade classroom, for example, height might be represented by x and weight by y. The height of the first child would be represented by x_1 and the weight of the first child by y_1; the height of the second child would be x_2, the weight y_2, and so on. If there were 30 children in the class, the last child measured would have a height of x_{30} and a weight of y_{30}.

For the purpose of describing the characteristics of a set of data the statistician calculates certain statistics. For example, the average height and average weight for the 30 children would probably be considered useful. One common type of average is found by dividing the sum of the measurements by the number of measurements. Thus, to calculate the average height of 30 children, one must first sum the measurements — that is, find $x_1 + x_2 + x_3 + x_4$ and so forth up through x_{30}. This sum can be written in shorter form,

$$x_1 + x_2 + x_3 + \cdots + x_{30}$$

The average height could, therefore, be represented symbolically as

$$\frac{x_1 + x_2 + x_3 + \cdots + x_{30}}{30}$$

This notation is still considered somewhat unwieldy. It can be simplified further by the introduction of *summation notation* by replacing $(x_1 + x_2 + x_3 + \cdots + x_{30})$ by

$$\sum_{i=1}^{30} x_i$$

The symbol Σ is a Greek capital s, representing the word "sum," and the i is called the index. The summation notation literally says "sum the expression x_i from $i = 1$ to $i = 30$." The use of the letter i is arbitrary; that is, the expression

$$\sum_{j=1}^{30} x_j$$

says exactly the same thing as

$$\sum_{i=1}^{30} x_i$$

The average height can now be represented succinctly as

$$\frac{\sum_{i=1}^{30} x_i}{30}$$

Since many statistical formulas involve extensive use of summation, we shall define summation notation in general and investigate some properties of the *summation operator*,

$$\sum_{i=1}^{n}$$

Definition 1.8:   Given that n is a positive integer,

$$\sum_{i=1}^{n} x_i = x_1 + x_2 + x_3 + \cdots + x_n$$

We shall understand this definition to mean that the summation operator tells us to replace the index *wherever it occurs after the operator* by one, then by two, and so on through n, separating each resulting term by a plus (+) sign. It follows from Definition 1.8, therefore, that we can write such things as

$$\sum_{i=1}^{2} (x_i - y_i) = (x_1 - y_1) + (x_2 - y_2)$$

$$\sum_{j=1}^{3} 2y_j^3 = 2y_1^3 + 2y_2^3 + 2y_3^3$$

$$\sum_{k=3}^{5} (x_k + 10) = (x_3 + 10) + (x_4 + 10) + (x_5 + 10)$$

Three important properties of the summation operator are stated below as theorems and then proved. (Reasons for each step are included in parentheses along the right-hand margin.)

Theorem 1.1:

$$\sum_{i=1}^{n} (x_i + y_i) = \sum_{i=1}^{n} x_i + \sum_{i=1}^{n} y_i$$

Proof:

$$\sum_{i=1}^{n} (x_i + y_i) = (x_1 + y_1) + (x_2 + y_2) + \cdots + (x_n + y_n) \qquad \text{(Def. 1.8)}$$

$$= (x_1 + x_2 + \cdots + x_n) + (y_1 + y_2 + \cdots + y_n) \quad \text{(Regroup)}$$

$$= \sum_{i=1}^{n} x_i + \sum_{i=1}^{n} y_i \qquad \text{(Def. 1.8)}$$

We shall assume without proof a more generalized version of Theorem 1.1 which states that the summation operator may be distributed over any finite series of terms which are being added to or subtracted from each other. That is, such statements as the following are legitimate.

$$\sum_{i=1}^{n} (x_i^2 - 2y_i) = \sum_{i=1}^{n} x_i^2 - \sum_{i=1}^{n} 2y_i$$

$$\sum_{j=1}^{n} (x_j^3 + y_j - 2) = \sum_{j=1}^{n} x_j^3 + \sum_{j=1}^{n} y_j - \sum_{j=1}^{n} 2$$

This generalized version of Theorem 1.1 may be called the *distributive property for summation.*

Theorem 1.2:   Given that c is a constant*

$$\sum_{i=1}^{n} c x_i = c \sum_{i=1}^{n} x_i$$

Proof:

$$\sum_{i=1}^{n} c x_i = c x_1 + c x_2 + \cdots + c x_n \qquad \text{(Def. 1.8)}$$

$$= c(x_1 + x_2 + \cdots + x_n) \qquad \text{(Factor out } c\text{)}$$

$$= c \sum_{i=1}^{n} x_i \qquad \text{(Def. 1.8)}$$

*A constant c may be thought of as being a variable x which always has the same value, namely c. That is, $x_1 = c$, $x_2 = c$, $x_3 = c$, and so forth. (Note that because c represents a constant, no subscript is used.)

It follows from Theorem 1.2, for example, that

$$\sum_{k=1}^{3} 5y_k = 5 \sum_{k=1}^{3} y_k$$

Theorem 1.3:   Given that c is a constant,

$$\sum_{i=1}^{n} c = nc$$

Proof:

$$\sum_{i=1}^{n} cx_i = cx_1 + cx_2 + \cdots + cx_n \tag{Def. 1.8}$$

If we let each $x_i = 1$, then substituting above we get

$$\sum_{i=1}^{n} c = c + c + \cdots + c = nc$$

In other words, if a constant c is added n times, the sum equals the product nc. For example,

$$\sum_{i=1}^{3} 15 = 3(15) = 45.$$

The preceding theorems serve to simplify summation problems such as the following.

Problem: Find the value of

$$\sum_{i=1}^{2} (x_i^2 + 2y_i - 5),$$

given that $x_1 = 2$, $x_2 = -1$, $y_1 = 1$, and $y_2 = 5$.

Solution:

$$\sum_{i=1}^{2} (x_i^2 + 2y_i - 5) = \sum_{i=1}^{2} x_i^2 + \sum_{i=1}^{2} 2y_i - \sum_{i=1}^{2} 5 \tag{Dist. Prop. for Σ}$$

$$= \sum_{i=1}^{2} x_i^2 + 2 \sum_{i=1}^{2} y_i - \sum_{i=1}^{2} 5 \tag{Thm. 1.2}$$

$$= \sum_{i=1}^{2} x_i^2 + 2 \sum_{i=1}^{2} y_i - 10 \tag{Thm. 1.3}$$

$$= [2^2 + (-1)^2] + 2 [1 + 5] - 10 \tag{Subst.}$$

$$= 7$$

Problem: If

$$\sum_{i=1}^{50} x_i^2 = 56.2$$

and

$$\sum_{i=1}^{50} x_i = 14.8$$

find the value of

$$\sum_{i=1}^{50} (x_i - 2)^2$$

Solution:

$$\sum_{i=1}^{50} (x_i - 2)^2 = \sum_{i=1}^{50} (x_i^2 - 4x_i + 4)$$

$$= \sum_{i=1}^{50} x_i^2 - \sum_{i=1}^{50} 4x_i + \sum_{i=1}^{50} 4 \qquad \text{(Dist. Prop. for } \Sigma\text{)}$$

$$= \sum_{i=1}^{50} x^2 - 4 \sum_{i=1}^{50} x_i + 200 \qquad \text{(Thms. 1.2 and 1.3)}$$

$$= 56.2 - 4(14.8) + 200 \qquad \text{(Subst.)}$$

$$= 197.0$$

Exercises

1. Expand each of the following summations using Definition 1.8.

    (a) $\displaystyle\sum_{i=1}^{5} y_i$ (b) $\displaystyle\sum_{k=3}^{6} x_k^2$ (c) $\displaystyle\sum_{j=1}^{7} x_j y_j$

    (d) $\displaystyle\sum_{i=1}^{3} (2x_i + 5)$ (e) $\displaystyle\sum_{j=2}^{4} x_j^2 f_j$

2. Write each of the following sums in summation notation form.
    (a) $y_1^2 + y_2^2 + y_3^2 + \cdots + y_{17}^2$
    (b) $2z_1 + 2z_2 + 2z_3 + \cdots + 2z_{100}$
    (c) $x_2 y_2^2 + x_3 y_3^2 + x_4 y_4^2 + x_5 y_5^2$
    (d) $(x_1^2 + 5)^3 + (x_2^2 + 5)^3 + (x_3^2 + 5)^3$

3. Given that $x_1 = 0$, $x_2 = 2$, $x_3 = -1$, $x_4 = 3$ and $x_5 = 1$, find the indicated sums.

    (a) $\displaystyle\sum_{i=1}^{5} x_i^2$ (b) $\displaystyle\sum_{i=3}^{5} x_i$

    (c) $\displaystyle\sum_{i=1}^{5} (2x_i + 1)^2$ (d) $\displaystyle\sum_{i=1}^{5} x_i(x_i - 1)$

4. Given that $x_1 = 2$, $x_2 = 5$, $x_3 = -2$, $y_1 = 3$, $y_2 = -1$, and $y_3 = 1$, find:

(a) $\displaystyle\sum_{k=1}^{3} x_k^2\, y_k$ (b) $\displaystyle\sum_{k=1}^{3} (4x_k + y_k)$

(c) $\displaystyle\sum_{k=1}^{3} (x_k + y_k)^2$ (d) $\displaystyle\left(\sum_{k=2}^{3} x_k\right)\left(\sum_{k=2}^{3} y_k^2\right)$

5. Given that

$$\sum_{j=1}^{10} x_j^3 = 91, \qquad \sum_{j=1}^{10} x_j^2 = 25$$

and

$$\sum_{j=1}^{10} x_j = 7$$

find the value of

$$\sum_{j=1}^{10} (5x_j^3 - 3x_j^2 + x_j + 22)$$

Summary

The field of modern statistics is a rapidly growing branch of applied mathematics which can be thought of as a combination of two subfields called *descriptive statistics* and *statistical inference*. The former deals with the collection, tabulation, and presentation of data and the calculation of measures which describe the data. Statistical inference, on the other hand, involves the use of data and descriptive measures as a basis for making predictions or decisions relating to unobserved events. A thorough knowledge of the basic concepts of *probability* is a prerequisite to success in understanding statistical inference. *Set notation* is basic to the development of probability theory. *Summation notation* is a short-cut language used extensively in statistical work.

part 1

descriptive statistics

2

Describing Ungrouped Data

2.1 Sample vs. Population

It is a common situation in which an investigator wishes to answer certain questions relating to the characteristics of a large group (set) of objects or people (elements). For convenience we call this large set the *population*. Often the population is of such a nature that it is virtually impossible to measure each element on the variables of interest. In these situations we do the next best thing—we select a subset of the population and measure only those elements. Such a subset is called a *sample*.

For, example, suppose we wish to know the IQ of the typical fifth grade student in the United States. Rather than attempt to measure the IQs of all fifth graders in the U.S., the limitations of time and money would almost certainly force us to select a sample for investigation. Similarly, a manufacturer of light bulbs might wish to know how many hours of light can be expected from a certain type of bulb he produces. Obviously, he cannot afford to burn out an entire stock of bulbs to get the information he needs. Thus, he selects a sample of bulbs from his total stock (population), measures how long it takes each bulb in the sample to burn out, and uses his findings as an indication of what he could expect from the total stock.

In both examples, as in any other instance where a population is measured *indirectly* through sampling, the information gained from the sample must be considered an *estimate* of the characteristics or properties of the parent population. Estimates are invariably in error to some degree, since they have resulted from only a partial look at the population. One of the major tasks of the statistician is to provide the investigator with the methods necessary to minimize sampling errors.

In this chapter we begin to investigate various methods of calculating descriptive statistics from sample data. That is, we shall assume that the decision to select a sample has been made, the sample has been selected, and measurements on the sample have been taken.

2.2   Measures of Central Location

One important way of describing a set of sample data is to give some indication of "central location" (average) in the distribution of measurements. We shall discuss and compare three different measures of central location: the *mean*, the *median*, and the *mode*.

Before we define these statistics, consider the following sample data, which we shall use for demonstration purposes in subsequent sections. Let the variable y represent the number of children in a family. Suppose that 15 families have been randomly selected, and the number of children in each was found to be

$$2, 6, 3, 2, 1, 2, 7, 5, 4, 1, 4, 0, 5, 2, 4$$

Thus, we have sample data composed of 15 measurements (counts), each a value of variable y. We begin to describe this data by calculating measures of central location.

2.2.1   The Sample Mean

The sample mean is the statistic most commonly associated with the term "average." To find it we divide the sum of the measurements by the number of measurements.

Definition 2.1: If $x_1, x_2, x_3, \ldots, x_n$ are the numerical measurements on variable x from a sample of n elements, then the *sample mean* of x, denoted by $\bar{x}$, is

$$\bar{x} = \frac{\sum\limits_{i=1}^{n} x_i}{n}$$

For the sample data the sum of the 15 measurements is 48. Therefore, the mean of varible y is 48 divided by 15, which equals 3.2. Thus, we say that $\bar{y} = 3.2$, or that the mean number of children for that sample of 15 families is 3.2.

2.2.2 The Sample Median

A second type of "average" is called the sample median. This statistic is a measure of central location in the sense that if you order the n measurements according to size, there are just as many measurements below the median as above it. A more precise definition appears below.

> **Definition 2.2:**   When the n measurements in a sample are listed in rank order, the *sample median* is the measurement with rank $(n + 1)/2$.

Using the same sample data we reorder the 15 measurements according to size.

Measurement: 0, 1, 1, 2, 2, 2, 2, 3, 4, 4, 4, 5, 5, 6, 7
        Rank: 1 2 3 4 5 6 7 8 9 10 11 12 13 14 15

Since $n = 15$, $(n + 1)/2 = (15 + 1)/2 = 8$. Therefore, the median is the measurement with rank equal to eight. In this case the median is three. Note that the mean and median are not equal in this example, even though both statistics attempt to measure central location.

Note also that Definition 2.2 appears to be inapplicable whenever there is an even number of measurements in the sample. For example, if $n = 20$, then $(n + 1)/2 = 10.5$. Strictly speaking, there is no measurement with rank 10.5 since all of our ranks are whole numbers. However, we can avoid the problem by agreeing that whenever $(n + 1)/2$ is a number of the form "$k.5$," we find the median by adding the measurements with ranks k and $k + 1$ and dividing the result by two. Therefore, if $(n + 1)/2 = 10.5$, we add the measurement with rank 10 to the measurement with rank 11 and divide the result by two. Suppose, for example, a sample contained the following measurements.

29, 7, 3, 13, 22, 18

Reordering according to rank, we get

3, 7, 13, 18, 22, 29

Since $n = 6$, $(n + 1)/2 = (6 + 1)/2 = 3.5$. Therefore, the median of this sample is $(13 + 18)/2 = 15.5$.

2.2.3 A Sample Mode

Definition 2.3: A *sample mode* is a measurement which occurs most frequently in the sample. (It is possible for two or more modes to exist in one sample if two or more different measurements tie as most frequent.)

For our sample data, the sample mode is two, since more families have two children than any other specific number. Note the distribution of frequencies below.

Number of Children: 0 1 2 3 4 5 6 7
　　　　Frequency: 1 2 ④ 1 3 2 1 1

The set of data: 2, 8, 5, 5, 3, 4, 5, 8, 9, 8 is said to be *bimodal*; that is, it has two modes: 5 and 8.

2.2.4 Mean, Median, and Mode—Advantages and Disadvantages

We have observed that the "average" number of children per family can be expressed three ways: mean = 3.2, median = 3, and mode = 2. Each statistic describes more or less a central location in the sample. Each statistic has its advantages and disadvantages depending upon the situation and purpose of the investigation.

The sample mean is certainly the most used measure of central location, probably because of its favorable properties: (a) the mean is unique—that is, there can be only one mean for a set of numerical data, (b) it is reliable, compared to other measures of central location, in the sense that means of several samples taken from the same population tend to vary less than do other measures, (c) its mathematical properties are such that further statistical manipulation is convenient (see Section 2.2.5, page 21), and (d) it takes into account the size of each individual measurement—that is, a change in one value changes the mean value.

The sample median is also unique, but is not as reliable as the mean, does not lend itself easily to further statistical manipulation, and is not necessarily affected by a change in value of one individual measurement. This last property is considered by some to be an advantage in certain situations. It can be argued that one or two extremely large or small measurements which are atypical with respect to the other measurements in the sample should not be allowed to significantly affect the value of a statistic designed to measure "average" value in the sample. Since the value of the median is not affected by the *size* of extreme values, the median is in some applied situations considered superior to the mean in measuring central location. Similarly, in situations where one or two errors occur in the taking or recording of measurements, the median has less chance of being affected by such errors than does the mean. Another advantage of the median is its computational ease, although for large samples the job of ranking can become rather tedious.

The sample mode also has the advantage of computational simplicity. However, its disadvantages make it a very unsatisfactory measure of central location. Perhaps the greatest drawback is that the mode is not unique; there may be several modes in a given set of data. In fact, if all n measurements are different, according to our definition we would have n modes!* In addition, the mode is relatively unreliable. In short, the sample mode is generally a rather poor measure of central location for *quantitive* data. Its major utility, however, lies in its use with *qualitative* data. For example, if we "measure" a sample of people on religious preference, we essentially divide the set into subsets such as: Protestant, Catholic, Jew, and so on. The mode would be the religion of the subset containing the most elements, and "average" would be interpreted as typical rather than central. For obvious reasons, the mean and median are of no use in describing a sample of people on the basis of religious preference.

2.2.5 The Weighted Mean

Occasionally it is desirable to give different weights to individual measurements in the process of finding a mean. For example, suppose that in a college course the professor gives three hour exams and one two-hour final exam. And since the final exam is twice as long as each hour exam, the professor wishes to give the final exam score twice as much weight as each hour exam score. He can do this by treating the final exam score as *two* scores, and then calculating the mean of the resulting five test scores. Suppose that a particular student had hour exam grades of 75, 90, and 82, and a final exam score of 86. The "weighted mean" of these scores (counting the final exam score twice) is

$$\frac{75 + 90 + 82 + 86 + 86}{5} = 83.8$$

Note that the calculation above can be written

$$\frac{(1)\ 75 + (1)\ 90 + (1)\ 82 + (2)\ 86}{5} = 83.8$$

Note also that the sum of the numbers in parentheses is five, the denominator. In other words, each score is first multiplied by an appropriate *weight*, the products are added, and the sum is divided by the sum of the weights. This example gives rise to the following definition.

---

*Some authors prefer to define the mode as nonexistent when all measurements are different.

Definition 2.4: Given measurements x_1, x_2, x_3, . . . , x_n which have weights w_1, w_2, w_3, . . . , w_n, respectively, the *weighted mean* of x is

$$\bar{x}_w = \frac{\sum\limits_{i=1}^{n} w_i x_i}{\sum\limits_{i=1}^{n} w_i}$$

Note that the definition of $\bar{x}_w$ above reduces to the usual definition of the sample mean when the measurements are weighted equally. For example, if each $w_i = 5$, then, using Definition 2.4 and Theorems 1.2 and 1.3, we get

$$\bar{x}_w = \frac{\sum\limits_{i=1}^{n} 5x_i}{\sum\limits_{i=1}^{n} 5} = \frac{5 \sum\limits_{i=1}^{n} x_i}{(n)(5)} = \frac{\sum\limits_{i=1}^{n} x_i}{n} = \bar{x}$$

The values to be used as weights in an applied problem are determined by the conditions in that particular problem; often the decision is arbitrary. For example, the professor might have decided to weight the final exam three times as much as each hour exam. There is a special application of Definition 2.4, however, for which the weights are determined by Definition 2.1. This application involves finding the mean of a set of data from the means of various subsets of the data. Consider the following example. Mr. Wilson goes on a fishing trip for seven days. Let x_i be the number of fish caught on the ith day. For the first five days of his trip the mean number of fish he caught per day was 4.4. However, during the last two days the mean per day was 6.5 fish. What is the mean number of fish caught per day for the seven-day trip? According to Definition 2.1 the mean number of fish caught per day is

$$\bar{x} = \frac{\sum\limits_{i=1}^{7} x_i}{7}$$

$$= \frac{x_1 + x_2 + x_3 + x_4 + x_5 + x_6 + x_7}{7}$$

$$= \frac{(x_1 + x_2 + x_3 + x_4 + x_5) + (x_6 + x_7)}{7}$$

$$= \frac{\sum\limits_{i=1}^{5} x_i + \sum\limits_{i=6}^{7} x_i}{7} = \frac{5 \dfrac{\sum\limits_{i=1}^{5} x_i}{5} + 2 \dfrac{\sum\limits_{i=6}^{7} x_i}{2}}{7}$$

Since

$$\frac{\sum\limits_{i=1}^{5} x_i}{5} = 4.4 \qquad \text{and} \qquad \frac{\sum\limits_{i=6}^{7} x_i}{2} = 6.5$$

$$\bar{x} = \frac{5(4.4) + 2(6.5)}{7} = 5.0$$

Note from the above equation that the mean for seven days is essentially a "weighted mean" based on the two given means (4.4 and 6.5), the weights being the number of measurements which determine each of the given means. In other words, we have developed the following theorem.

Theorem 2.1: If a set of n measurements is divided into k mutually exclusive subsets where $\bar{x}_i$ is the mean of the measurements in subset i, and n_i is the number of measurements in subset i $(n_1 + n_2 + n_3 + \cdots + n_k = n)$, then the overall mean $\bar{x}$ of the n measurements may be expressed as the weighted mean of the k subset means $\bar{x}_1, \bar{x}_2, \ldots, \bar{x}_k$; that is,

$$\bar{x} = \frac{\sum\limits_{i=1}^{k} n_i \bar{x}_i}{\sum\limits_{i=1}^{k} n_i} = \frac{\sum\limits_{i=1}^{k} n_i \bar{x}_i}{n}$$

Theorem 2.1 is important in that it allows us to calculate the sample mean for a set of n measurements directly from the means of subsets *without going back to the original data.* By way of additional illustration, consider the following example.

Problem: On a football team's starting lineup, the mean weight for the two ends is 200 lb., for the other five linemen is 240 lb., and for the four backfield men is 180 lb. What is the mean weight for the starting eleven?

Solution: $\bar{x}_1 = 200 \qquad n_1 = 2$
$\bar{x}_2 = 240 \qquad n_2 = 5$
$\bar{x}_3 = 180 \qquad n_3 = 4$

Therefore, according to Theorem 2.1, the mean weight for the starting eleven is

$$\bar{x} = \frac{2(200) + 5(240) + 4(180)}{2 + 5 + 4} = \frac{2320}{11} = 211 \text{ (to the nearest pound)}$$

Note that in this example we have found the mean weight of eleven players without knowing the eleven individual weights.

Exercises

1. Find the mean, median, and mode for the following sample data: 8, 2, 22, 16, 9, 10, 12, 8, 7.

2. Bill competed in the 100 yard dash 10 times last season. His measured times, in seconds, were 10.2, 10.4, 10.0, 10.2, 10.0, 9.8, 10.0, 9.7, 9.9, 9.8. Find the mean, median, and mode for these data.

3. The college grade-point averages for 16 graduating seniors are

2.6	2.7	3.0	3.2
3.2	3.5	3.2	2.5
3.8	2.3	2.9	2.6
3.3	2.9	2.7	3.6

    Find the mean, median, and mode for these data.

4. In a sample of 103 working men, 100 have annual salaries of exactly $8500. The other men have salaries of $100,000, $250,000, and $300,000.
    (a) Find the mean, median, and modal salary.
    (b) Why does the mean differ so much from the median and mode?
    (c) Which of the three is the best measure of "average" in this case?
    (d) If the problem had read "100 have a *mean* annual salary of $8500" rather than "100 have annual salaries of exactly $8500," what would be the mean, median, and modal salary of the 103 men?

5. A baseball player's "batting average" (number of hits divided by number of times at bat) is 0.350 for his first 20 times at bat, 0.200 for his next 100 times at bat, and 0.500 for his next 30 times at bat. What is his overall batting average?

6. In a learning experiment a rat runs a maze 20 times with a mean time of 10.25 seconds. In 10 additional runs through the maze the same rat has a mean time of 9.65 seconds. What is the rat's mean time for the 30 runs?

7. A measure of central location called the *geometric mean* is sometimes used. This statistic is defined as the nth root of the product of n measurements. (i.e., $\sqrt[n]{x_1 \cdot x_2 \cdot x_3 \cdot \cdots \cdot x_n}$). Find the geometric mean of
    (a) 12 and 75
    (b) 0.16, 0.50, and 0.80

8. Another measure of central location is called the *harmonic mean*. For a sample of size n the harmonic mean is n divided by the sum of the reciprocals of the measurements. In symbolic notation, it is

    $$n \Big/ \sum_{i=1}^{n} \frac{1}{x_i}$$

    (a) Find the harmonic mean of 20, 30, and 60.
    (b) A race driver travels a two-mile course at 100 mph., then repeats the run at 120 mph. Verify that his mean rate of speed over the four miles can be found by calculating the harmonic mean of 100 and 120.

2.3 Measures of Dispersion

For the statistician there is one very important characteristic of sample data, other than central location, which differentiates one sample from another. Consider the following data from two different samples.

Sample *A*:   2, 5, 1, 9, 12, 5, 5, 1
Sample *B*:   5, 3, 5, 4, 5, 7, 6, 5

Note that the three measures of central location—mean, median, and mode—are all equal to five for both samples. That is, measures of central location do not differentiate Sample *A* from Sample *B*. If we list the measurements for each sample in rank order,

Sample *A*:   1, 1, 2, 5, 5, 5, 9, 12
Sample *B*:   3, 4, 5, 5, 5, 5, 6, 7

we note that the two samples do in fact differ in one important way; there is clearly more *dispersion* in Sample *A*. By "dispersion" we mean "spread" or "variation".

Suppose that *A* and *B* represent dive bomber pilots. Each has just taken a run over eight practice targets, dropping one bomb on each. The data represent distances in yards from the target to the point of bomb impact. On the "average" it appears that the two pilots are equally good at their work. However, a comparison of the two samples on the basis of dispersion shows that pilot *B* is much more *consistent* with his aim, though *A* has the three most accurate shots. If it were true that a bomb must hit within eight yards of a target to assure complete destruction of that target, then pilot *B* would probably be considered superior to pilot *A*, since he completely destroyed eight targets; *A* destroyed only six. Consistent performance is preferred to erratic performance (other things equal) in many real-life situations. On the other hand, suppose that *A* and *B* are professional golfers and the data represent place finishes in eight consecutive tournaments in which they both played. Although the "average" place finish for each is the same, player *A*, the *less consistent* of the two, must be considered the more successsful golfer from a financial standpoint, due to the disproportionately large purses which go to winners and runners-up.

The important point in these illustrations is not whether dispersion is good or bad but that dispersion is an important characteristic of sample data and one which varies independently of central location. Therefore, we need ways of measuring dispersion; that is, we need statistics which are sensitive to the amount of dispersion existing in sample data.

2.3.1 The Sample Range

Perhaps the most obvious and simplest measure of dispersion in sample data is the number obtained by subtracting the smallest measurement from the largest; more specifically,

> **Definition 2.5:** If x_l is the largest measurement in the sample data and x_s is the smallest, then $x_l - x_s$ is called the *sample range*.

Applying Definition 2.5 to the two samples in the previous section, we find that A and B have ranges of 11 and 4, respectively. The larger range, of course, indicates the greater dispersion existing in Sample A — a result consistent with our earlier observations.

It is true, however, that the sample range, as a measure of dispersion, is not generally very useful. Although it is easy to calculate and understand, it suffers from a serious drawback–namely, that it is not sensitive to dispersion *between* the extreme values. The three samples below serve to illustrate this weakness. Each sample has a range of 11 (12 minus 1), but simple observation reveals obvious differences in dispersion among the three. Since the range is a somewhat inadequate measure of overall sample dispersion for most purposes, we usually employ more reliable measures.

Sample 1: 1, 2, 3, 4, 5, 6, 7, 8, 9, 10, 11, 12
Sample 2: 1, 5, 6, 6, 6, 6, 7, 7, 7, 7, 8, 12
Sample 3: 1, 1, 1, 1, 2, 2, 2, 3, 3, 4, 7, 12

2.3.2 The Sample Mean Deviation

It makes sense to reason that sample data which tends to bunch closely around a central location in the sample would have small dispersion, and sample data in which many values are distributed at considerable distances from a central location would have large dispersion. Therefore, if we use the sample mean, $\bar{x}$, as our measure of central location, then $x_i - \bar{x}$ represents the distance (deviation) of x_i from $\bar{x}$. It seems reasonable that the mean of all such distances, i.e.,

$$\frac{\sum_{i=1}^{n} (x_i - \bar{x})}{n}$$

should be a statistic which is *sensitive* to the dispersion in the sample data; that is, a statistic which is large when the dispersion is large and small when the dispersion is small. Unfortunately, this line of reasoning breaks down because the numerator of the proposed statistic is always zero!

Theorem 2.2: Given a sample of size n with measurements x_1, x_2, x_3, . . . , x_n, it is always true that

$$\sum_{i=1}^{n} (x - \bar{x}) = 0$$

Proof:

$$\sum_{i=1}^{n} (x_i - \bar{x}) = \sum_{i=1}^{n} x_i - \sum_{i=1}^{n} \bar{x} \qquad \text{(Dist. Prop. for } \Sigma\text{)}$$

$$= \sum_{i=1}^{n} x_i - n\bar{x} \qquad \text{(Thm. 1.3)}$$

$$= \sum_{i=1}^{n} x_i - n \frac{\sum_{i=1}^{n} x_i}{n} \qquad \text{(Def. 2.1)}$$

$$= \sum_{i=1}^{n} x_i - \sum_{i=1}^{n} x_i \qquad \text{(Cancel the } n\text{s)}$$

$$= 0$$

Of course, the reason why the summation of the distances, $(x_i - \bar{x})$, is zero is that $x_i - \bar{x}$ is *positive* when x_i is greater than $\bar{x}$ and *negative* when x_i is smaller than $\bar{x}$. Thus, in summing all the distances, the negative and positive values cancel each other out. But do we want to differentiate between positive and negative distance from the mean? Is it not true that if $\bar{x} = 50$, then the measurements of 45 and 57 are at distances of five and seven units, respectively, from the mean? And would we not want to consider the mean distance from $\bar{x}$ as six units $((5 + 7)/2)$, rather than one unit $((-5 + 7)/2)$? What we need to do, therefore, is to ignore the negative sign when $x_i - \bar{x}$ is negative. In other words, we really want to calculate the *absolute value* of $x_i - \bar{x}$, denoted by $|x_i - \bar{x}|$. This reasoning leads us to a definition of *mean deviation*, which is a legitimate measure of dispersion.

Definition 2.6: Given a sample of size n with measurements x_1, x_2, x_3, . . . , x_n, the *sample mean deviation* is defined as

$$\frac{\sum_{i=1}^{n} |x_i - \bar{x}|}{n}$$

From the summations in Table 2.1 we can verify Theorem 2.2 and show that Definition 2.6 is a reasonable statistic for measuring dispersion, using the data from samples A and B. Note that

$$\sum_{i=1}^{n} (x_i - \bar{x}) = 0$$

for both samples, as expected (Theorem 2.2), and the mean deviation for Sample A (2.75) is larger than the mean deviation for Sample B (0.75), which is consistent with our visual expectations and the values we obtained earlier for the ranges. The mean deviation, therefore, appears to be sensitive to dispersion with the added advantage (compared to the range) that every measurement affects its value. However, the mean deviation has one unfortunate drawback; due to the use of absolute values the mathematical properties of the mean deviation are such that theoretical treatment of the statistic is difficult. Fortunately, there are other alternatives.

Table 2.1. Summations Needed to Calculate the Mean Deviations for Samples A and B

	Sample A $(\bar{x} = 5)$			Sample B $(\bar{x} = 5)$	
x_i	$(x_i - \bar{x})$	$\|x_i - \bar{x}\|$	x_i	$(x_i - \bar{x})$	$\|x_i - \bar{x}\|$
1	−4	4	3	−2	2
1	−4	4	4	−1	1
2	−3	3	5	0	0
5	0	0	5	0	0
5	0	0	5	0	0
5	0	0	5	0	0
9	4	4	6	1	1
12	7	7	7	2	2
Σ	0	22	Σ	0	6

Sample A: $\dfrac{\text{Mean}}{\text{Deviation}} = \dfrac{|x_i - \bar{x}|}{8} = \dfrac{22}{8} = 2.75$ Sample B: $\dfrac{\text{Mean}}{\text{Deviation}} = \dfrac{|x_i - \bar{x}|}{8} = \dfrac{6}{8} = 0.75$

2.3.3   The Sample Variance

Thinking back for a moment to our earlier discussion, we recall that originally the absolute value was introduced so that negative deviations could be made positive, thus avoiding the problem caused by negative and positive deviations from the mean summing to zero. There are, however, other ways of making negative numbers positive. We could use, for example, the fact that *the square of a negative number is positive.* That is, instead of finding the average deviation from the mean, we could find the *average squared deviation* from the mean. Such a statistic is called the *sample variance.*

Definition 2.7: Given a sample of size n with measurements x_1, x_2, x_3, . . . , x_n, the *sample variance*, denoted by s_x^2, is

$$s_x^2 = \frac{\sum_{i=1}^{n} (x_i - \bar{x})^2}{n - 1}$$

Note that $n - 1$ is used in the denominator of $s_x{}^2$, rather than n (as one might have expected). The explanation for this cannot truly be appreciated at this point in the text (see page 231). Intuitively, however, we may still think of $s_x{}^2$ as a sort of *average* squared deviation from $\bar{x}$. Using Definition 2.7 and the summation of $(x_i - \bar{x})^2$ from Table 2.2, we get for Sample A

$$s_x{}^2 = \frac{106}{8-1} = \frac{106}{7} = 15.14$$

It is left to the reader to verify that the variance for Sample B is 1.43. We see again that the data in Sample A has greater dispersion.

Table 2.2. Summation Necessary for the Calculation of $s_x{}^2$ for Sample A

x_i	$(x_i - \bar{x})$	$(x_i - \bar{x})^2$
1	−4	16
1	−4	16
2	−3	9
5	0	0
5	0	0
5	0	0
9	4	16
12	7	49
Σ		106

2.3.4 *The Sample Standard Deviation*

Though the sample variance is used extensively in statistical work, its positive square root, called the *sample standard deviation*, is perhaps the best known measure of dispersion due to its use in a rather wide variety of applications.

Definition 2.8: Given a sample of size n with measurements $x_1, x_2, x_3, \ldots,$ x_n, the *sample standard deviation*, denoted by s_x, is the *positive* square root of the sample variance; that is,

$$s_x = \sqrt{\frac{\sum_{i=1}^{n} (x_i - \bar{x})^2}{n-1}}$$

Since the variance cannot be negative (see Exercise 5, page 32), the positive square root of $s_x{}^2$ always exists. It should be noted that by taking the square root of the variance to find the standard deviation, we are in a sense compensating for the fact that we originally *squared* the deviations to find the variance. Thus, the standard deviation is a measure comparable to the mean deviation, though they are rarely equal. Table 2.3 compares the range, mean deviation, variance, and standard deviation by summarizing the values obtained for Samples A and B. It is clear from the table and from the definitions that these four statistics are indeed sensitive to dispersion existing in sample data. Large values for these statistics correspond to large dispersion in the sample data, and vice versa. For the time being, this shall be the extent to which we interpret the values we get for the various measures of dispersion. At this point, for example, a sample standard deviation of 5.3 tells us little unless we can compare it to the standard deviation of a different sample. Later we shall discuss interpretations which can be made with respect to a single sample.

Table 2.3. Measures of Dispersion for Samples A and B

	Size	Range	Mean Deviation	Variance (s_x^2)	Standard Deviation (s_x)
Sample A:	8	11	2.75	15.14	3.89
Sample B:	8	4	0.75	1.43	1.20

2.3.5 Calculation Formula for s_x^2 (and s_x)

The calculation of s_x^2 from Definition 2.7 often involves considerable work, especially when the size of the sample is large and when $\bar{x}$ does not come out an integer (consider the work that would have been involved in the earlier calculation of s_x^2 for Sample A if $\bar{x}$ had been 5.125 instead of 5). The following theorem gives us a formula for s_x^2 which is equivalent to the definition but which can be used to save calculation time and effort.

Theorem 2.3: Calculation Formula

$$s_x^2 = \frac{n \sum_{i=1}^{n} x_i^2 - \left(\sum_{i=1}^{n} x_i \right)^2}{n(n-1)}$$

Proof:

$$s_x^2 = \frac{1}{n-1} \left[\sum_{i=1}^{n} (x_i - \bar{x})^2 \right] \qquad \text{(Def. 2.7)}$$

$$= \frac{1}{n-1} \left[\sum_{i=1}^{n} (x_i^2 - 2x_i\bar{x} + \bar{x}^2) \right] \qquad \text{(Square } x_i - \bar{x})$$

$$= \frac{1}{n-1} \left[\sum_{i=1}^{n} x_i^2 - \sum_{i=1}^{n} 2x_i\bar{x} + \sum_{i=1}^{n} \bar{x}^2 \right] \qquad \text{(Dist. Prop. for } \Sigma)$$

$$= \frac{1}{n-1} \left[\sum_{i=1}^{n} x_i^2 - 2\bar{x} \sum_{i=1}^{n} x_i + n\bar{x}^2 \right] \qquad \text{(Thms. 1.2 and 1.3)}$$

$$= \frac{1}{n-1} \left[\sum_{i=1}^{n} x_i^2 - 2 \left(\frac{\sum_{i=1}^{n} x_i}{n} \right) \sum_{i=1}^{n} x_i + n \left(\frac{\sum_{i=1}^{n} x_i}{n} \right)^2 \right] \qquad \text{(Def. 2.1)}$$

$$= \frac{1}{n-1} \left[\sum_{i=1}^{n} x_i^2 - \frac{2}{n} \left(\sum_{i=1}^{n} x_i \right)^2 + \frac{1}{n} \left(\sum_{i=1}^{n} x_i \right)^2 \right]$$

$$= \frac{1}{n-1} \left[\sum_{i=1}^{n} x_i^2 - \frac{1}{n} \left(\sum_{i=1}^{n} x_i \right)^2 \right]$$

$$= \frac{n}{n} \cdot \frac{1}{n-1} \left[\sum_{i=1}^{n} x_i^2 - \frac{1}{n} \left(\sum_{i=1}^{n} x_i \right)^2 \right]$$

$$\therefore s_x^2 = \frac{n \sum_{i=1}^{n} x_i^2 - \left(\sum_{i=1}^{n} x_i \right)^2}{n(n-1)}$$

Note that this formula allows us to find s_x^2 without calculating $\bar{x}$ or squaring any deviations; we simply need to know n, the summation of x_i, and the summation of x_i^2. To verify this formula we use it to calculate the variance for Sample A.

x	x^2
1	1
1	1
2	4
5	25
5	25
5	25
9	81
12	144
Σ 40	306

$$\text{Variance} = s_x^2 = \frac{8(306) - (40)^2}{8(7)} = 15.14$$

Note that 15.14 is the same value we arrived at earlier using Definition 2.7. Except for occasional differences due to rounding errors, the definition of s_x^2 and the Calculation Formula yield identical answers. For Sample A there is little gained in using the Calculation Formula. However, in more instances than not, the Calculation Formula is much easier to use and, in fact, is more accurate.* A table of squares is provided in the Appendix to facilitate these calculations (see Table A).

Since the standard deviation (s_x) is the positive square root of the variance (s_x^2), it follows that the formula

$$s_x = \sqrt{\frac{n \sum_{i=1}^{n} x_i^2 - \left(\sum_{i=1}^{n} x_i \right)^2}{n(n-1)}}$$

is a legitimate alternative to that found in Definition 2.8. A square root table is also provided in the Appendix (see Table B).

---

*If $\bar{x}$ is rounded before it is applied in Definition 2.7, each squared deviation contributes a small error in the result. No such rounding errors can occur with the Calculation Formula.

Exercises

1. Calculate the range, mean deviation, variance, and standard deviation for the sample data in Exercise 1, page 24.

2. Calculate the sample variance of x, where the measurements are 23, 29, 25, 24, 25, using
   (a) Definition 2.7;
   (b) Theorem 2.3 (Calculation Formula).

3. Find the variance and standard deviation for the college grade-point averages in Exercise 3, page 24.

4. A particular stock was quoted at the following prices during the month of April.

10.22	11.09	11.61	13.00	11.95
10.51	10.82	12.26	12.98	12.01
10.66	10.92	12.87	12.90	12.22
10.75	11.55	12.63	11.88	12.18
10.64	11.54	12.82	11.89	12.31

   Find the mean, variance, and standard deviation of these data.

5. (a) Can the sample variance be negative?
   (b) Under what circumstances, if any, could the sample variance be zero?
   (c) Would you answer (a) and (b) the same way if we were talking about the standard deviation rather than the variance?

6. Show by expanding the summations that
   $$\sum_{i=1}^{3} x_i^2 \neq \left(\sum_{i=1}^{3} x_i \right)^2$$

7. If the variance for 10 measurements is 32.6 and the variance for 20 other measurements is 36.2, find the overall variance for the 30 measurements.

8. If the standard deviation for five measurements is 8.0 and the standard deviation for 12 other measurements is 7.5, find the overall standard deviation for the 17 measurements. Assume that the mean for the 5 measurements and for the 12 measurements are equal.

9. Over a two-month period a housewife collected data on the price per pound (x) of porterhouse steak in three different food stores—10 observations in each. The sample mean and sample variance for each store are listed below. If the housewife is willing to wait for a bargain price on porterhouse steak, from which store is she most likely to get it?

	$\bar{x}$	s_x^2
Store #1:	$1.35	$0.20
Store #2:	1.34	0.24
Store #3:	1.36	0.32

2.4   Transforming the Data

Occasionally we have the opportunity to transform the data in order to simplify our statistical calculations. For example, the data presented earlier for Sample B might be transformed by subtracting three from each measurement. That is, the original data $(x_i s)$

3, 4, 5, 5, 5, 5, 6, 7

would now become new data ($u_i s$, where $u_i = x_i - 3$)

0, 1, 2, 2, 2, 2, 3, 4

Consider the effect this *transformation* would have on the sample mean, the sample variance, and the sample standard deviation.

u_i	u_i^2
0	0
1	1
2	4
2	4
2	4
2	4
3	9
4	16
Σ 16	42

$$\bar{u} = \frac{16}{8} = 2.00$$

$$s_u^2 = \frac{8(42) - (16)^2}{8(7)} = \frac{80}{56} = 1.43$$

$$s_u = \sqrt{1.43} = 1.20$$

For this transformation ($u_i = x_i - 3$) $\bar{u}$ is three less than $\bar{x}$, $s_u^2 = s_x^2$, and $s_u = s_x$. In other words, the subtraction of three from each measurement in the sample subtracts three from the mean but does not affect the variance or standard deviation. We can see that this makes sense by looking at Figure 2.1 which compares the original data to the transformed data on the number line. We can see that central location shifts to the left three units, but dispersion is unaffected.

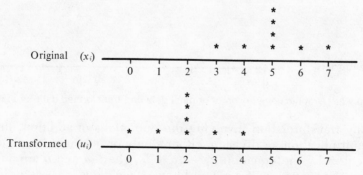

Figure 2.1. Graphic comparison of original data and transformed data for $u_i = x_i - 3$.

Consider a different transformation; suppose we were to multiply each measurement in the original Sample B data by two. The result would be the transformed data (u_is, where $u_i = 2x$)

6, 8, 10, 10, 10, 10, 12, 14

The values of the mean, variance, and standard deviation of this transformed data reveal some interesting effects.

u_i	u_i^2
6	36
8	64
10	100
10	100
10	100
10	100
12	144
14	196
Σ 80	840

$$\bar{u} = \frac{80}{8} = 10.00$$

$$s_u^2 = \frac{8(840) - (80)^2}{8(7)} = \frac{320}{56} = 5.71$$

$$s_u = \sqrt{5.71} = 2.39$$

Note that, except for some error introduced in rounding, the effect of multiplying every measurement in the original data by two is to increase the mean and standard deviation by a factor of two and increase the variance by a factor of $(2)^2$. Thus, multiplying each measurement by a constant affects the standard deviation and variance as well as the mean. These effects can be observed in Figure 2.2; central location clearly shifts to the right, and the dispersion is obviously greater in the transformed data.

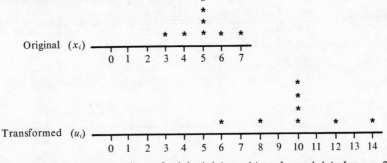

Figure 2.2. Graphic comparison of original data and transformed data for $u_i = 2x_i$.

Any transformation involving the operations of addition, subtraction, multiplication, or division, either singly or in combination, can be represented by the equation $u_i = ax_i + b$, where a and b are any real numbers. For instance, if $a = 1$ and $b = -3$, we get $u_i = x_i - 3$; or if $a = 2$ and $b = 0$, we get $u_i = 2x_i$. Mathematicians call a transformation of the form $u_i = ax_i + b$ a *linear transformation*. The following theorem summarizes the effects which a linear transformation has on the mean, variance, and standard deviation.

Theorem 2.4: Given a sample of n measurements x_1, x_2, x_3, . . . , x_n, if $u_i = ax_i + b$, then $\bar{u} = a\bar{x} + b$, $s_u^2 = a^2 s_x^2$, and $s_u = |a|s_x$.

Proof:

$$\bar{u} = \frac{\sum_{i=1}^{n} u_i}{n} \qquad \text{(Def. 2.1)}$$

$$= \frac{1}{n} \sum_{i=1}^{n} (ax_i + b) \qquad (u_i = ax_i + b, \text{ given})$$

$$= \frac{1}{n} \left[\sum_{i=1}^{n} ax_i + \sum_{i=1}^{n} b \right] \qquad \text{(Dist. Prop. for } \Sigma)$$

$$= \frac{1}{n} \left[a \sum_{i=1}^{n} x_i + nb \right] \qquad \text{(Thms. 1.2 and 1.3)}$$

$$= a \cdot \frac{\sum_{i=1}^{n} x_i}{n} + \frac{nb}{n} \qquad \text{(Mult. by } 1/n)$$

$$\therefore \quad \bar{u} = a\bar{x} + b \qquad \text{(Def. 2.1)}$$

$$s_u^2 = \frac{\sum_{i=1}^{n} (u_i - \bar{u})^2}{n - 1} \qquad \text{(Def. 2.7)}$$

$$= \frac{1}{n - 1} \sum_{i=1}^{n} \left[(ax_i + b) - (a\bar{x} + b) \right]^2 \qquad \begin{array}{l} (u_i = ax_i + b, \\ \text{given, and} \\ \bar{u} = a\bar{x} + b, \\ \text{just proved)} \end{array}$$

$$= \frac{1}{n - 1} \sum_{i=1}^{n} \left[ax_i + b - a\bar{x} - b \right]^2$$

$$= \frac{1}{n - 1} \sum_{i=1}^{n} \left[ax_i - a\bar{x} \right]^2$$

$$= \frac{1}{n - 1} \sum_{i=1}^{n} \left[a(x_i - \bar{x}) \right]^2$$

$$= \frac{1}{n - 1} \sum_{i=1}^{n} a^2 (x_i - \bar{x})^2$$

$$= a^2 \left[\frac{1}{n - 1} \sum_{i=1}^{n} (x_i - \bar{x})^2 \right] \qquad \text{(Thm. 1.2)}$$

$$\therefore \quad s_u^2 = a^2 s_x^2 \qquad \text{(Def. 2.7)}$$

$$s_u = \sqrt{s_u^2} \qquad \text{(Def. 2.8)}$$

$$= \sqrt{a^2 s_x^2} \qquad (s_u^2 = a^2 s_x^2, \text{ just proved)}$$

$$= \sqrt{a^2} \cdot \sqrt{s_x^2}$$

$$= \sqrt{a^2} \cdot s_x \qquad \text{(Def. 2.8)}$$

$$\therefore \quad s_u = |a|s_x$$

In summary, Theorem 2.4 tells us that a transformation of the form $u_i = ax_i + b$ (1) affects the mean ($\bar{x}$) in exactly the same way it affects each measurement (x_i), (2) affects the variance (s_x^2) by a factor of a^2, and (3) affects the standard deviation by a factor of $|a|$.

Probably the most useful transformations involve the subtraction and/or division by a positive constant, since these operations tend to reduce the numerical values in the data and thus simplify statistical calculations. In a typical situation, an appropriate transformation is chosen, its application generates new data, sample statistics are calculated for the new data, and then these values are adjusted (using Theorem 2.4) to find the values of the corresponding statistics for the original data.

For example, we may wish to find the mean, variance, and standard deviation of the following data (x_is):

95, 85, 95, 90, 90, 100, 85, 90

By subtracting 90 from each measurement and then dividing each result by 5, we get the transformed data (u_is):

1, −1, 1, 0, 0, 2, −1, 0

This transformation: $u_i = (x_i - 90)/5$ can be written $u_i = 1/5\, x_i + (-18)$; that is, it can be expressed as $u_i = ax_i + b$, where $a = 1/5$ and $b = -18$. Therefore, it is a linear transformation and its effect on the mean, variance, and standard deviation of x is expressed by Theorem 2.4. Next, we calculate the mean, variance, and standard deviation for the new data.

u_i	u_i^2	
−1	1	$\bar{u} = \dfrac{2}{8} = .25$
−1	1	
0	0	
0	0	$s_u^2 = \dfrac{8(8) - (2)^2}{8(7)} = \dfrac{60}{56} = 1.07$
0	0	
1	1	
1	1	$s_u = \sqrt{1.07} = 1.03$
2	4	
Σ 2	8	

But we want to find the mean, variance, and standard deviation of x, not u. So, we apply Theorem 2.4 to get

$$\bar{u} = \frac{1}{5}\bar{x} + (-18), \qquad s_u^2 = \left(\frac{1}{5}\right)^2 s_x^2, \qquad \text{and} \qquad s_u = \frac{1}{5}s_x$$

Substituting for $\bar{u}$, s_u^2, and s_u, we find

$$\bar{x} = 91.25, \qquad s_x^2 = 26.75, \qquad \text{and} \qquad s_x = 5.15$$

Exercises

1. Express the following transformations in linear transformation form (i.e., in the form $u_i = ax_i + b$).
   (a) $u_i = x_i - 50$ (b) $u_i = (10x_i + 25) \div 5$
   (c) $u_i = \dfrac{x_i - 2}{10}$

2. Given $\bar{x} = 50$ and $s_x^2 = 16$, find $\bar{u}$, s_u^2, and s_u, where $u_i =$
   (a) $2x_i + 5$ (b) $x_i - 25$
   (c) $\dfrac{x_i}{10}$ (d) $\dfrac{x_i - 50}{4}$

3. Transform the data in Exercise 2, page 32, by subtracting 25 from each of the measurements, and then again find the variance of x.

4. Calculate $\bar{x}$, s_x^2, and s_x for the following data using a convenient transformation: 426, 424, 425, 430, 422, 425, 424, 428.

5. Variable y is measured on a certain sample and the data collected is transformed by first multiplying each measurement by 2 and then subtracting 50 (i.e., $u_i = 2y_i - 50$). For the new data $\bar{u} = 2$ and $s_u^2 = 1.44$. Find the mean, variance, and standard deviation of y.

6. In the previous exercise suppose that the transformed variable u has median $= 3$, mode $= 6$, and range $= 8$. Find the median, mode, and range of y.

7. Given a sample of size n with measurements $x_1, x_2, x_3, \ldots, x_n$. Suppose that for each measurement we subtract $\bar{x}$, and then divide the result by s_x. That is, we transform the data according to the equation

$$u_i = \frac{x_i - \bar{x}}{s_x}.$$

   Find $\bar{u}$, s_u^2, and s_u. (Hint: Look carefully at Exercise 2(d) above.)

8. The average daily temperatures (centigrade) were recorded for the month of May. For this sample of 31 measurements, the mean is 18 and the standard deviation is 5. Find the mean, variance, and standard deviation of these temperatures on the *farenheit* scale.
   [Farenheit (F) and centigrade (C) are equated by the relation: $F = \frac{9}{5}C + 32$.]

Summary

Typically, the investigator wishes to answer questions relating to the characteristics of a large group of objects or people, called the *population*. When an exhaustive examination of every element in the population is either unfeasible or impossible, a subset of the population, called a *sample*, is selected for examination. Two important characteristics of a sample are its *central location* and its *dispersion*. Sample statistics which measure central location include the *mean, median,* and *mode*. The *range, mean deviation, variance,* and *standard deviation* are statistics designed to measure dispersion. Whenever feasible, a *transformation* of the sample data may be employed to simplify calculations.

3

Methods of Grouping Data

3.1 Grouped vs. Ungrouped Data

Consider the sample data in Table 3.1. The measurements represent weight-loss in pounds for 90 women who have been on a special diet for 60 days. A casual glance at this collection of data does not readily reveal the major characteristics of the sample. For example, it is not immediately obvious: (a) what weight-loss would be near the average, (b) what the extreme values are, nor (c) whether spread between the extremes is relatively uniform or concentrated at the center, or some other location. Data in this form is called *ungrouped* data.

The tabulation displayed in Table 3.2 suggests a grouping of the above data. This grouping is probably more informative to the observer than the original ungrouped data. For example, we see that nine women lost from 5 to 9 pounds, 19 women lost 10 to 14 pounds, and so on. We can quickly see that the central location in the sample is probably somewhere between 15 and 19 pounds, that the extremes are (at the outside) 5 and 44, and that the spread is not uniform throughout but concentrated in and around the 15 to 19 interval.

Table 3.1.   Weight-Loss Data—"Ungrouped"

15	(8)	12	18	44	30	15	18	23	(6)
23	16	20	17	21	12	12	23	25	13
19	17	17	28	13	17	17	28	18	16
20	(7)	14	(8)	15	27	10	19	13	15
18	10	(8)	11	16	40	18	21	14	27
15	32	28	22	10	(9)	18	12	25	25
18	20	21	18	18	16	(9)	(8)	21	17
29	23	14	14	25	15	12	10	20	16
24	19	15	11	21	12	15	(8)	17	19

Table 3.2.   Weight-Loss Data—"Grouped"

Weight-Loss	Tally	Frequency
5–9	ʃʋʅ ////	9
10–14	ʃʋʅ ʃʋʅ ʃʋʅ ////	19
15–19	ʃʋʅ ʃʋʅ ʃʋʅ ʃʋʅ ʃʋʅ ʃʋʅ ///	33
20–24	ʃʋʅ ʃʋʅ ʃʋʅ	15
25–29	ʃʋʅ ʃʋʅ	10
30–34	//	2
35–39		0
40–44	//	2
		90

Although it is true that grouping the data in this fashion is clearly more useful for purposes of descriptive presentation, there is the disadvantage that *some information has been lost* in the process. For example, the grouped distribution tells us that nine women lost from five to nine pounds, but it does not tell us the *exact* weight-loss for each of them. Therefore, it is fair to say that the grouped data above provides useful descriptive information (on sight) at the expense of some specific details. When ungrouped data is grouped, the resulting tabulation is called a *frequency distribution*.

Although the method of grouping data just demonstrated involves some loss of information, it is not true that information must always be lost when ungrouped data is grouped into a frequency distribution. For example, a sample of 60 families is investigated and the number of dependent children in each family is recorded (see Table 3.3). In Tables 3.4 and 3.5 the data is grouped in two distinct ways. The distribution in Table 3.4 tells us, for example, that 13 families have either four or five dependent children. Table 3.5, however, has preserved the original information that exactly eight families have four children and exactly five have five children.

Table 3.3. Number of Dependent Children for 60 Families

2	4	3	1	2	6	3	1	3	4	2	2
0	2	2	3	0	3	5	3	1	2	4	3
3	3	1	2	3	4	3	2	4	0	3	1
6	1	3	5	3	7	1	5	2	3	1	4
3	3	4	3	5	2	4	1	2	3	5	3

Table 3.4. Frequency Distribution One (information lost)

Interval	Frequency
0–1	12
2–3	32
4–5	13
6–7	3
	60

Table 3.5. Frequency Distribution Two (no information lost)

Interval	Frequency
0	3
1	9
2	12
3	20
4	8
5	5
6	2
7	1
	60

Grouping in intervals (as in Table 3.4) is not a unique operation; that is, the length and number of intervals are arbitrarily chosen from several possibilities. It is clear, however, that interval-grouping loses information.* On the other hand, we can always choose to group data without losing information (as in Table 3.5) by listing all of the different measurements that occur and the frequencies with which they occur. Grouping in this manner is no real advantage, however, when x takes on a great number of different values since the resulting table would have too many entries to be useful. When this situation occurs, interval-grouping is an acceptable and popular alternative, even though some information is lost.

3.2 Constructing a Frequency Distribution

Since interval-grouping involves several arbitrary choices, we proceed in this section to outline the ground rules for making these choices.

---

*By "loses information" we mean that in a sample of n measurements we no longer know (from the frequency distribution) the exact values of the n measurements.

3.2.1 Basic Procedure

To proceed from the ungrouped data (Table 3.1) to the frequency distribution (Table 3.2), we follow a few simple steps:

(a) *Find the sample range;* that is, we find the smallest and largest measurements in the sample and subtract the former from the latter. For the weight-loss data we get $44 - 6 = 38$.

(b) *Decide how many grouping intervals we want and how wide each should be.* To make these decisions we follow two rules of thumb: (1) *the number of intervals should be no less than 6 and no more than 12* (the decision depends upon the size of the sample and the length of interval desired), and (2) *the interval length should be the same for each class* *—a whole number, if possible.* If we divide the range (38) by 6 we get 6.33, whereas 38 divided by 12 is 3.17. Therefore, if the *number* of intervals is to be between 6 and 12 (inclusive), then the *length* of each interval must be a number between 3.17 and 6.33. The whole number five is chosen arbitrarily as the length of each interval. Since $38/5 = 7.6$, we need at least eight intervals of length five to group the given data.

(c) *List the intervals, tally the number of measurements which fall into each interval, and tabulate the results.* Rules of thumb are (1) *the first interval must include the smallest measurement,* (2) *the interval limits are expressed to the same degree of accuracy as the measurements themselves,* and (3) *there must be no gaps nor overlaps between intervals.* For example, for the weight-loss data, we begin with the interval 5–9 (6 is the smallest measurement). The limits, 5 and 9, represent weight-loss to the nearest pound (same as all the measurements). The second interval is 10–14, not 9–13 (overlapping) nor 11–15 (leaving a gap).

We say that the length of each interval in this frequency distribution is five. To justify this statement we note that an *actual* weight-loss x, such that $4.5 \le x < 5.5$ would be recorded as "5" (if measured to the nearest whole number), an *actual* weight-loss x, such that $5.5 \le x < 6.5$ would be measured as "6," etc. Thus, the interval 5–9 really represents weight-loss x such that $4.5 \le x < 9.5$, and the difference between 9.5 and 4.5 is 5—the length of the interval 5–9. The same reasoning can be applied to the other intervals.

---

*We insist on equal interval lengths (1) to eliminate visual misinterpretations of the data, and (2) to simplify statistical calculations (Chapter 4).

3.2.2 Basic Terminology

The grouping intervals shall be called *classes*. The numbers used to express each class are called *upper* and *lower class limits*. The length of each class is called the *class length* (equal for each class). The value at the midpoint of each class is called the *class mark* for that class. For a particular class, the class mark is found easily by calculating the mean of the two class limits. Each class has two *class boundaries*. The *lower* class boundary is found by subtracting half of the class length from the class mark; the *upper* class boundary equals the class mark plus half the class length.

Figure 3.1 shows the relationship among the terms "class limit," "class mark," and "class boundary," using the weight-loss example. Note that the difference between any successive pair of class boundaries, any successive pair of class marks, any successive pair of upper class limits, and any successive pair of lower class limits equals the class length.

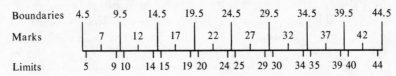

Figure 3.1. Class boundaries, limits, and marks for the weight-loss data.

Table 3.6 is an extended version of Table 3.2. Remember that the weight-loss data is expressed to the nearest whole pound. The class limits are therefore expressed in whole pounds. If the sample had been

Table 3.6. Frequency Distribution° for the Weight-Loss Data

Class Boundaries	Class Limits	Class Mark (x_i)	Frequency (f_i)	Percentage
4.5–9.5	5–9	7	9	10.0
9.5–14.5	10–14	12	19	21.1
14.5–19.5	15–19	17	33	36.7
19.5–24.5	20–24	22	15	16.7
24.5–29.5	25–29	27	10	11.1
29.5–34.5	30–34	32	2	2.2
34.5–39.5	35–39	37	0	0.0
39.5–44.5	40–44	42	2	2.2
			90	100.0

°Actually the "Class Limits" and "Frequency" columns alone constitute a frequency distribution. The other columns, representing additional information, can be generated from the limits and frequencies.

measured to the nearest *tenth* of a pound, the class limits would have been 5.0–9.9, 10.0–14.9, 15.0–19.9, etc., and therefore the class boundaries would be 4.95, 9.95, 14.95, and so on. If the data had been measurements to two decimal places (i.e., to the nearest hundredth of a pound), the class limits would have been 5.00–9.99, 10.00–14.99, 15.00–19.99, and so on, and the class boundaries: 4.995, 9.995, 14.995, etc.

It should be noted also that due to the above mentioned restrictions it is impossible for the class length to be expressed to a greater degree of precision than the original measurements. For the original weight-loss sample (measured to the nearest whole pound) we could not have used, for example, a class length of 4.8.

An additional column appears in Table 3.6 labelled "Percentage." The percentage for each class is found by dividing the frequency (f_i) by the sample size (90, in this sample) and then multiplying by 100. For the class with limits 20–24, percentage = $(15/90) \times 100 = 16.7$, which means that 16.7 percent of the women in the sample lost anywhere from 20 to 24 pounds.

A variation on the frequency distribution is the *cumulative frequency distribution*. The *cumulative frequencies* are simply accumulations of frequencies from either low values of the variable to high (Table 3.7), or vice versa (Table 3.8). From Table 3.7, for example, we learn that 76 women (84.4 percent) lost *less than* 25 pounds. And from Table 3.8 we learn that 62 women (68.9 percent) lost *more than* 14 pounds.[*] The percentages are calculated the same way as in a frequency distribution.

Table 3.7.   Cumulative Frequency Distribution for Weight-Loss Data (less than)

Weight-Loss	Cumulative Frequency	Percentage
less than 5	0	0.0
less than 10	9	10.0
less than 15	28	31.1
less than 20	61	67.8
less than 25	76	84.4
less than 30	86	95.6
less than 35	88	97.8
less than 40	88	97.8
less than 45	90	100.0

[*] Note that the phrase "less than 5" in Table 3.7 could be replaced by "4 or less," "less than 10" by "9 or less," etc., with the same cumulative frequencies and percentages remaining. Similarly in Table 3.8 more than 4 could be replaced by 5 or more, and so forth.

Table 3.8. Cumulative Frequency Distribution
for Weight-Loss Data (more than)

Weight-Loss	Cumulative Frequency	Percentage
more than 4	90	100.0
more than 9	81	90.0
more than 14	62	68.9
more than 19	29	32.2
more than 24	14	15.6
more than 29	4	4.4
more than 34	2	2.2
more than 39	2	2.2
more than 44	0	0.0

3.3 Graphical Representations

If we were to use the information given in Table 3.6 to *graph* class marks (x_i) against frequencies (f_i) in the usual sense, the graph shown in Figure 3.2 would result. However, the graphical representation known as the *frequency histogram* is a more common, and perhaps a more eye-catching, method of representing frequency distributions. We construct a frequency histogram from the basic graph by simply placing a rectangle on the horizontal axis (x_i) for each class in such a way that the height of each rectangle equals f_i, the base equals the class length, and the midpoint of each base is at the class mark. In addition, for simplicity, we usually label only the class boundaries on the x_i axis, rather than the class marks or class limits. The frequency histogram for the weight-loss data is pictured in Figure 3.3.

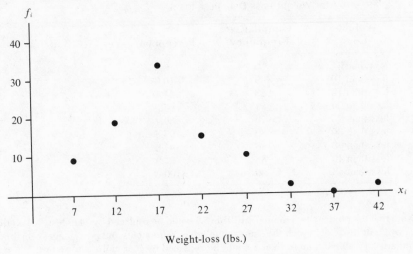

Weight-loss (lbs.)

Figure 3.2. Graph of weight-loss (x_i) vs. frequency (f_i).

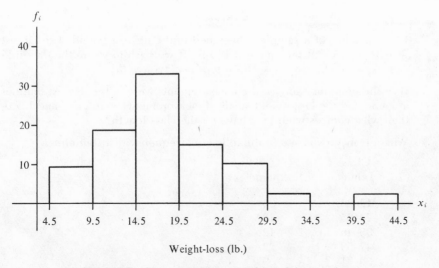

Figure 3.3. Frequency histogram for weight-loss data.

By representing a frequency distribution by a frequency histogram we are essentially viewing the total distribution in terms of *area*. Each rectangular area is proportional to the percentage of sample measurements which fall within the limits of that class. This area concept will prove very useful to us in later chapters.

A variation on the histogram is called the *frequency polygon*. It too may be constructed directly from the basic graph. We simply connect the points of the graph with straight lines. In addition, the two end points are connected to points on the x_i axis which are distances of one class length from the two extreme class marks. Figure 3.4 shows a frequency polygon for the weight-loss data.

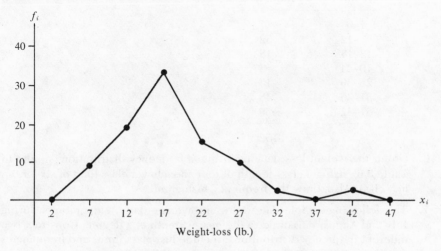

Figure 3.4. Frequency polygon for weight-loss data.

Exercises

1. Each member of a sample is weighed to the nearest pound. The largest weight is 215 and the smallest is 127. If we wish to group the data into classes, what choices do we have for the *class length*?

2. If in the previous exercise the measurements were to the nearest tenth of a pound and the largest and smallest measurements were 215.0 and 127.0, then what choices would we have for the class length?

3. What errors do you see in the following frequency distribution?

Class Limits	Frequencies
12–23	5
24–35	10
35–46	12
47–58	22
60–71	9
72–85	3
	61

4. A sample of measurements to the nearest hundredth of an inch are grouped in a frequency distribution which has class boundaries of: 15.995, 22.995, 29.995, 36.995, 43.995, 50.995, and 57.995. Find the class limits, class marks, and class length for this distribution.

5. For the given frequency distribution:
   (a) Describe approximately the central location, the extremes, and the dispersion, without the aid of statistical calculations.
   (b) List the class boundaries and class marks, and find the class length.

Class Limits	Frequency
1–6	7
7–12	52
13–18	18
19–24	3
25–30	15
31–36	48
37–42	9
	152

6. Group the weight-loss data again into a frequency distribution (similar to Table 3.6), using a class length of four and a lower class limit of six for the first class. Also, draw the frequency histogram.

7. The following sample data represent average daily temperatures during July and August measured to the nearest tenth of a degree. Construct two different frequency distributions (include just class limits and frequencies

in each case), one with a class length of 2.0 and the other with a class length of 3.5. In each case, make the lower class limit in the first class 72.0.

80.1	78.2	75.5	86.1	80.5	76.0	83.7	95.4	81.8	78.1	90.7
74.6	72.6	81.2	80.7	82.9	79.1	87.1	91.1	83.1	79.0	89.3
78.9	78.5	80.2	82.2	83.1	84.2	86.5	81.9	83.7	85.2	
85.0	73.1	89.0	80.1	82.5	84.3	80.1	85.4	87.4	80.7	
83.1	82.9	87.7	90.9	77.8	83.3	92.2	85.7	82.6	85.1	
84.4	80.2	88.4	87.7	75.2	82.7	90.8	84.0	82.1	88.0	

8. The diameters of 54 ball bearings are measured to the nearest thousandth of an inch. Construct a frequency distribution, including boundaries, limits, marks, frequencies, and percentages, in which the limits for the first class are 0.425–0.427. What is the class length?

.440	.434	.435	.434	.435	.437	.430	.436	.427
.432	.438	.426	.434	.441	.439	.433	.435	.430
.427	.437	.431	.429	.440	.435	.434	.427	.437
.432	.434	.441	.435	.430	.438	.425	.428	.434
.441	.426	.433	.445	.436	.437	.432	.435	.438
.444	.436	.442	.431	.432	.439	.428	.437	.439

9. The weekly salaries in dollars for a sample of 64 factory workers are listed below. Construct a frequency distribution, including boundaries, limits, marks, frequencies, and percentages.

96.56	110.15	101.50	116.15	151.22	99.10	110.94	104.44
106.25	92.35	132.15	108.12	128.80	138.52	98.60	107.47
98.50	106.20	121.00	109.72	114.45	122.61	118.21	101.00
98.50	102.86	95.01	131.60	118.21	108.26	115.06	134.62
124.22	88.96	109.95	91.20	95.55	106.12	108.76	120.89
100.50	115.28	101.50	98.74	112.60	107.72	102.41	118.23
106.32	141.20	97.75	100.05	105.50	111.62	104.78	114.06
150.65	99.10	128.01	119.85	113.12	109.25	123.75	96.81

10. For the frequency distributions in Exercises 5 and 8:
    (a) Construct a "less than" and a "more than" cumulative frequency distribution (including the percentage columns).
    (b) Draw the frequency histogram.
    (c) Draw the frequency polygon.

11. We have introduced the frequency histogram and frequency polygon in terms of interval-grouped distributions only. We can, however, construct these same graphical representations for frequency distributions such as the one in Table 3.5 (in which no information has been lost). In such cases we simply think of each value of x as the "class mark" of a very short class interval and proceed logically from there. Draw the frequency histogram and frequency polygon for the frequency distribution in Table 3.5. (Label the class marks rather than the class boundaries in the histogram.)

Summary

Ungrouped data may be grouped into a *frequency distribution*. This process can always be done without losing information; in most cases, however, the construction of a useful frequency distribution invariably involves grouping into intervals, or *classes*, and thus the exact values of the original measurements are lost. Although the grouping process is somewhat arbitrary in practice, certain basic procedures and rules of thumb are outlined. Basic to the structure of an interval-grouped frequency distribution are its *class limits, class boundaries, class marks,* and *class length*. The most common graphical representations of frequency distributions are the *frequency histogram* and the *frequency polygon*.

4

Describing Grouped Data

We have discussed ways of describing sample data in its ungrouped form, and have developed procedures for grouping data into frequency distributions. We noted that grouping can be accomplished without loss of information, but more often than not, it is more convenient to group into intervals (classes)—a procedure which does lose some information. We shall now consider methods for describing data which have been grouped into a frequency distribution. As we might expect, if the data have been grouped with no loss of information, we can use the same formulas (or equivalent versions) defined in Chapter 2 to describe central location and dispersion. If, on the other hand, we group into interval classes, thus losing information, we need to develop statistical measures which when applied to the frequency distribution will give us values that closely approximate the corresponding statistical descriptions of the ungrouped data. In other words, it is still our goal to describe the original (ungrouped) data, but we want to do this indirectly by using only the information provided by the frequency distribution. If no information has been lost in the process

49

of grouping, there is no problem—the old formulas still apply. If some information has been lost, however, we need new or altered formulas that will give us close approximations to the statistical descriptions which we are unable to calculate directly from the ungrouped data. In this chapter we shall concentrate on the most frequently used sample statistics—namely, the median, mean, variance, and standard deviation.

4.1   When No Information Is Lost

Let us consider the following sample data for the variable x in both its ungrouped form, 17, 14, 13, 15, 17, 17, 15, 13, 18, 17, 15, 17, 16, 15, 16; and in the following grouped form,

x_i	f_i
13	2
14	1
15	4
16	2
17	5
18	1
	15

Since no information has been lost in this grouping, we can apply the formulas for the various measures of central location and dispersion (defined in Chapter 2) to either form of the data and get identical results. It is true, however, that for some sample statistics (specifically the mean, variance, and standard deviation) the working formulas introduced earlier can be altered into forms consistent with the frequency distribution format, thus giving us more convenient tools for the job.

4.1.1   Measures of Central Location

We need no new formulas for calculating the sample *mode* and sample *median* from the frequency distribution above. The mode is clearly 17, since it occurs more frequently (five times) than any other value of x. To find the median we note that the total frequency (total number of measurements in the sample) is 15, and therefore the median is the $((15 + 1)/2)$th, or eighth, measurement in rank order (Definition 2.2). We see from the grouped data that a total of seven measurements have values of 15 or less. Therefore, the eighth measurement must be 16, the median.

Similarly, we need no new formulas to calculate the sample *mean* from the frequency distribution. Definition 2.1 tells us that we find $\bar{x}$ by adding the 15 values of x and then dividing the sum by 15. From the frequency distribution we know that we have two values of 13, one value of 14, four values of 15, and so on, and therefore

$$\bar{x} = (13 + 13 + 14 + 15 + 15 + 15 + 15 + 16 + 16 + 17 + 17 + 17$$
$$+ 17 + 17 + 18)/15 = 15.67$$

At this point, however, we see that it is possible to shorten the work, noting that $13 + 13 = (13)(2)$, $14 = (14)(1)$, $15 + 15 + 15 + 15 = (15)(4)$, etc., and thus

$$\bar{x} = [(13)(2) + (14)(1) + (15)(4) + (16)(2) + (17)(5) + (18)(1)]/15$$
$$= 15.67$$

Note that we have essentially applied the following formula to the frequency distribution.

$$\bar{x} = \frac{\sum_{i=1}^{6} x_i f_i}{15}$$

In other words, we found the total sum of the measurements by finding the value "$x_i f_i$" for each of the six different values of x, and then adding the results. This observation leads us to an alternate form of Definition 2.1, to be applied to frequency distribution in which no information has been lost.

Definition 4.1:   In a sample of n measurements on variable x, the *sample mean* of x is

$$\bar{x} = \frac{\sum_{i=1}^{k} x_i f_i}{n}$$

where k is the number of *different* values of x in the sample, each x_i represents a different value of x, and each f_i represents the number of times x_i occurs in the sample.

The reader should compare Definition 4.1 to Definition 2.1, noting that the only difference in the formula is the substitution of

$$\sum_{i=1}^{k} x_i f_i \quad \text{for} \quad \sum_{i=1}^{n} x_i$$

Remember, in the ungrouped situation we sum over all n measurements of x, but when the data is grouped, we sum over the k different values of x (each multiplied by its own frequency).

4.1.2 Measures of Dispersion

Referring to the same grouped data (page 50), we see that the sample *range* is easily found to be 5 (18 minus 13) and sufficient information is given for us to find the other measures of dispersion, using the formulas in Definition 2.7 (sample *variance*) and Definition 2.8 (sample *standard deviation*).* However, for simplicity in calculation, we can express each formula in an equivalent form applicable directly to the grouped data.

Definition 4.2:   In a sample of n measurements on variable x, where k is the number of different values of x, each x_i represents a different value of x ($i = 1, 2, 3, \ldots, k$), and each f_i represents the number of times x_i occurs,

$$\text{Sample Variance} = s_x^2 = \frac{\sum_{i=1}^{k} (x_i - \bar{x})^2 f_i}{n - 1}$$

$$\text{Sample Standard Deviation} = s_x = \sqrt{s_x^2} \text{ (as always)}$$

As before (Theorem 2.3), the formula for sample variance has a form more suitable for calculation purposes. Note the similarity between Theorem 2.3 and the following theorem.

Theorem 4.1:

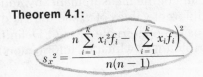

$$s_x^2 = \frac{n \sum_{i=1}^{k} x_i^2 f_i - \left(\sum_{i=1}^{k} x_i f_i \right)^2}{n(n-1)}$$

Proof: Left to the reader.

In applying Theorem 4.1 to grouped data to find the sample variance, it is customary to do the work in tabular form, as shown in Table 4.1. Therefore,

$$s_x^2 = \frac{15(3715) - (235)^2}{15(14)}$$

$$= 2.38$$

The reader should verify that Theorem 2.3 (applied to the *ungrouped* data) also gives us $s_x^2 = 2.38$.

Table 4.1.   Sums Needed for Calculation of s_x^2

x_i	f_i	$x_i f_i$	$x_i^2 f_i$
13	2	26	338
14	1	14	196
15	4	60	900
16	2	32	512
17	5	85	1445
18	1	18	324
Σ	15	235	3715

*Since the sample *mean deviation* is not very often used in practice, we shall not bother to discuss it further in this text, except on occasion in the exercises.

Exercises

1. For the frequency distribution in Table 3.5, let x represent the number of dependent children in the family. Then, using the definitions in this chapter, find the
   (a) mean, median, and mode of x.
   (b) range, variance, and standard deviation of x.

2. Group the following sample data into a frequency distribution without losing information, and then calculate the mean, median, mode, range, variance, and standard deviation from the grouped data. The data represent time (x) in seconds that it takes each of 56 boys to run 50 yards.

6.6	7.6	7.4	7.0	7.1	7.5	6.7
6.8	6.9	7.3	7.1	7.2	7.3	6.8
7.2	7.0	7.5	7.0	7.1	7.0	7.0
6.8	7.1	6.9	7.5	7.1	6.8	7.2
7.0	7.2	7.1	7.2	7.4	7.1	7.1
7.3	7.0	7.2	6.8	6.9	7.0	7.4
6.7	6.9	7.5	7.0	7.3	7.1	6.9
6.9	7.4	7.4	7.2	7.1	7.2	7.0

3. Find the sample variance of x (s_x^2) for the following frequency distribution, using (a) Definition 4.2, and (b) Theorem 4.1.

x_i	f_i
5.5	4
6.0	5
6.5	6
7.0	2
7.5	2
8.0	1
	20

4. The sample *mean deviation* of x for grouped data is defined as

$$\frac{\sum_{i=1}^{k} |x_i - \bar{x}| f_i}{n}$$

where the symbols have the same meaning as in Definitions 4.1 and 4.2.
   (a) Find the sample mean deviation for the data in Exercise 1.
   (b) Which of the other statistics calculated in Exercise 1 comes closest to measuring the same thing as the mean deviation?

4.2 When Information Is Lost (Interval Grouping)

We now turn to the problem of describing the original ungrouped data, using only the information supplied by a frequency distribution which has been constructed in such a way that information has been lost. That is, the data are grouped into intervals so that the exact values of the n measurements cannot be determined from the frequency distribution.

Reflecting back on the various formulas already defined for describing sample characteristics, we see that not a single one of these formulas can be directly applied to this type of frequency distribution. The information which has been lost is exactly the information called for in the formulas! Not only can we not legitimately apply the formulas already defined, but no matter what formulas we ultimately decide to apply to the frequency distribution (for the purpose of describing the original ungrouped data), the statistical values thus obtained must be thought of as *estimates* of values defined on the ungrouped data.

Before developing these new formulas, we should note that with the advent of the electronic computer age, any discussion of the calculation of statistical measures from frequency distributions is considered by some authors as hardly worth the trouble. They argue that the computer can quickly and efficiently calculate the desired statistics from the ungrouped data (which is, after all, what we are after), and thus avoid the effort of grouping the data and eliminate the chance that the measures calculated from the grouped data are not close enough to the exact values that we are after. On the contrary, this author believes that the discussion in this section is worth the effort because (a) electronic computers are not always available, (b) sometimes the data we want to analyze comes to us already "grouped" (in which case the ungrouped data is not available at all), and (c) as we shall soon see, the "estimates" that we calculate from a frequency distribution are in most cases very close approximations to values we could attain if we were to calculate the corresponding descriptive measures directly from the ungrouped data.

4.2.1 Measures of Central Location

To estimate the *mode* (or modes) from an interval-grouped frequency distribution, we simply find the class mark of the class (or classes) with the highest frequency. For example, the frequency distribution for the weight-loss data (Table 3.6) shows that the class with limits "15–19" has the highest frequency (33). Therefore, our mode estimate is 17, the class mark. In addition, we call the class with the highest frequency the *modal class*.

To estimate the sample *median* from an interval-grouped frequency distribution, we try to locate in the frequency histogram a line perpendicular to the x_i axis which divides the total area of the histogram exactly in half. The median estimate is then defined as *the value of x_i at the point which that perpendicular line intersects the x_i axis*. To estimate the sample median we apply Definition 4.3.

Definition 4.3: The sample *median*, estimated from an interval-grouped frequency distribution, is

$$b + \frac{d}{f}(l)$$

where $b =$ the lower class boundary of the class containing the sample median estimate,

$f =$ the frequency of that same class,

$l =$ the class length,

$d =$ the difference between $n/2$ (n is the sample size) and the sum of the frequencies below b.

To apply Definition 4.3 we need to know which class contains the median estimate. In a frequency histogram the area is directly proportional to the frequency f_i. That is, for example, a class containing 25 percent of the sample is represented by a rectangle whose area is 25 percent of the total area of the histogram. Therefore, since half the area of the frequency histogram lies to the left of the point on the x_i axis known as the "median estimate," it must follow that this half-area represents one-half of the total frequency, or $n/2$.* Thus, the first step in applying Definition 4.3 is to find $n/2$. In the weight-loss frequency distribution we have 90 measurements, so that $n/2 = 45$. Since a total of 28 measurements fall in the first two classes and 61 in the first three classes, it is clear that the median estimate must lie in the third class (see Figure 4.1). The lower boundary for the third class is $b = 14.5$. The other three values needed for Definition 4.3 are $f = 33$, $l = 5$, and $d = 17$ (45 minus 28). Therefore, the estimated sample median for the weight-loss data is

$$14.5 + \frac{17}{33}(5) = 17.08$$

Note that this estimate is only slightly higher than the actual median (17), which can be found directly from the ungrouped data on page 39.

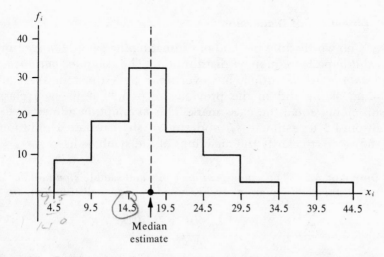

Figure 4.1. Position of median estimate in the weight-loss frequency histogram.

*Implicit in this reasoning is the assumption that measurements are uniformly distributed throughout each class.

To estimate the sample *mean* from an interval-grouped frequency distribution, we make the following assumption: *in each class all measurements equal the class mark.* That is, we assume for every class (although we know it is not true) that each of the f_i measurements in the *i*th class is equal in value to the class mark (x_i). With this assumption, we can essentially use Definition 4.1 to find an estimate of $\bar{x}$, although we need to restate the definition here because the symbols used now have slightly different meanings.

Definition 4.4: The sample *mean* of *x*, estimated from an interval-grouped frequency distribution, is

$$\hat{\bar{x}} = \frac{\sum\limits_{i=1}^{k} x_i f_i}{n}$$

where k = the number of classes,
x_i = the class mark of the *i*th class,
f_i = the frequency of the *i*th class,
n = the total frequency (sample size).

Although Definitions 4.4 and 4.1 appear on the surface to be identical, it should be carefully noted that (a) Definition 4.1 applies only to frequency distributions in which no information has been lost, while Definition 4.4 applies only to interval-grouped frequency distributions, (b) the symbols k, x_i, f_i, and n have slightly different meanings in the two definitions, and (c) Definition 4.1 provides an exact value for the sample mean, while Definition 4.4 gives only an estimate (although the estimate is usually quite good).

4.2.2 Measures of Dispersion

There is no worthwhile method of estimating the sample *range* from an interval-grouped frequency distribution. The sample *variance* and sample *standard deviation*, however, may be estimated. We start by assuming, as we did in the previous section, that in each class all measurements equal the class mark. This assumption allows us to use Definition 4.2 to estimate s_x^2 and s_x. We shall restate Definition 4.2 here, however, to clarify the meanings of the symbols used.

Definition 4.5: The sample *variance* of *x* and sample *standard deviation* of *x*, estimated from an interval-grouped frequency distribution, are

$$\hat{s}_x^2 = \frac{\sum\limits_{i=1}^{k} (x_i - \bar{x})^2 f_i}{n-1} = \frac{n \sum\limits_{i=1}^{k} x_i^2 f_i - \left(\sum\limits_{i=1}^{k} x_i f_i \right)^2}{n(n-1)}$$

$$\hat{s}_x = \sqrt{\hat{s}_x^2} \text{ (as always)}$$

where k = the number of classes,
x_i = the class mark of the ith class,
f_i = the frequency of the ith class,
n = the total frequency (sample size).

Compare Definitions 4.5 and 4.2 as you did with Definitions 4.4 and 4.1 in the previous section. Note that the Calculation Formula for $\hat{s}_x^2$ is included in Definition 4.5 for convenience.

In Table 4.2 the weight-loss frequency distribution is used to demonstrate the use of Definitions 4.4 and 4.5 in calculating estimates of the mean, variance, and standard deviation. Therefore,

$$\hat{\bar{x}} = \frac{1600}{90} = 17.78$$

$$\hat{s}_x^2 = \frac{90(32840) - (1600)^2}{90(89)} = 49.39$$

$$\hat{s}_x = \sqrt{49.39} = 7.03$$

These estimates are quite close to the actual values of $\bar{x}$, s_x^2 and s_x, which can be calculated directly from the ungrouped data (page 39). The actual values are $\bar{x} = 17.70$, $s_x^2 = 47.45$, and $s_x = 6.89$.

Calculations like those involved in the above example can be simplified by the use of appropriate transformations (discussed earlier in Chapter 2). In the next section we discuss a very special transformation used to simplify the calculation of the mean, variance, and standard deviation for an interval-grouped frequency distribution.

Before we go on, it should be noted that in this text we have used the term "estimate" in two different but related contexts. First, any sample statistic, calculated from either grouped or ungrouped data, is

Table 4.2. Sums Needed to Calculate $\hat{s}_x^2$ and $\hat{s}_x$ from the Grouped Weight-Loss Data

x_i	f_i	$x_i f_i$	$x_i^2 f_i$
7	9	63	441
12	19	228	2736
17	33	561	9537
22	15	330	7260
27	10	270	7290
32	2	64	2048
37	0	0	0
42	2	84	3528
Σ	90	1600	32840

thought of as an "estimate" of a corresponding measure in the population from which the sample was selected. This idea is discussed in Section 2.1, page 18. Secondly, we think of statistics that are calculated from interval-grouped frequency distributions as estimates of the corresponding statistics defined on ungrouped data. For example, the sample mean, $\bar{x}$, calculated from ungrouped data is considered an estimate of the mean of the population from which the sample was drawn. If the sample data are grouped into intervals, then $\hat{\bar{x}}$ (the sample mean calculated from the resulting interval-grouped frequency distribution) is considered an estimate of $\bar{x}$. However, $\hat{\bar{x}}$ is also an estimate of the true population mean. Both $\bar{x}$ and $\hat{\bar{x}}$ are estimates of the population mean; however, $\bar{x}$ is a better (more reliable) estimate than $\hat{\bar{x}}$. Although both $\bar{x}$ and $\hat{\bar{x}}$ are subject to random error (the sample is randomly selected from the population), $\hat{\bar{x}}$ is also subject to error introduced in the process of grouping the data into classes. This error is often almost negligible, especially in large samples; that is, $\bar{x}$ and $\hat{\bar{x}}$ are usually close in value for a particular sample. It is important, however, that the reader understand the distinction between $\bar{x}$ and $\hat{\bar{x}}$ (s_x^2 and $\hat{s}_x^2$, etc.) and the dual use of the term "estimate."

4.2.3   A Special Transformation

To simplify the calculations in the previous section we transform the variable x_i to the variable u_i according to the equation

$$u_i = \frac{x_i - m}{l}$$

where *m is any class mark* we wish to choose, and *l is the class length*. Note that this transformation is *linear* (refer back to page 34) since it can be expressed in the form

$$u_i = \frac{1}{l} x_i + \frac{-m}{l}$$

Next we estimate the mean, variance, and standard deviation of u_i; that is, we calculate $\hat{\bar{u}}$, $\hat{s}_u^2$, and $\hat{s}_u$ using Definitions 4.4 and 4.5. Finally, we make the necessary "adjustments" on $\hat{\bar{u}}$, $\hat{s}_u^2$, and $\hat{s}_u$ to give us $\hat{\bar{x}}$, $\hat{s}_x^2$, and $\hat{s}_x$. To make these adjustments we need to know the relationships between $\hat{\bar{x}}$ and $\hat{\bar{u}}$, $\hat{s}_x^2$ and $\hat{s}_u^2$, and $\hat{s}_x$ and $\hat{s}_u$. These relationships are expressed in Theorem 4.2.

Theorem 4.2:   Given an interval-grouped frequency distribution for the variable x, if x is transformed into u according to the equation $u_i = (x_i - m)/l$, where m is any class mark and l is the class length, then $\hat{\bar{x}} = l\hat{\bar{u}} + m$, $\hat{s}_x^2 = l^2\hat{s}_u^2$, and $\hat{s}_x = \sqrt{\hat{s}_x^2} = |l|\hat{s}_u$.

Proof: Left to the reader.

Note the similarity between Theorem 4.2 (applied to grouped data) and Theorem 2.4 (applied to ungrouped data, page 35). We see that this linear transformation has the same effect on grouped data as it would on ungrouped data with respect to the mean, variance, and standard deviation. The real saving afforded by this special transformation is obvious only with reference to actual experience. Therefore, we shall calculate $\hat{\bar{x}}$, $\hat{s}_x^2$, and $\hat{s}_x$ again for the weight-loss frequency distribution, this time using the transformation $u_i = (x_i - 17)/5$, where 17 is an arbitrarily chosen class mark, and five is the class length. Therefore, using the sums from Table 4.3, we get

$$\hat{\bar{u}} = \frac{\sum\limits_{i=1}^{8} u_i f_i}{n} = \frac{14}{90} = \frac{7}{45}$$

$$\hat{s}_u^2 = \frac{n \sum\limits_{i=1}^{8} u_i^2 f_i - \left(\sum\limits_{i=1}^{8} u_i f_i\right)^2}{n(n-1)}$$

$$= \frac{90(178) - (14)^2}{90(89)} = 1.9755$$

$$\hat{s}_u = \sqrt{1.9755} = 1.41$$

And consequently, according to Theorem 4.2,

$$\hat{\bar{x}} = 5\left(\frac{7}{45}\right) + 17 = 17.78$$

$$\hat{s}_x^2 = 5^2(1.9755) = 49.39$$

and

$$\hat{s}_x = \sqrt{49.39} = 7.03 \quad (\text{or } |5| \cdot 1.41 = 7.05^*)$$

Table 4.3. Sums Needed to Calculate $\hat{\bar{u}}$, $\hat{s}_u^2$ and $\hat{s}_u$ from the Grouped Weight-Loss Data

Class Mark x_i	Frequency f_i	$u_i = \dfrac{x_i - 17}{5}$	$u_i f_i$	$u_i^2 f_i$
7	9	-2	-18	36
12	19	-1	-19	19
17	33	0	0	0
22	15	1	15	15
27	10	2	20	40
32	2	3	6	18
37	0	4	0	0
42	2	5	10	50
Σ	90		14	178

*Rounding error is introduced since $\hat{s}_u$ was rounded to two decimal places.

This special transformation clearly saves calculation effort. It essentially converts class marks (x_is) to *small integers* (u_is), which make most of the multiplication simple. The choice for m is arbitrary, although choosing the class mark corresponding to the largest frequency (as we did) usually saves the most work. Why? The reader should do the calculations, using a different class mark for m in the transformation (see Exercise 3 below). The results should be identical.

Exercises

1. Find estimates of the mean, variance, and standard deviation of x for the following frequency distribution, using Definitions 4.4 and 4.5.

Class Mark x_i	Frequency f_i
4	6
7	13
10	20
13	15
16	7
19	5
22	2
25	2

2. Rework Exercise 1, using the special transformation (Theorem 4.2).

3. Rework the estimates of $\bar{x}$, s_x^2 and s_x for the weight-loss frequency distribution, using the special transformation with $m = 12$ (rather than 17).

4. Find estimates of the mean, median, mode, variance, and standard deviation of y for the following frequency distribution, using the special transformation where applicable.

Class Limits	f_i
10–19	5
20–29	12
30–39	24
40–49	35
50–59	18
60–69	7

   101

5. Find estimates of the mean, median, mode, variance, and standard deviation, using the special transformation where applicable, for the interval-grouped frequency distributions found in
   (a) Exercise 7, page 46.
   (b) Exercise 8, page 47.
   (c) Exercise 9, page 47.

6. It can be said that the estimate of $\bar{x}$, calculated from an interval-grouped frequency distribution, is in fact a weighted mean of the class marks. Explain this statement with reference to Theorem 2.1 and Definition 4.4.

4.3 Other Measures of Location

Intuitively, the sample median of x is, in essence, a value of x such that half the measurements have values less than x and half have values greater than x. This intuitive definition of the sample median is consistent with our formal definition for ungrouped data (Definition 2.2) and our definition for an estimate of the sample median from interval-grouped data (Definition 4.3). Our best way of visualizing the position of the sample median in the data is to draw a frequency histogram and label that point on the x axis through which a perpendicular line divides the total area in half (look again at Figure 4.1). If a student takes an exam and his score is at the median, he knows "intuitively" that 50 percent of the class got scores lower than his. When students take standardized examinations, such as the College Board examinations, their results are typically reported in terms of *percentile*. If a person's score is at the 65th percentile, this means (intuitively) that 65 percent of the scores (measurements in the sample) were lower than his. A score at the 95th percentile tells the person that 95 percent of the scores are lower than his. The concept of "percentile" therefore gives us a means of talking about location in the sample data other than central location. The median becomes a special case of the more general concept of percentile; that is, the median is the 50th percentile. Whereas we can imagine the median, with one perpendicular line, dividing the histogram into two equal parts, we can also imagine the various percentile points, with a perpendicular line through each, dividing the histogram into 100 equal parts. Since it takes only one perpendicular line to divide a histogram into two parts, it takes only 99 perpendicular lines to divide a histogram into 100 parts. Thus, there are 99 percentiles, and we shall designate them by $P_1, P_2, P_3, \ldots, P_{99}$. This means that the median is P_{50}.

Unlike the median, the various percentiles (other than P_{50}) were not defined for ungrouped data. The reason for this is that unless the sample is quite large, no single definition of percentile gives us useful measures. In fact, in some small sample cases certain percentiles would be impossible to find. For example, in a sample of 15 measurements, the largest being 17.8, what would be a reasonable value for P_{98}? Only 93.3 percent of the sample is less than 17.8 (assuming that 17.8 occurs only once), and 100 percent of the sample has values less than any point above 17.8. So P_{98} does not exist in this sample.

Thus we shall define percentile exclusively in the context of the interval-grouped distribution.

Definition 4.6:   The rth percentile ($r = 1, 2, 3, \ldots , 99$) of an interval-grouped frequency distribution, denoted by P_r, is

$$P_r = b + \frac{d}{f}\,(l)$$

where $b =$ the lower class boundary of the class containing P_r,
$\quad\quad f =$ the frequency of the same class,
$\quad\quad l =$ the class length,
$\quad\quad d =$ the difference between $(r/100)n$, (n is the sample size), and the
$\quad\quad\quad$ sum of the frequencies below b.

Note the similarity between Definition 4.6 and Definition 4.3 (page 54). If we let $r = 50$ in Definition 4.6, we get Definition 4.3! Remember that the median estimate is the 50th percentile.

To apply Definition 4.6 we first find the lower class boundary of the class containing P_r. For example, to calculate P_{37} (the 37th percentile) for the weight-loss frequency distribution, we first find $(.37)(90)$, which gives us 33.3 — *the proportion of measurements below P_{37}.* Looking at the cumulative frequencies, we see that a total of 28 measurements fall in the first two classes and a total of 61 in the first three classes. Therefore, P_{37} must lie in the third class, which means that $b = 14.5$ (the lower class boundary of the third class). In addition, it follows that $f = 33$, $l = 5$, and $d = 33.3 - 28 = 5.3$. Therefore,

$$P_{37} = 14.5 + \frac{5.3}{33}(5) = 15.30$$

Certain subsets of the 99 percentile points are given special names. The percentiles P_{25}, P_{50}, and P_{75} are called the first, second, and third *quartiles*, respectively. The three quartiles separate the distribution into four equal parts. The percentiles $P_{10}, P_{20}, P_{30}, \ldots , P_{90}$ are called *deciles* (P_{10} is the first decile, P_{20} is the second, and so on). These nine percentiles separate the distribution into 10 equal parts.

Exercises

1.   Give four names for P_{50}.

2.   Find P_6, P_{82}, the third decile, and the first quartile for the weight-loss frequency distribution.

3.   The following frequency distribution represents scores for 150 sociology students on a final examination. The professor decides to assign letter grades according to the following rule: A (scores $\geq P_{90}$), B ($P_{70} \leq$ scores $< P_{90}$), C ($P_{30} \leq$ scores $< P_{70}$), D ($P_{10} \leq$ scores $< P_{30}$), and E (scores $< P_{10}$). What are the cut off scores for each grade?

Scores	Frequency
40–49	3
50–59	18
60–69	38
70–79	56
80–89	27
90–99	8

4. Is it true that $\dfrac{P_{30} + P_{70}}{2} = P_{50}$? Explain.

4.4 Symmetry and Skewness

The reader has undoubtedly noticed the considerable variation which exists with regard to the over all shape of different frequency distributions — especially with respect to the general position of the major concentration of measurements. Figure 4.2 shows histograms which represent three distinct types of frequency distributions. We say that frequency distributions are either *symmetric* or *skewed* (asymmetric). If a frequency distribution is skewed, it is either *positively skewed* (skewed to the right) or *negatively skewed* (skewed to the left).

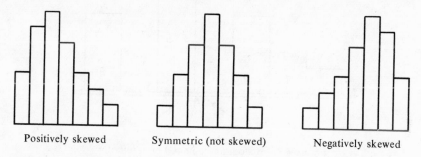

Positively skewed   Symmetric (not skewed)   Negatively skewed

Figure 4.2. Symmetry and skewness depicted by histograms of different shapes.

Although for convenience we picture symmetry and skewness with the histogram (which represents grouped data), we formally define symmetry with respect to the original data.

Definition 4.7: Given a sample of n measurements on variable x, we say that x is *symmetrically distributed* if the sample mean equals the sample median.

Referring again to Figure 4.2, we note that for a symmetrically distributed variable x, the major concentration of measurements is centered on the horizontal axis, with equal frequencies for classes that are equidistant from the center. For the two skewed types, measurements are concentrated either to the left or right of center. For symmetric distributions the mean and median are at the same point on the x axis. For a positively skewed distribution the mean is larger than (to the right of) the median, and in a negatively skewed distribution the mean is smaller than (to the left of) the median.

To aid our intuitive understanding of the above statements, it is helpful to think of the mean as the *balancing point* along the x axis of a histogram. That is, imagine that the rectangles of the histogram are piles of boxes which are piled on a long steel beam (the x axis), and the beam is perched on a fulcrum (see Figure 4.3). If the beam balances on the fulcrum, the point at which the fulcrum touches the beam is the *mean*.

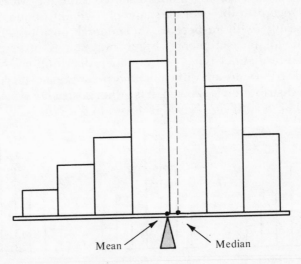

Figure 4.3. Histogram balanced on a fulcrum.

We recall from elementary physics, or even the backyard seesaw, that the point of balance depends not only on the *weight* on each side of the point, but also on the *distance* over which the weight is distributed on each side. (For example, a light person can balance a heavy person on a seesaw by sitting further from the fulcrum than the heavy person.) If the fulcrum is at the median, the weight on each side of the fulcrum is equal (i.e., equal area in the histogram). Thus, in a symmetric distribution the balance point (the mean) is at the median (i.e., mean = median). For a skewed distribution, however, the median is not a balance point, because although "weight" is equal on either side of the median, the distance on each side over which the "weight" is distributed is *not* equal. If the distribution is negatively skewed, the distance to the left of the median is longer; thus, the point of balance (the mean) must be to the left of the median. Using the same reasoning, it follows that for a positively skewed distribution the mean is larger than the median.

So that we may measure the extent and direction to which a distribution is skewed, we define the following statistic.

Definition 4.8: In a sample of n measurements on variable x, where m is the median of x, the *coefficient of skewness*, denoted by K, is

$$K = \frac{3(\bar{x} - m)}{s_x}$$

Clearly, for a symmetric distribution $K = 0$ (by Definition·4.7). When the distribution is skewed to the left (negatively skewed), K is negative. Why? When the distribution is skewed to the right, K is positive. The larger the absolute value of K, the greater the skewness. Although $|K|$ can have a value as high as 3.00, experience has shown that $|K|$ rarely exceeds 1.00.[*] In the weight-loss sample

$$K = \frac{3(17.70 - 17)}{6.89} = 0.30$$

Since K is positive, but small, we know that the weight-loss data is skewed slightly to the right. We can observe this skewness in Figure 3.3. Our calculated value of K is, in essence, an estimate of the K we could get if we substituted the exact values of the mean, median, and standard deviation of the *population* (from which the sample was selected) into the formula for K. Thus, K gives us an estimate of the degree of skewness existing in the population.

Exercises

1. The following numbers are values of the coefficient of skewness (K). List them in order of degree of skewness (greatest skewness first) and indicate in each case whether the skewness is to the right or left.

    2.16, 0.09, −1.72, −0.05, −2.46, 0, 0.81

2. For each of the following exercises guess the sign of K by *observation*, and then calculate K.
    (a) Exercise 1, page 53
    (b) Exercise 3, page 53
    (c) Exercise 1, page 60 (use estimates of $\bar{x}$, m, and s_x)

3. In 2(c) above, we use estimates of $\bar{x}$, m, and s_x. Why? How does 2(c) differ from 2(a) and 2(b) in this respect?

[*]See Croxten, Cowden and Klein (1967) for more detail regarding the derivation of K and for information concerning other coefficients of skewness.

Summary

The descriptive characteristics of ungrouped data may be either (a) calculated directly from the corresponding grouped data, or (b) estimated from the grouped data, depending upon how the data has been grouped. If sample data is grouped in such a way that no information is lost, then the sample statistics which we defined for ungrouped data may be calculated without error directly from the frequency distribution, using the same definitions, or equivalent definitions which fit the frequency distribution format. However, if the sample data is grouped into classes (i.e., some information lost), we use new definitions to calculate (from the frequency distribution) *estimates* of the sample statistics.

A special linear transformation is used to simplify the calculations involved in estimating the values of $\bar{x}$, s_x^2 and s_x from interval-grouped frequency distributions.

Other measures of location are defined on grouped data, called *percentiles* and symbolized by P_r, $(r = 1, 2, 3, \ldots, 99)$. Generally speaking, r percent of the measurements in a sample have values smaller than P_r. Certain subsets of the percentiles are called *quartiles* and *deciles*.

The concepts of *symmetry* and *skewness* help to describe the overall shape of a distribution. If the sample mean of x equals the sample median of x, we say that the variable x is *symmetrically distributed*. If the two sample statistics are not equal, then the distribution of x is *skewed* — either *positively* or *negatively*.

5

Regression
and Correlation

In the preceding chapters we have discussed only those descriptive statistics which deal with a single variable. That is, each of the sample statistics defined thus far describes some characteristic associated with a single variable only. In this chapter we shall investigate methods by which we can describe relationships between two or more variables. Before plunging directly into this discussion, we must first establish some basic terminology and notation with regard to graphing equations in two variables.

5.1 Graphing Linear Equations

Mathematical equations may be used to express the relationship between two variables. For example, the equation $w = 2z^2 - 1$ expresses a mathematical relationship between w and z. Similarly, the equation $m = 3p + 4$ represents a mathematical relationship between m and p. The latter example is called a *linear equation*.

Definition 5.1:   Any equation in two variables, x and y, which can be expressed in the form $y = ax + b$, where a and b are constants (any real numbers), is called a *linear equation*.

Thus, according to Definition 5.1, $m = 3p + 4$ is a linear equation. On the other hand, the equation $w = 2z^2 - 1$ is not linear.

All equations in two variables can be represented by a graph drawn on a plane determined by two coordinate axes placed at right angles. The graph of an equation in two variables, x and y, is simply a set of points in the plane, each point representing an ordered pair of numbers (x,y) which satisfies the equation. When the ordered pairs of numbers which satisfy a *linear* equation are translated into points on the graph, the points all lie in a *straight line*. On the other hand, points graphed for *nonlinear* equations, such as $w = 2z^2 - 1$, do not lie in a straight line. Figure 5.1 shows several ordered pairs which satisfy the equation $w = 2z^2 - 1$, represented by points on the graph. Note that it is necessary to plot a few well-spaced points on the graph, and then the graphing job is completed by sketching a curved line through the plotted points.* Similarly, every ordered pair of real numbers of the form (p,m) which satisfies $m = 3p + 4$ is represented by a point on the graph of that equation. Since the graph of $m = 3p + 4$ is a straight line, and since it takes only two points to determine a straight line, we need to find only two ordered pairs which satisfy the equation $m = 3p + 4$, such as $(1,7)$ and $(-2,-2)$. Then we simply plot the two points and draw a straight line through them. It is understood that every point on the straight line corresponds to an ordered pair which satisfies the equation.

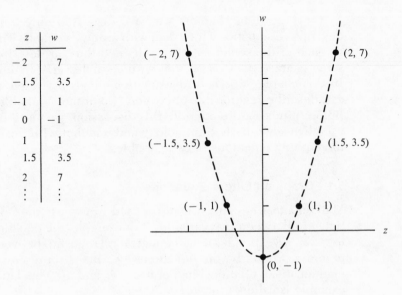

Figure 5.1.   Graph of $w = 2z^2 - 1$.

*This assumes, of course, as we shall assume throughout this section, that the domain of the dependent variable is the set of all real numbers.

Definition 5.2:   For any linear equation $y = ax + b$, a is called the *slope* and b is called the *y-intercept*.

The terms "slope" and "y-intercept" refer to characteristics of the graph of $y = ax + b$. The y-intercept tells us the point where the graph crosses the y-axis; namely, at $(0,b)$. The slope reflects the angle at which the line is situated relative to the x-axis. The larger the slope, the more vertical the line. A slope of zero indicates a horizontal line. In addition, if the slope is positive, the line runs from lower left to upper right (as in Figure 5.2). If the slope is negative, the line runs from upper left to lower right. Figure 5.3 shows some basic variations in the graph of $y = ax + b$ for different as and bs.

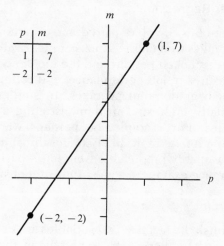

p	m
1	7
-2	-2

Figure 5.2.   Graph of $m = 3p + 4$.

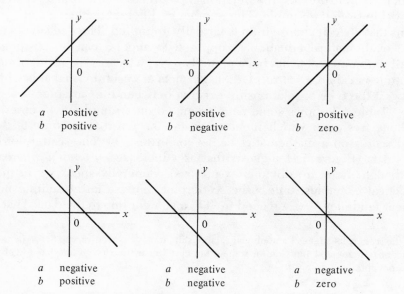

Figure 5.3.   Variations in the graph of $y = ax + b$ for different as and bs.

Exercises

1. Indicate which of the following mathematical statements are *linear* equations and rewrite each of them in the form $y = ax + b$. (Assume x can be any real number.)

   (a) $y = x + 1$ (b) $y = (x - 1)^2$ (c) $y = \dfrac{5 - 2x}{2}$

   (d) $y = 8x$ (e) $y = 5$ (f) $y = 2/x$

2. Graph each of the equations in Exercise 1.

3. Determine the slope and y-intercept for each of the linear equations in Exercise 1.

5.2 Simple Linear Regression

Many statistical studies involve paired measurements; that is, two different variables measured on the same sample. Examples include the number of cigarettes smoked per day and the incidence of throat cancer, IQ scores and measures of job performance, and college entrance exam scores and freshman grade-point averages. In studies of this sort the investigator is usually interested in demonstrating a relationship between the variables. For example, he perhaps wants to show that people who smoke a lot are more likely to get throat cancer, or that job performance is reasonably predictable from one's basic intelligence, or that scores on entrance exams are a good indicator of success in college.

5.2.1 The Scatterplot

If we wish to establish the fact that two variables are in some way related or dependent upon each other, we try to find the mathematical equation which best expresses the relationship, given the information (data) available from the sample. The number of different mathematical equations that we can choose from is virtually unlimited. The decision to use one mathematical equation as opposed to another can be postponed until the sample data has been plotted on graph paper. The result of such plotting is called a *scatterplot*. Scatterplots are useful visual aids which suggest the type of relationship existing between the variables.

Table 5.1 shows some data generated on a sample of 15 entering college freshmen which includes scores on a verbal aptitude test (x) and scores on a mathematics achievement test (y). The scatterplot for this data (Figure 5.4) suggests that as x increases, y tends to increase (although there are obvious exceptions). Generally speaking, in most practical examples, unless the scatterplot shows a fairly definite non-linear tendency,* we proceed to "fit" a straight line to the data. That is,

---

*If the scatterplot suggests a nonlinear relationship, it may be necessary to fit a hyperbolic function, exponential function, or some other curvilinear function to the data to get the best possible fit.

we assume that the relationship is *linear* and we proceed to find the equation of a specific straight line which "best fits" the data. This best-fitting line is called the *regression line.*[*]

Table 5.1. Verbal Aptitude Scores (*x*) and Mathematics Achievement Scores (*y*) for a Sample of 15 Individuals

	x_i	y_i
1	80	60
2	75	62
3	95	78
4	50	40
5	65	58
6	55	65
7	58	52
8	87	72
9	76	70
10	40	50
11	68	78
12	35	40
13	55	48
14	74	80
15	98	88

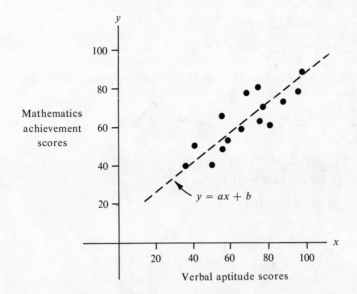

Figure 5.4. Scatterplot: verbal aptitude (*x*) vs. mathematics achievement (*y*).

[*]The term "regression" is somewhat of a misnomer. See Walker and Lev (1969, page 210) for an explanation of the origin of the term.

5.2.2 The Least Squares Method

Note in Figure 5.4 that a dotted line has been sketched in to represent the regression line for the given data. Intuitively, we interpret "best fit" to mean that the swarm of points in the scatterplot is as closely grouped around the line as possible — so that the distance between each point and the line is on the average as small as possible. To find this straight line of best fit we use the method of *least squares*. This procedure defines best fit to mean that the *total sum of squares of the vertical distances from the points to the straight line must be as small as possible.* We use the least squares method to find the *slope "a"* and the *y-intercept "b"* of the regression line $y = ax + b$. This method involves calculus which unfortunately makes a thorough explanation of the procedure beyond the scope of this text.* However, the least squares method provides us with calculation formulas for the slope and the y-intercept of the regression line.

Definition 5.3: Given n paired measurements on variables x and y, the equation of the *y-on-x regression line* is $y = ax + b$, where

$$a = \frac{n\left(\sum x_i y_i\right) - \left(\sum x_i\right)\left(\sum y_i\right)}{n \sum x_i^2 - \left(\sum x_i\right)^2}$$

$$b = \frac{\sum y_i - a \sum x_i}{n}$$

$$\left(\text{``}\sum\text{''} \text{ replaces } \sum_{i=1}^{n} \text{ for simplicity}\right)$$

Table 5.2 shows the calculation of the four sums needed to find a and b, according to Definition 5.3. These sums are

$$\sum x_i, \quad \sum y_i, \quad \sum x_i^2 \quad \text{and} \quad \sum x_i y_i$$

Therefore,

$$a = \frac{15(66{,}693) - (1011)(941)}{15(73{,}063) - (1011)^2} = 0.664$$

$$b = \frac{941 - 0.664(1011)}{15} = 17.98$$

and the *y-on-x* regression equation is

$$y = 0.664x + 17.98$$

*For a more thorough explanation see Gemignani (1970, page 271).

The reader should graph this equation on the scatterplot in Figure 5.4 to see how close it comes to the dotted line (which was a visual guess). Moreover, graphing the regression line on the scatterplot is a wise procedure, since it will readily reveal any gross errors in calculation. If either a or b is significantly in error, the line will not closely fit the scatterplot—a situation which can be spotted visually without difficulty.

It is important that we think of a and b as statistics (in the same sense that $\bar{x}$, s_x^2, and s_x are statistics). The slope and y-intercept are calculated from sample data. If we were to do the same analysis on a different sample of 15 students, no doubt the values of a and b would be different (although probably not much). The point is, we are attempting here to estimate the linear relationship between variables x and y *in the population*; we shall represent this "true" linear relationship by the equation "$y = \alpha x + \beta$." Since our available information is confined to sample data, we calculate a and b, which are estimates of α and β. Thus, we can think of the equation $y = ax + b$ (generated from the sample) as an estimate of the equation $y = \alpha x + \beta$ which is our representation for the hypothetical linear relationship between x and y existing in the population.

Table 5.2. Four Sums Needed to Find
a and b

x_i	y_i	x_i^2	$x_i y_i$
80	60	6400	4800
75	62	5625	4650
95	78	9025	7410
50	40	2500	2000
65	58	4225	3770
55	65	3025	3575
58	52	3364	3016
87	72	7569	6264
76	70	5776	5320
40	50	1600	2000
68	78	4624	5304
35	40	1225	1400
55	48	3025	2640
74	80	5476	5920
98	88	9604	8624
Σ 1011	941	73,063	66,693

5.2.3 *Prediction*

Whenever we can show that two variables are mathematically related to each other, our next thought is a practical one — namely, could a given value of one variable be used to *predict* a value of the other variable? Using the same sample, if we know that a particular person received a 72 on the verbal aptitude test, can we predict his mathematics achievement? If we knew nothing regarding the relationship between the two variables, then our best prediction would have to be the sample mean for the mathematics achievement scores, or $\bar{y} = 62.70$. However, since we have estimated a linear relationship, namely the equation $y = 0.664x + 17.98$, we substitute 72 for x in this equation and solve for y, giving us $y = (0.664)(72) + 17.98 = 65.79$. We are more confident with the prediction $y = 65.79$ than we are of the prediction $y = 62.70$ (the value of the sample mean of y), since the former takes advantage of more information — namely, that x and y are linearly related. Thus, the y-on-x regression equation, as defined in this section, has two uses: (a) it expresses a linear relationship between two variables, and (b) it provides a means for predicting values of y for given values of x.

If, on the other hand, we wished to predict values of x for given values of y, we would need the equation of the regression line of x-*on*-y, i.e., $x = ay + b$, where the formulas for a and b are the same as those given in Definition 5.3, but with the xs and ys reversed. The regression lines, "y-on-x" and "x-on-y", are different lines. As a general rule, if we wish to predict values on one variable from given values of another, we automatically think of the former variable as "y" and the latter as "x" and proceed, using Definition 5.3, to find a regression equation of the form "$y = ax + b$".

One last word of caution is in order. It is dangerous to predict the value of y for a value of x which is outside the range of the actual measurements collected on variable x. For example, for the data in Table 5.1 we should not attempt to predict a mathematics achievement score for a hypothetical person who has a verbal aptitude less than 40 or greater than 88. Such predictions can be very unreliable, and in some instances ridiculous (see Exercise 6, page 75), the reason being that the relationship between the two variables outside the range of actual sample data is, strictly speaking, still unknown to us.

Exercises

1. Given the following set of paired measurements, generated by measuring the height (h) and weight (w) of 10 people,
    (a) find a linear equation which can be used to predict a person's height, given his weight.
    (b) construct the scatterplot and plot the equation on the scatterplot.
    (c) predict the height of a person who weighs 175 lb.

Persons	Height (inches)	Weight (lb.)
1	60	120
2	72	180
3	70	200
4	70	160
5	74	190
6	67	148
7	68	155
8	76	220
9	61	130
10	65	145

2. Find the x-on-y regression equation for the verbal-aptitude–mathematics-achievement data (Table 5.1).

3. A bacteriologist observes the frequency of a malignant cell growth in a particular organism over a period of time and he records the number of malignant cells (c) existing at various moments in time (t). Using this data (below),
   (a) find the regression equation which can be used to predict the number of malignant cells in existence at a given time.
   (b) construct a scatterplot and plot the equation on it.
   (c) predict the total number of cells existing at the end of 20 hours.

t:	3	6	9	12	15	18	21	24	27	30	33	36
c:	7	17	21	33	43	58	78	93	115	143	178	230

4. The equation $y = 15.2x - 2.6$ is a regression equation (y-on-x). What is the predicted value of y, given that $x = 2.0$? What is the predicted value of x, given that $y = 42.6$?

5. Using the following data:

x:	1	1	0	0	−1	−1	−2	−2
y:	0	−2	0	−1	−1	1	1	2

   (a) Find both the "y-on-x" and "x-on-y" regression equations.
   (b) Construct the scatterplot and graph both equations on it.

6. Eight students were asked how many hours they had studied prior to a particular hour examination. The responses were paired with the scores on an examination which had a maximum score of 100.

Hours:	0.5	1.0	1.5	2.5	3.0	3.0	3.5	4.0
Scores:	70	60	65	85	80	90	85	90

   (a) Find the regression equation which will predict scores, given hours of study.
   (b) What is the predicted score for a student who studies 10 hours?
   (c) In view of the fact that 100 is given as the maximum possible examination score, how do you explain your answer to (b)?

5.3   The Coefficient of Correlation

5.3.1   *Analysis of* Q_R

In the previous section we learned to fit a straight line, called a regression line, to a two-variable scatterplot by the "least squares" method. This method determines the equation of that particular straight line $y = ax + b$ which best fits the scatterplot in the sense that the total sum of squares of the vertical distances between actual and predicted values of y for given values of x is a *minimum*. If we let Q_R be that total sum of squared deviations, then

$$Q_R = \sum [y_i - (ax_i + b)]^2$$

where a and b are determined by Definition 5.3. This means that any other choice of a or b (other than those determined by Definition 5.3) would give us the equation of a line different from the regression line, *and* that the total sum of squares of vertical deviations for such a line would be *larger* than Q_R.

Thus, we know how to express mathematically an estimate of the linear relationship between two variables. What we have not yet done is determine the *extent* to which the two variables are linearly related. Except in very unusual circumstances, the least squares method will always produce a linear regression line regardless of the extent to which the two variables are in fact related in a linear way. It would seem obvious that when the two variables do not have a strong linear relationship, that even the best-fitting straight line would have some rather large vertical deviations, giving us a relatively large Q_R value. On the other hand, if the two variables do have a strong linear relationship, we could expect the regression line to "fit" the scatterplot rather closely, thus making Q_R relatively small. Figure 5.5 shows two different scatterplots: one (A) which suggests only a slight linear relationship between the variables, the other (B) a strong linear trend. Since the vertical deviations are obviously larger in scatterplot A than in B, we would expect Q_R *(the total sum of squared vertical deviations from the y-on-x regression line)* to be larger for A than for B. Thus, it appears that a small Q_R is indicative of a strong linear relationship, and vice versa.

The previous conclusion, although appealing, is not necessarily true if (a) the two scatterplots being compared have an unequal number of paired measurements, or if (b) the unit of measure is not the same for both scatterplots. If one scatterplot has a great many more paired measurements (and therefore many more deviations), it would be expected

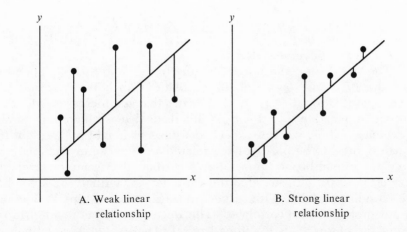

Figure 5.5.   Two scatterplots – showing "weak" and "strong" linear relationship according to the size of the vertical deviations from the regression line.

to have a larger Q_R whether a strong linear relationship exists or not. Or
if the two scatterplots have equal numbers of measurements and represent two samples measured on the same pair of variables (and therefore
display approximately the same degree of linearity), but one is measured
in feet and the other in inches, we can expect the latter one to yield a
larger Q_R. Therefore, Q_R is sensitive to "goodness of fit" (with respect
to the regression line), but it is not a reliable statistic for the purpose of
measuring the strength of linear relationship.

5.3.2   Definition of r_{xy}

To measure the strength of linear relationship between x and y on a
sample of n paired measurements on the two variables, we use a statistic
called the *sample linear correlation*, symbolized by r_{xy}.

> **Definition 5.4:**   Given a sample of n paired measurements on variables
> x and y, the statistic
>
> $$r_{xy} = \frac{n \sum x_i y_i - \left(\sum x_i\right)\left(\sum y_i\right)}{\sqrt{\left[n \sum x_i^2 - \left(\sum x_i\right)^2\right] \cdot \left[n \sum y_i^2 - \left(\sum y_i\right)^2\right]}}$$
>
> is called the *sample linear correlation* between x and y, (sometimes called
> the *Pearson* product-moment coefficient of correlation*).

*Named after English statistician, Karl Pearson (1857–1936).

Before we discuss the relationship between r_{xy} and Q_R, we shall discuss some properties of r_{xy} and then calculate it for the verbal-aptitude–mathematics-achievement data.

The statistic r_{xy} can range in value from -1.00 to 1.00 depending upon the *type* and *degree* of linear relationship between x and y. A positive correlation ($0 < r_{xy} \leq 1$) indicates that as x increases, y tends to increase. A negative r_{xy} ($-1 \leq r_{xy} < 0$) tells us that as x increases, y tends to decrease. Thus, we can expect measures of height and weight (for human beings) to be positively correlated and we can expect measures of weight and agility to be negatively correlated. A sample correlation of zero ($r_{xy} = 0$) indicates that the variables are virtually independent of each other (at least, with respect to *linear* relationship). Weight and IQ are probably zero-correlated. The *degree* of linear relationship is greater the closer r_{xy} is to either 1 or -1. That is, *low* correlations are those near zero (either positive or negative) and *high* correlations are those near 1 or -1. We say, for example, that $r_{xy} = -0.57$ is a higher correlation than $r_{xy} = 0.35$, and that $r_{xy} = -0.95$ is a higher correlation than $r_{xy} = -0.15$. In other words, the sign of r_{xy} indicates the *type* of linear relationship, but the size of the absolute value of r_{xy} indicates the *degree* of relationship.

Definition 5.4 formulates r_{xy} for calculation purposes. To use the formula we need to know

$$\sum x_i, \quad \sum y_i, \quad \sum x_i^2, \quad \sum y_i^2, \quad \sum x_i y_i$$

and the number of paired measurements (n). Note that all of these pieces of information, except $\sum y_i^2$, are needed in the calculation of a and b for the regression equation. Referring again to the calculations on page 72, it can be established by the reader that $\sum y_i^2 = 62{,}137$. Thus, the sample linear correlation between verbal aptitude and mathematics achievement is

$$r_{xy} = \frac{15(66{,}693) - (1{,}011)(941)}{\sqrt{[15(73{,}063) - (1{,}011)^2] \cdot [15(62{,}137) - (941)^2]}}$$

$$= \frac{49{,}044}{\sqrt{(73{,}824)(46{,}574)}}$$

$$= 0.84$$

This correlation indicates that the linear relationship between verbal aptitude and mathematics achievement is positive and relatively strong.

We have been assuming, of course, throughout this chapter that the paired data is ungrouped. There exist methods for grouping paired data into a two-way frequency distribution, and it is possible to estimate r_{xy} from the grouped data (see Walker and Lev, 1969, page 217).

5.3.3 Relationship Between Q_R and r_{xy}

We observed earlier that Q_R is sensitive to how well the regression line fits the scatterplot, but as a measure of correlation strength it has two disadvantages — namely, it is also sensitive to (a) the *number* of paired measurements in the sample, and (b) the *units* used for measuring y. Therefore, we use r_{xy} (Definition 5.4) rather than Q_R to measure linear correlation. It is not clear from Definition 5.4 just how r_{xy} overcomes the disadvantages of Q_R. The following equation (which can be derived mathematically from Definition 5.4) clarifies the relationship between r_{xy} and Q_R and, upon close scrutiny, intuitively justifies r_{xy} as a reasonable statistic for the measurement of linear correlation.

$$r_{xy}^2 = 1 - \frac{Q_R}{Q_M}, \qquad \text{where} \qquad Q_M = \sum (y_i - \bar{y})^2$$

The equation says essentially that the square of r_{xy} depends not on Q_R alone, but on a comparison between Q_R and $\sum (y_i - \bar{y})^2$. Why does this make sense? The reader may recall an earlier statement (page 74) which suggests that a reasonable way to predict a y value when no linear relationship between x and y has been established is to calculate the sample mean of y, that is, $\bar{y}$. In other words, if we had no y-on-x regression equation available, the best estimate of y, for *any* given value of x would be $\bar{y}$! Thus, rather than use the regression equation $y = ax + b$ to estimate y, we would use the equation $y = \bar{y}$. The graph of the equation $y = \bar{y}$ is a straight line parallel to the axis (slope is zero). The y-intercept is $\bar{y}$. Figure 5.6 shows the line $y = \bar{y}$ and the regression line $y = ax + b$ "fitted" to the *same* scatterplot. The vertical deviations (differences between ys and predicted ys) for each line shows a visual comparison of "fit." The total sum of squared deviations from the regression line is Q_R, and from the line $y = \bar{y}$ is $\sum (y - \bar{y})^2$, which we shall call Q_M. *If there is a significant linear trend, we would expect the regression line to fit the scatterplot more closely than the line $y = \bar{y}$* (i.e., $Q_R < Q_M$). If, however, there is no linear trend in the data, we would not expect the regression line to fit any better than the line $y = \bar{y}$ (i.e., $Q_R = Q_M$). Therefore, when no linear trend exists, Q_R/Q_M is 1, which makes $r_{xy}^2 = 1 - 1 = 0$, giving us $r_{xy} = 0$. When there is a linear trend, $Q_R/Q_M < 1$, which makes $r_{xy}^2 >$ zero, giving us $r_{xy} > 0$ (if the slope of the regression line is positive) or $r_{xy} < 0$ (if the slope is negative). In essence, *r_{xy} tells us the extent to which the regression line is a better fit to the data than the line $y = \bar{y}$.*

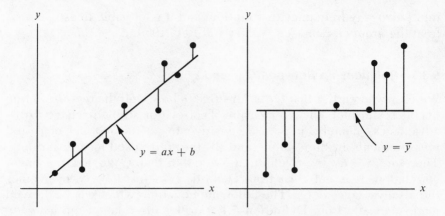

Figure 5.6. Vertical deviations from $y = ax + b$ and $y = \bar{y}$ compared for the same data.

5.3.4 Effect of Data Transformation on r_{xy}

The most useful data transformation for the purpose of simplifying the calculation of r_{xy} is the subtraction of a constant c_1 from each x measurement and/or a constant c_2 from each y measurement. Theorem 5.1 expresses the effect of such a transformation on the correlation between x and y.

> **Theorem 5.1:** Given a sample of n paired measurements on variables x and y, if $u_i = x_i - c_1$ and $v_i = y_i - c_2$, then $r_{uv} = r_{xy}$.

> **Proof:** Left to the reader (see Exercise 6, page 82).

In other words, if we were to subtract a constant from each measurement on x and subtract a constant from each measurement on y, the resulting set of pairs (u_i, v_i) would have the *same* correlation as the original data.

Let us test Theorem 5.1 on the verbal-aptitude–mathematics-achievement data. We shall let $u_i = x_i - 67$ and $v_i = y_i - 63$. The values 67 and 63 are chosen because they approximate to the nearest whole number the means of x and y ($\bar{x} = 67.4$ and $\bar{y} = 62.7$). These choices tend to minimize the absolute values of the transformed measurements, and therefore reduce significantly the calculation effort. Table 5.3 shows the transformed data and the new sums. Therefore,

$$r_{xy} = r_{uv} = \frac{15(3{,}268) - (6)(-4)}{\sqrt{[15(4{,}924) - (6)^2] \cdot [15(3{,}106) - (-4)^2]}}$$

$$= \frac{49{,}044}{\sqrt{(73{,}824)(46{,}594)}}$$

$$= 0.84$$

Table 5.3. Recalculation of Sums Using Transformations

x_i	y_i	$(x_i - 67)$ u_i	$(y_i - 63)$ v_i	u_i^2	v_i^2	$u_i v_i$
80	60	13	−3	169	9	−39
75	62	8	−1	64	1	−8
95	78	28	15	784	225	420
50	40	−17	−23	289	529	391
65	58	−2	−5	4	25	10
55	65	−12	2	144	4	−24
58	52	−9	−11	81	121	99
87	72	20	9	400	81	180
76	70	9	7	81	49	63
40	50	−27	−13	729	169	351
68	78	1	15	1	225	15
35	40	−32	−23	1024	529	736
55	48	−12	−15	144	225	180
74	80	7	17	49	289	119
98	88	31	25	961	625	775
Σ		6	−4	4924	3106	3268

Note that we get the exact same value for r_{uv} that we calculated earlier for r_{xy} (page 78). Thus, the transformation does not affect the original correlation between x and y. Other types of transformations are possible but not usually productive in practice (see Exercise 5).

Exercises

1. For each of the following sets of paired data:
   (a) Plot the data.
   (b) Calculate the y-on-x regression equation.
   (c) Graph the regression equation on the scatterplot.
   (d) Calculate r_{xy}, using Definition 5.4.

i.

x	y
2	−1
5	1
4	2
2	0
1	−3

ii.

x	y
2	2
3	0
1	1
0	2
−1	4

2. In Exercise 1, in each case calculate Q_R and Q_M, and show that the equation

$$r_{xy} = \pm \sqrt{1 - \frac{Q_R}{Q_M}}$$

gives the same answer for r_{xy}. (The sign of r_{xy} is the same as the sign of the slope of the regression line.)

3. Find the sample linear correlation for the data in:
    (a) Exercise 1, page 74.
    (b) Exercise 3, page 75.
    (c) Exercise 6, page 75.

4. Do Exercise 3(c) again using the transformation $u_i = x_i - 2.5$ (x represents hours) and $v_i = y_i - 75$ (y represents scores).

5. Given a sample of n paired measurements on variables x and y, if we apply *linear* transformation $u_i = ax_i + b$ and $v_i = cy_i + d$, then $r_{uv} = r_{xy}$ if a and c have the same sign and $r_{uv} = -r_{xy}$ if a and c have opposite signs. Do Exercise 3(c) again using the transformations:

$$u_i = \frac{x_i - 2.5}{0.5} \quad \text{and} \quad v_i = \frac{y_i - 75}{5}$$

6. Prove Theorem 5.1.

5.4   Interpretation of r_{xy}

We know that $r_{xy} = 0$ suggests that the two variables are linearly independent in the sense that knowing a value of one variable is of no help in predicting a corresponding value for the other variable. We know that a positive correlation implies that high values of one variable *tend* to accompany high values of the other (and vice versa). And we know that a negative correlation implies that high values of one variable *tend* to accompany low values of the other (and vice versa). The last two statements suggest that when $r_{xy} \neq 0$, values of one variable can be used to predict values of the other. The accuracy of the prediction (i.e., the strength of the linear relationship) is determined by the size of $|r_{xy}|$. The higher the value of $|r_{xy}|$, the more accurate the prediction. If $|r_{xy}| = 1$ (i.e., if $r_{xy} = 1$ or $r_{xy} = -1$), prediction is perfect.

5.4.1 r_{xy}^2 as a Measure of Variation in y

Usually r_{xy} is not zero, nor is it plus or minus one. What can we say about a value such as 0.84, which we found earlier to be the sample linear correlation between verbal aptitude and mathematics achievement? Is it sufficient to interpret this correlation as strong evidence that the higher a person's verbal aptitude is, the higher his mathematics achievement will be? We can say more, and to do so we refer again to Q_R and Q_M. Note, first of all, that Q_M is in essence a measure of the amount of *error* involved in predicting y with $\bar{y}$. Similarly, Q_R is a measure of the amount of error involved in predicting y with "$ax + b$." If x and y are not linearly related, then $Q_R = Q_M$. Hopefully, x is linearly related to y to some extent so that $Q_R < Q_M$. Thus, $Q_M - Q_R$ is the reduction in error due to predicting y with $ax + b$ as opposed to predicting y with $\bar{y}$. It follows, therefore, that $(Q_M - Q_R)/Q_M$ is the proportional reduction in error. But $(Q_M - Q_R)/Q_M = 1 - Q_R/Q_M$, which we have already stated is equal to $r_{xy}{}^2$ (page 79). Therefore, $r_{xy}{}^2$ *represents the proportion of the variation (error) in y which can be attributed to its linear relationship with x*. For example, if $r_{xy} = 0.80$, then $(0.80)^2 = 0.64$; therefore, we say that 64 percent of the variation in y can be attributed to its linear relationship with x. Thus, 36 percent is still unaccounted for. Therefore, a correlation of 0.80 is not twice as strong as a correlation of 0.40. In fact, $r_{xy} = 0.80$ represents *four* times the strength of $r_{xy} = 0.40$, because the relative strength of r_{xy} is proportional to $r_{xy}{}^2$ (in this case, 0.64 and 0.16).

5.4.2   Mathematical vs. Causal Relationship

The user of correlation statistics must keep in mind at all times the fact that r_{xy} represents a *mathematical* relationship between x and y, and nothing more. The researcher who uses correlation is often trying to show a *causal* relationship between two variables. For example, he may hypothesize that cigarette smoking causes lung cancer. Suppose that for a sample of 100 people he finds $r_{xy} = 0.75$, where $x =$ the number of cigarettes smoked per day, and y is a measure of lung cancer growth. What legitimate conclusions can he make based solely on the correlation 0.75? He may conclude that the two variables are mathematically correlated,* but he cannot assume from this fact alone that either one of the variables has a *causal* effect upon the other. Certainly, if causality exists, he should expect to find a healthy mathematical relationship between x and y. Finding that mathematical relationship, however, does not prove the causality hypothesis; it simply *fails to disprove* the hypothesis. In the absence of a more thorough investigation there is always the possibility that a third variable may cause people to want to smoke and also causes cancer to grow in lung tissue.†

The existence of other variables which may be causing variations in the principal variables is not an uncommon occurrence. For example, if we were to correlate the number of catholic priests and the number of cases of whiskey consumed per year in a random sample of 50 Massachusetts cities, we could expect to find a high positive correlation, not necessarily because the priests are consuming the whiskey, but more likely because of a third variable, population size. A city with a large

---

*Assuming that $r_{xy} = 0.75$ is a good estimate of the true correlation in the population. Remember that r_{xy} is a descriptive statistic calculated from a sample and used to estimate the *true* relationship between x and y in the *population*. (See Section 7.7.3.)

†These remarks are not intended as an endorsement for cigarette smoking. On the contrary, the evidence, based on much more than a few correlations, is overwhelmingly in support of the hypothesis that cigarette smoking causes lung cancer. (See the committee report to the Surgeon General, *Smoking and Health*, U.S. Government Printing Office, 1964.)

population would tend to have a large number of priests *and* a large whiskey consumption compared to cities with smaller populations. In Section 5.6.2 we discuss *partial correlation* and learn how to adjust r_{xy} to remove the effect of other variables.

5.4.3 Non-linear Correlation

The correlation user must also remember that r_{xy} measures only the *linear* relationship between x and y. Even though r_{xy} may be small, it is possible that better predictions could be made if a nonlinear (curvilinear) relationship were investigated. Nonlinear methods are beyond the scope of this text. An excellent reference for curvilinear regression and correlation is Armore (1966, page 463).

Exercises

1. List the following correlations in order of strength, the weakest first: -0.52, 0.09, -0.95, 0.27, 0.67, -0.29.

2. Which correlation indicates more predictive strength, -0.50 or 0.20? How much more strength?

3. Given a y-on-x regression analysis, If $Q_R = 45$ and $Q_M = 60$, then what is the value of r_{xy}? How much of the variation of y is due to y's linear relationship to x?

4. Is the y-on-x regression equation (predicting y from x) the same as the x-on-y regression equation? Would Q_R and Q_M for y-on-x regression be the same values as Q_R and Q_M for the x-on-y regression? Does $r_{xy} = r_{yx}$? Does Q_R/Q_M for the y-on-x regression equal Q_R/Q_M for the x-on-y regression?

5.5 Multiple Regression and Correlation

Up to this point in our discussion we have considered only the two-variable case; that is, we have learned to measure the linear correlation between two variables x and y, and to use that relationship for the purpose of predicting y from x. Actually, we have been restricting ourselves to the simplest case of the more general model involving the prediction of y from n variables: $x_1, x_2, x_3, \ldots, x_n$.[*]

For example, suppose that we wish to predict success at John Doe College, using the college freshman grade-point index as the criterion (y). Reasonable predictor variables might be high school average (x_1), the College Board Scholastic Aptitude Test score (x_2), and the score on

---

[*]Note of clarification: in this section the notation $x_1, x_2, x_3, \ldots, x_n$ stands for n *different variables* — *not* n measurements on variable x. If it became necessary for us to represent individual measurements on x_1, for example, we would use a double subscript; that is, $x_{11}, x_{12}, x_{13}, \ldots, x_{1k}$ would represent k measurements on variable x_1.

the John Doe College entrance examination (x_3). Assuming that the measurements on x_1, x_2, and x_3 are a requirement for admission to John Doe, data on x_1, x_2, x_3, and y would be available on any sample of freshmen selected at the end of the freshman year. Table 5.4 shows how such data might look in part.

Table 5.4. Partial Data on Four Variables on College Freshmen

Freshman	y	x_1	x_2	x_3
1	2.35	82.6	1090	80
2	1.80	73.0	960	75
3	2.24	78.5	1140	84
4	3.56	85.7	1350	93
5	2.71	90.2	1280	96
.	.	.	.	.
.	.	.	.	.
.	.	.	.	.

So that we may predict y, given a knowledge of x_1, x_2, and x_3, we "fit" the given data to the equation

$$y = a_1 x_1 + a_2 x_2 + a_3 x_3 + b$$

where a_1, a_2, a_3, and b are real numbers. Again, the procedure for finding the "best fit" is supplied by the least squares method, which gives us formulas for a_1, a_2, a_3, and b in terms of various summations obtained from the sample data. The prediction equation above is called the *sample multiple linear regression equation.*

The sample multiple linear correlation may be defined by a formula similar to the one used to express r_{xy}, namely

$$\pm \sqrt{1 - \frac{Q_R}{Q_M}}$$

where

$$Q_R = \sum [y_i - (a_1 x_{1i} + a_2 x_{2i} + a_3 x_{3i} + b)]^2,$$
$$Q_M = \sum (y_i - \bar{y})^2, \text{ as before.}$$

In other words, the sample multiple linear correlation is a measure of the comparison between variation due to predicting y from $\bar{y}$ (Q_M) and variation due to predicting y from $a_1 x_1 + a_2 x_2 + a_3 x_3 + b$ (Q_R). That is, sample linear correlation is *always* a measure of the proportion of the original variation in y (the criterion) which is explained or accounted for by the predictor variables, regardless of how many predictors we use.

When we have more than one predictor variable, we can study the effect on the correlation of adding new predictors to the regression equation. For example, we may find that the correlation between y and x_1 is 0.52 and between y and x_2 is 0.45, whereas the multiple correlation between y and the best combinations of x_1 and x_2 might be 0.64. If we throw x_3 into the analysis, we might be able to increase 0.64 to 0.70. These increases depend upon the intercorrelations among the predictors themselves, as well as the correlations between each predictor and y. Generally speaking, if two variables are highly correlated with each other and one is already being used in the regression equation to predict y, then the addition of the other variable to the regression equation will not significantly increase the multiple correlation.

In addition to giving the admissions director of John Doe College a formula (i.e., the regression equation itself) for predicting the freshman grade-point indices for the following year, this type of analysis can also show the relative contribution of each x variable to the prediction of y, thus telling the college what types of predictive information are worth collecting and what types of abilities and aptitudes are most likely to lead to success at John Doe.

Multiple regression and correlation analysis is relatively complicated with respect to the calculations involved. Such analyses are best suited for the electronic computer. We shall not attempt to present detailed formulas and examples here. Appropriate references such as Fryer (1966), McNemar (1962), and Garrett (1966) are included in the bibliography.

5.6 Other Measures of Correlation

5.6.1 Rank Correlation

In Section 5.3.2 we defined the sample linear correlation r_{xy} and used the verbal-aptitude–mathematics-achievement data to demonstrate the rather tedious process of calculating that statistic. It is possible, however, to simplify the process somewhat by replacing the *measurements* by their appropriate *ranks* before calculating the correlation. Of course, such a departure from the normal procedure loses some information (i.e., introduces some error) so that the end result must be thought of as an estimate of r_{xy}, though a very close estimate in most cases. Take, for example, the same data: x (verbal aptitude) and y (mathematics achievement) and replace the individual measures for each column with numbers representing rank order within that column. This procedure produces two columns of ranks denoted by variables x' and y'. The largest measurement in a column gets a rank of "1," the next largest a rank of "2," and so forth. If two or more measurements are identical, each should be assigned a value equal to the mean of the rank order positions occupied by those measurements.

For example (see Table 5.5), the largest y measurement (88) is assigned a rank of one, the second largest (80) has rank two, the third largest measurement is 78, which occurs twice. Therefore, each 78 is assigned a rank of $(3 + 4)/2 = 3.5$. The next smaller measurement gets a rank of 5, and so forth. After all ranks have been assigned for both columns, the rank correlation r_{xy}' may be found by applying Definition 5.4 to the two sets of ranks (columns headed by x_i' and y_i'). However, it is simpler to find r_{xy}' by applying the formula in Definition 5.5.

Definition 5.5: Given a set of n paired rankings (x_i', y_i') for a sample of n paired measurements on variables x and y, the statistic

$$r_{xy}' = 1 - \frac{6 \sum_{i=1}^{n} d_i^2}{n(n^2 - 1)} \qquad \text{where } d_i = x_i' - y_i' \text{ for each } i$$

is called the *sample rank correlation* between x and y.

For our example, therefore,

$$r_{xy}' = 1 - \frac{6(112.5)}{15(224)} = 1 - 0.20 = 0.80$$

Note that the value $r_{xy}' = 0.80$ (calculated above) is close to the value $r_{xy} = 0.84$ which we calculated earlier. In most cases, especially when n is large, r_{xy}' will give us a good estimate of r_{xy}.

Table 5.5. Conversion to Ranks and Calculation of the d_i^2 Summation

x_i	y_i	x_i'	y_i'	d_i	d_i^2
80	60	4	9	−5	25
75	62	6	8	−2	4
95	78	2	3.5	−1.5	2.25
50	40	13	14.5	−1.5	2.25
65	58	9	10	−1	1
55	65	11.5	7	4.5	20.25
58	52	10	11	−1	1
87	72	3	5	−2	4
76	70	5	6	−1	1
40	50	14	12	2	4
68	78	8	3.5	4.5	20.25
35	40	15	14.5	.5	.25
55	48	11.5	13	−1.5	2.25
74	80	7	2	5	25
98	88	1	1	0	0
Σ					112.50

5.6.2 Partial Correlation

In Section 5.4.2 we discussed the fact that the correlation coefficient r_{xy} measures a mathematical relationship, rather than a *causal* one, and we noted that sometimes a high correlation between two variables may be "caused" by a third variable. In the example used earlier, the correlation between the number of Catholic priests (x) and the number of cases of whiskey consumed (y) is no doubt affected by population size (z). Let us suppose that $r_{xy} = 0.80$, $r_{xz} = 0.90$ and $r_{yz} = 0.85$. It is possible for us to adjust r_{xy} by mathematically removing from it the effect of variable z. That is, we can recalculate r_{xy}, assuming that z is fixed (constant). This adjustment is accomplished by application of the following definition.

> **Definition 5.6:** Given three variables x, y, and z, and the sample correlations r_{xy}, r_{xz}, and r_{yz}, the quantity
>
> $$r_{xy \cdot z} = \frac{r_{xy} - r_{xz} r_{yz}}{\sqrt{(1 - r_{xz}{}^2)(1 - r_{yz}{}^2)}}$$
>
> is called the *sample partial correlation of x and y with the effect of z removed* (i.e., with z held fixed).

Applying Definition 5.6 to our example we get

$$r_{xy \cdot z} = \frac{0.80 - (0.90)(0.85)}{\sqrt{[1 - (0.90)^2][1 - (0.85)^2]}} = 0.15$$

which tells us that for a *fixed population size*, x and y are not very highly correlated. That is, if we were to do the experiment over again using data only from cities with approximately equal population (z held constant) we could expect to find r_{xy} much closer to 0.15 than to 0.80. Similarly, we could examine the relationship between x and z, with y held constant ($r_{xz \cdot y}$), and the relationship between y and z, with x held constant ($r_{yz \cdot x}$), using Definition 5.6. How would you expect these partial correlations to relate to r_{xz} and r_{yz} in this case? (See Exercise 6, page 90.) Sometimes the partial correlation between two variables is larger than the simple correlation (see Exercise 7, page 90).

This example should reinforce the notion that great care must be exercised in the interpretation of correlation coefficients. Partial correlation is useful for determining the extent to which apparent relationship between variables can be explained by the effects of outside variables. The partial correlation technique may be generalized to measure simultaneously the effects of more than one outside variable; see Garrett (1966, page 411).

5.6.3 Correlation for Dichotomous-Categorical Variables

We have been assuming in this chapter that the variables studied by the statistician are always *numerical* in nature, such as measures of height, weight, intelligence, and family size. Measurements on these variables can be easily and meaningfully translated into numbers. There are, however, many variables of interest to researchers which are considered *categorical* in nature rather than numerical. For example, if we were to study the relationship between army rank (x) and leadership ability (y), the y variable could be a numerical score on a leadership examination and the x variable would be categorical (PFC, corporal, captain, major, etc.). Obviously, no correlation technique that we have yet described could be used to measure the relationship between x and y — or between *any* pair of variables — one or both of which are categorical in nature. There do exist, however, some statistical techniques which give us correlation coefficients between two variables in situations where one or both of the variables have exactly *two* categories. We shall call these variables *dichotomous-categorical* variables.

For example, we might study the relationship between annual income and completion of a college degree, with annual income reported as numerical data and the other variable measured on two categories — "yes" and "no." Or we might want to study the correlation between two dichotomous-categorical variables: "church attendance" (yes or no) versus "belief in God" (yes or no).

There are four correlation coefficients which are used to describe the linear relationship between two variables, at least one of which is dichotomous-categorical: the *point biserial*, the *biserial*, the *phi coefficient*, and the *tetrachoric r*. An excellent reference for this material is Roscoe (1969, page 84).

Exercises

1. Find the rank correlation (r_{xy}') for the data in Exercise 1, page 74, and compare your answer to the corresponding value for r_{xy} calculated in Exercise 3(a), page 82.

2. The high school averages (x) and college freshman grade-point averages (y) for a sample of 20 subjects in a particular university are listed below. Find r_{xy}' for the sample and interpret your results.

Subject	x	y	Subject	x	y
1	92	2.50	11	86	3.15
2	83	1.95	12	78	2.75
3	80	2.60	13	94	3.28
4	95	3.55	14	75	2.20
5	82	2.95	15	85	3.36
6	90	3.40	16	78	3.08
7	74	1.62	17	88	3.15
8	88	3.60	18	82	3.02
9	82	2.82	19	82	2.56
10	74	1.98	20	74	2.42

3. Calculate r_{xy}' for the following data.

x:	186	225	676	87	121	158	492	302
y:	28	39	15	62	51	48	40	36

4. Ten prize hogs are submitted for competition at the county fair. Three judges study each entry and then list the ten in rank order (rank of one for the one judged best, two for second-best, etc.). Judges' results are as follows:

		Ranks	
Hog #	Judge A	Judge B	Judge C
1	9	9	6
2	3	1	3
3	7	10	8
4	5	4	7
5	1	3	2
6	6	8	4
7	2	5	1
8	4	2	5
9	10	7	9
10	8	6	10

Calculate the appropriate rank correlations to help you answer the following questions:
(a) Which pair of judges agree the most?
(b) Which pair of judges disagree the most?

5. If $r_{xy} = 0.38$, $r_{xz} = 0.62$, and $r_{yz} = 0.91$, find the partial correlation between x and y, with z held fixed.

6. In the text example (page 88) calculate $r_{xz \cdot y}$ and $r_{yz \cdot x}$ and interpret your results.

7. Suppose that we have sample data for a group of college sophomores on three variables (1) freshman year index, (2) intelligence, and (3) number of hours studied per week on the average during freshman year. We find that $r_{12} = 0.50$, $r_{13} = 0.40$, and $r_{23} = -0.30$. Find $r_{12 \cdot 3}$, $r_{13 \cdot 2}$, and $r_{23 \cdot 1}$, and interpret your results.

Summary

The *y-on-x linear regression line*, which has an equation of the form $y = ax + b$, provides a means for (a) describing a mathematical relationship between two variables, and (b) predicting the value of y for a given value of x.

The *Pearson product-moment correlation coefficient* measures the extent to which the regression line fits the given data. The coefficient, r_{xy}, varies in the interval from -1 to $+1$ depending upon the strength and direction of the linear relationship. If x tends to increase as y decreases, then r_{xy} is negative. If x tends to increase as y increases, then r_{xy}

is positive. The relationship between x and y is considered weakest when $r_{xy} = 0$ and progressively stronger as r_{xy} approaches either -1 or $+1$. It is important to remember that (a) r_{xy} measures only *linear* relationship, and (b) that this relationship is strictly mathematical with no causal implications attached.

Multiple regression and multiple correlation techniques are available for measuring the relationship among three or more variables. Other measures of linear relationship include: *rank, partial, biserial, point biserial, phi,* and *tetrachoric* correlations.

part 2

probability

6

Fundamental Concepts
of Probability

6.1 Introduction

We have already established the importance of probability
in the field of statistics (Chapter 1, Section 1.2). In Part II
we explore in detail the basic concepts of probability.

6.1.1 Decision Making

Each day we face situations in which decisions must be
made. Typically the decision-making process involves the
selection of a course of action chosen from several possibilities. The decision is based on what we expect will occur
at some future time. Since most, if not all, of us cannot
accurately predict future events, we find ourselves making
decisions in the face of doubt or uncertainty. For example,
the decision to take (or not take) an umbrella to work is
based on the likelihood that rain may fall. The decision to
buy (or not buy) a certain stock is based on the likelihood
that the stock will be a good investment. The football
quarterback makes hundreds of decisions per game, each
time basing his decision on what he expects the defense
will do.

In most situations decisions are not made randomly. That is, the decision to take or not take an umbrella is not determined by a flip of a coin. The investor does not choose a stock by pointing randomly at the *New York Times* stock quotations. Nor does the quarterback randomly flip the pages of his play book to make his decisions on the field. Decisions in everyday life are based on past experiences involving similar courses of action under similar conditions. If the weather man says it is going to rain, and in the past he has been right more often than not, then it is likely that one would carry an umbrella to work. If the sky is overcast on a certain morning, and we recall that in the past overcast mornings turned into sunny days more often than not, the umbrella would probably remain at home. Investors base their stock purchases on past performance of the stocks relative to the characteristics of the companies involved and the overall conditions of the market. The quarterback studies films of his future opponents to determine how the defense is likely to respond to each of the plays he might call.

Since good decision making relies heavily upon past experiences, it is no surprise that experience is considered an important asset in any job situation. An experienced investor is more likely to make correct decisions, all other things equal, than an inexperienced one. A senior quarterback with plenty of game experience is often more successful than a sophomore candidate with no varsity experience, even though the sophomore may be bigger, faster, and able to throw a ball farther.

When dealing with everyday situations involving uncertainty, the layman uses terms such as "probably," "likely," and "possibly" to express that uncertainty. These terms are somewhat ambiguous since they do not mean exactly the same thing to everyone, nor do they give a clue as to the extent of uncertainty involved. The statistician prefers to be somewhat more precise and quantitative when dealing with events involving uncertainty. He avoids such terms as probably, likely, and possibly, and talks instead about the *probability of an event*. Before we attach any quantitative significance to the probability of an event, we shall first introduce some basic notation.

6.1.2 Notation

An action which may or may not (a) have occurred in the past, (b) be occurring now, or (c) occur in the future, may be described as an *event*. We shall use capital letters to represent events. For example, we could let R represent the event: "It will rain today." If we wish to talk about the probability that it will rain today, we use the notation $P(R)$.

Definition 6.1: Given that E represents an event, the notation $P(E)$ represents *the probability that E occurs*, or simply, *the probability of E*.

For simplicity we think of events as either occurring or not occurring. We represent the probability that E does *not* occur by the notation $P(\text{not-}E)$.

> **Definition 6.2:** Given event E, the notation $P(\text{not-}E)$ represents *the probability that E does not occur.*

If we consider two different events, E and F, then one, and only one, of the following possibilities is true;

(i) E and F both occur,
(ii) E occurs, but F does not,
(iii) F occurs, but E does not, or
(iv) neither E nor F occurs.

To represent the probability that E and F both occur, we write $P(E \text{ and } F)$. This notation expresses the likelihood that (i) is true, assuming that (i), (ii), (iii), and (iv) are all possible.

> **Definition 6.3** Given the events E and F, the notation $P(E \text{ and } F)$ represents *the probability that E and F both occur.*

To represent the probability that at least one of the events E and F occurs we write $P(E \text{ or } F)$. This notation expresses the likelihood that (i), (ii), or (iii) is true, assuming that (i), (ii), (iii), and (iv) are all possible.

> **Definition 6.4:** Given the events E and F, the notation $P(E \text{ or } F)$ represents *the probability that at least one of the events E and F occurs.*

Suppose that John and Bill attempt independently to decode a secret message. Let A represent the event that John will break the code. Let B represent the event that Bill will break the code. Then the probability that the code will be broken may be represented by $P(A \text{ or } B)$, since the code will be broken if either one or both of the men break it—that is, if *at least one* of them breaks it.*

Frequently in statistical work we have occasion to talk about the probability that one event occurs, given that another definitely occurs. To express the probability that E occurs, given that F occurs, we use the notation $P(E|F)$. This notation expresses the likelihood that (i) is true, assuming that (i) and (iii) *only* are possible!

---

*Note the similarity between "E and F" and "$A \cap B$," and between "E or F" and "$A \cup B$," where E and F are events and A and B are sets. We shall capitalize on this similarity shortly.

Definition 6.5:   Given any two events, E and F, the notation $P(E|F)$ represents *the probability of E, given* that F occurs*, and we say that such an expression denotes *conditional* probability.

A careful study of Definitions 6.3, 6.4, and 6.5 should convince the reader that $P(E$ and $F)$ and $P(F$ and $E)$ mean the same thing; $P(E$ or $F)$ and $P(F$ or $E)$ mean the same thing; *but* $P(E|F)$ and $P(F|E)$ are *not* equivalent notations.

Exercises

1. Let A represent the event that Bill will get cancer within the next ten years, and B the event that Bill will stop smoking within the same period of time. Write the following probability statements using the appropriate symbolism.
    The probability that within the next ten years Bill will:
    (a)   stop smoking,
    (b)   stop smoking and get cancer,
    (c)   not get cancer,
    (d)   either stop smoking or get cancer (or both),
    (e)   get cancer, given that he stops smoking.

2. A phone number is chosen randomly from the telephone book, and the number is called. Let E represent the event that the person who answers is a male, and F the event that the person who answers is a Republican. What probability statement does each of the following represent?
    (a) $P(\text{not-}E)$,   (b) $P(E$ and $F)$,   (c) $P(E$ or $F)$,
    (d) $P(E|F)$,   (e) $P(F|E)$.

3. Judging from the definitions in this section, what would be reasonable interpretations for the following symbolisms?
    (a) $P(A$ and B and $C)$ (b) $P(A$ or B or $C)$

6.2   The Empirical Approach

Now that we have defined some probability notation, we turn to the task of defining probability itself. That is, we want to be able to make statements such as, "The probability of event E is x," where x is a *number* (i.e., $P(E) = x$). We need to have a definition of probability in order to determine the value of x. There are basically two ways to approach a definition of probability—the *empirical* approach and the *classical* approach. The empirical approach depends upon the notation of *relative frequency*.

Consider the following example. Miss Smith is supposed to be at work at 8 o'clock every Monday morning. Let T represent the event that "Miss Smith will be on time (at or before 8:00 a.m.) Monday morning."

---

*If the term "given" is confusing, substitute the word "assuming."

Miss Smith's boss, Mr. Jones, has noticed that she does not always come to work on time on Monday morning. Therefore, if Mr. Jones wished to dictate a letter to Miss Smith at 8:00 a.m. on a particular Monday, he would feel more or less *uncertain* with regard to Miss Smith's availability at that time. That is, he would feel more or less uncertain with regard to the occurrence of event T. His *degree* of uncertainty would depend upon what proportion of times in the past Miss Smith arrived late to work on Monday morning. If Miss Smith were rarely late to work, let us say around five percent of the time, then Mr. Jones would feel very little uncertainty with regard to the occurrence of T (i.e., he would be almost certain that T would occur). If Miss Smith were often late, say about 95 percent of the time, again Mr. Jones would feel very little uncertainty with respect to the occurrence of T (i.e., he would be almost certain that T would *not* occur). On the other hand, if his secretary were late about half the time, Mr. Jones would feel a great deal of uncertainty with regard to the occurrence of T (i.e., he would consider the occurrence of T a "toss-up"). Thus, the degree of uncertainty regarding the occurrence of event T depends upon the proportion of times in the past that Miss Smith arrived at work on time. That proportion is called the *relative frequency* of event T.

> **Definition 6.6:** The number of times event E occurs in a certain time period divided by the number of times it had an opportunity to occur in that period is called the *relative frequency* of E for that period.

If Miss Smith had previously been on time 40 times out of 50 Mondays, then the relative frequency of T would be $40/50 = 0.80$.

Clearly, the size of the relative frequency of an event reflects one's feelings of uncertainty with regard to the occurrence of that event. A very small or very large relative frequency reflects little uncertainty; a relative frequency around 0.50 indicates *maximum* uncertainty. Thus the size of the relative frequency of T is reflective of Mr. Jones' feelings of uncertainty. However, Mr. Jones' feelings are affected also by the number of times he had an opportunity to observe Miss Smith arriving on Monday morning. If Mr. Jones observes her time of arrival on only five Mondays and finds her on time only twice (relative frequency = 0.40), it could be argued that in such a small sample of observations Mr. Jones did not get a true indication of his secretary's degree of punctuality. On the other hand, if Mr. Jones observes that his secretary is on time 20 times out of 50 Mondays, he would feel more confident that the relative frequency of T (still 0.40) is a *reliable* measure. Intuitively, it is clear that as the number of observations increases, the relative frequency becomes a more reliable measure. The ideal situation would be one in which the number of observations approaches infinite size! Statisticians refer to this theoretical situation as "the long run." We say theoretical because, in practice, it is impossible to make an infinite number of observations. However, since it is intuitively clear that (a) the higher the relative frequency of an event, the more "likely" or "probable" the event is, and (b) the greater the number of observations, the more reliable the relative frequency, we choose to define "probability" as a "long-run relative frequency."

> **Definition 6.7:   The Empirical Definition of Probability.** The *probability of event E*, denoted by $P(E)$, is the relative frequency of E *in the long run*.

It is obvious from Definition 6.7 that we cannot in fact calculate an exact value of $P(E)$ because we cannot achieve a "long run" situation in practice. However, we can use the relative frequency of E as an *estimate* of $P(E)$. The more observations we have for the calculation of the relative frequency of E, the more accurate is our estimate of $P(E)$. The probability of any event E, therefore, is an elusive, theoretical number which can be at best *estimated* by calculating the relative frequency of E.

Exercises

1. Listed below are the relative frequencies for five different events. Which one represents the greatest degree of uncertainty with respect to the occurrence of the event?
    (a) 1.00   (b) 0.46   (c) 0.12
    (d) 0.31   (e) 0.87

2. Let A be the event that the noon train will arrive on time. Assume that the train runs every day. In the past year (365 days) the train has been late only 13 times. Estimate the probability of A.

3. The probability that a baseball player will get a hit when he steps up to the plate is estimated by a relative frequency which baseball fans call a _____.

4. Flip a pair of pennies 100 times, and each time record the number of heads that occur. Let A = the event that two heads occur, B = the event that only one head occurs, and C = the event that no heads occur. Estimate $P(A)$, $P(B)$, and $P(C)$. How could you improve on the reliability of your estimates?

6.3   The Classical Approach

6.3.1   Historical Antecedent

Probability theory had its early beginnings back around the seventeenth century as an outgrowth of gambling games. Such games of chance included actions such as flipping coins, drawing playing cards, and throwing dice. The outcome of any flip, draw, or throw is, of course, uncertain. The development of a theory of probability received early impetus when gamblers began noticing that although individual out-comes are uncertain, there are "long-term" outcomes which are more or less predictable. Although the occurrence of a head is no more predictable than the occurrence of a tail on a single coin flip, it is true that when a coin is flipped many times, heads occur typically about half of the time. In a single draw from a deck of playing cards, a spade is no more likely to occur than any other suit, but in many draws from a deck of cards we can "predict" that about one-fourth of the selections will be spades. Early observations such as these were essentially a primitive attempt at the empirical approach, described in the previous section. From this point on mathematicians took over and within the same setting (i.e., games of chance) developed a classical approach to the definition of probability.

6.3.2 Chance Experiments, Sample Spaces, and Events

In statistical work the term "experiment" is generally used to denote a well-defined course of action which has two or more possible outcomes. Typically, the result of a given experiment is somewhat unpredictable; that is, there is an element of uncertainty with regard to the possible outcomes of the experiment. We shall begin our development of a classical definition of probability by examining some experiments associated with games of chance, such as flipping coins, tossing dice, and making selections from a group of objects. We shall call experiments of this type *chance experiments*.

> **Definition 6.8:** The set of all possible physical outcomes of a chance experiment is called the *sample space*, denoted by S, for that experiment.

For example, if the chance experiment involves flipping a single coin, then $S = \{h,t\}$. That is, the sample space is the set of all possible outcomes h (head) and t (tail). If the chance experiment involves flipping two coins, then $S = \{(h,h), (h,t), (t,h), (t,t)\}$. The notation "$(h,h)$" repre-sents the possibility that both coins could come up heads. Note that both of the outcomes (h,t) and (t,h) are included. Why? When we talk about two different sample spaces in the same sentence or paragraph, for the sake of clarity it is wise not to use the same symbol (S) to represent both sets. To avoid any possible ambiguity we could let $S_1 = \{h,t\}$ and $S_2 = \{(h,h), (h,t), (t,h), (t,t)\}$. Note, in reference to Definition 1.8 in Chapter 1, that $S_2 = S_1 \times S_1$ (i.e., the cartesian product of S_1 with S_1). We usually shorten $\{(h,h), (h,t), (t,h), (t,t)\}$ to $\{hh, ht, th, tt\}$ for convenience. When numbers are used, however, the shortened form can lead to con-fusion since, for instance, we think of 12 as "twelve" rather than "one-two." A single throw of a die gives us $S_3 = \{1, 2, 3, 4, 5, 6\}$ and a single throw of a pair of dice gives us

$$S_4 = S_3 \times S_3 = \{(1,1), (1,2), (1,3), \ldots , (6,6)\}$$

where (1,3), for example, is the outcome "one on the first die, and three on the second die." It would not be appropriate for us to use $\{11, 12, 13, \ldots , 66\}$ in place of $\{(1,1), (1,2), (1,3), \ldots , (6,6)\}$. How many different outcomes are possible when two dice are tossed?*

The reader should take specal note of the word "physical" in Definition 6.8. This term is included in the definition of sample space as a reminder that all possible *physically different* results must be represented separately in S. For example, one might be tempted to express the sample space for the two-coin chance experiment as

$$S = \{0,1,2\}$$

where 0 stands for the outcome "no heads," 1 for "one head," and 2 for "two heads." According to Definition 6.8 this set is not a legitimate sample space, since the element 1 actually represents two different physical outcomes. Similarly, one might try to express the sample space for the two-dice chance experiment as

$$S = \{2,3,4,5,6,7,8,9,10,11,12\}$$

where each number in set S represents one of the possible sums on the two dice. Again, this set is not a proper sample space according to Definition 6.8. Why?

Up to this point we have considered only *finite* sample spaces. It is possible to conceive of an *infinite* sample space, that is, one with an unlimited number of outcomes. For example, if in a chance experiment a coin were flipped until a head appeared, the sample space would be

$$S = \{h, th, tth, ttth, tttth, \ldots\}$$

Although infinite sample spaces are possible, for the purpose of defining probability we shall limit ourselves to *finite* sample spaces.

Definition 6.9: Every subset of the sample space represents an event, and therefore, any *event* associated with a chance experiment can be represented by a *subset* of the sample space for that experiment.

---

*It should be noted that n flips of a single coin and one flip of n identical coins are for all intents and purposes the same experiment with the same outcomes. In the former case, each outcome in S consists of the individual outcomes for flip one, flip two, and so on, while in the latter case, each outcome in S consists of the individual outcomes for coin one, coin two, etc. We can make a similar statement with respect to tossing one die n times versus tossing n identical dice once.

For example, let A represent the event that (in the two-dice experiment) the same number comes up on both dice. Therefore,

$$A = \{(1,1), (2,2), (3,3), (4,4), (5,5), (6,6)\}$$

Or let B represent the event that the sum of the numbers on the dice equals six. Then

$$B = \{(1,5), (2,4), (3,3), (4,2), (5,1)\}$$

If C represents the event that the sum on the dice is 15, then $C = \phi$, since a sum of 15 is not possible. (Note that the use of the empty set to represent the event C is legitimate since ϕ is a subset of S.)

Consider a different chance experiment. Suppose that we wish to randomly select one marble from a group of three marbles, two red and one black. Let r_1 and r_2 represent the red marbles and b the black one. Let R represent the event that the marble chosen is red. It follows, then, that

$$S = \{r_1, r_2, b\} \quad \text{and} \quad R = \{r_1, r_2\}$$

Now let us change the experiment slightly by making it a selection of two marbles instead of one. Let T be the event that at least one marble is black, and let M be the event that both marbles are the same color. Therefore,

$$S = \{br_1, br_2, r_1b, r_1r_2, r_2b, r_2r_1\}$$
$$T = \{br_1, br_2, r_1b, r_2b\}$$
$$M = \{r_1r_2, r_2r_1\}$$

In this example we are, of course, assuming that after the first selection is made, the marble selected cannot be chosen again on the second selection. That is, we make the first selection from three marbles and the second from the two remaining. This procedure is called *selection without replacement*.

On the other hand, we may wish to select the two marbles *with replacement*. In that case the sample space S and the events T and M would be

$$S = \{bb, br_1, br_2, r_1b, r_1r_1, r_1r_2, r_2b, r_2r_1, r_2r_2\}$$
$$T = \{bb, br_1, br_2, r_1b, r_2b\}$$
$$M = \{bb, r_1r_1, r_1r_2, r_2r_1, r_2r_2\}$$

Note that S, T, and M were all affected by replacement.

When we think about flipping a coin, we assume that the coin is perfectly balanced so that the outcomes on a single flip, h and t, are *equally likely* to occur. Similarly, we assume that the dice we talk about are perfect cubes with no irregularities such as occur in "loaded" dice. That is, we assume that each of the six possible outcomes on a die is *equally likely* to occur in a single throw. Any selections we make (with or without replacement) from a group of objects are made randomly so that on any particular selection each object is *equally likely* to be chosen. *If the outcomes in the basic experiment (i.e., outcomes resulting from one flip, one throw, one selection, and so on) are equally likely, then the outcomes in sample spaces resulting from multiple flips, throws, selections, etc., are also equally likely.*

We are now ready to define *probability* from the classical viewpoint. Note that the following definition applies only to chance experiments whose sample spaces are expressed as sets of *equally likely* outcomes.

Definition 6.10: The Classical Definition of Probability. Given a finite sample space S whose outcomes are equally likely, and an event E defined on S (i.e., E is a subset of S), *the probability of E, denoted by $P(E)$, is*

$$P(E) = \frac{n(E)}{n(S)}$$

Referring back to the two-dice experiment, the probability values assigned by Definition 6.10 to events A, B, and C are

$$P(A) = 6/36 = 1/6 \qquad P(B) = 5/36 \qquad \text{and} \qquad P(C) = 0/36 = 0$$

In the three-marble experiment (*without* replacement)

$$P(T) = 4/6 = 2/3 \qquad \text{and} \qquad P(M) = 2/6 = 1/3$$

In the three-marble experiment (*with* replacement)

$$P(T) = 5/9 \qquad \text{and} \qquad P(M) = 5/9$$

Note the effect that replacement has on the probability of events T and M.

6.3.3 *The Empirical vs. the Classical Approach*

Clearly the empirical and classical approaches are quite different. On the other hand, the two approaches are consistent with each other. That is, in both cases the probability of E is conceived as a number which represents the proportion of times E occurs in the *long run*. In the empirical approach we *estimate* $P(E)$ by actually repeating an experiment many times and recording the proportional occurrence of E. The major weakness of the empirical approach is the tediousness of the procedure. In the classical approach we think of E as a subset of a sample space S which we *assume* has equally likely outcomes. $P(E)$ is defined as a proportion—that is, the number of outcomes in E divided by the number of outcomes in S. A practical disadvantage of the classical approach is that many conceivable events, for which we may want to express probabilities, cannot be reasonably described as a subset of a sample space which has equally likely outcomes. The event that a radio tube will last more than 150 hours, and the event that a certain boy will graduate from a particular college are examples which cannot be legitimately forced into the classical framework. To express probabilities for events such as these, we seek empirical evidence.

Whereas the empirical approach deals with the actual manipulation of concrete physical objects, the classical approach operates strictly at a theoretical level. If we are interested in finding the probability, $P(E)$, that two coin flips will produce two heads, the empirical approach would involve actually flipping a pair of coins many times, recording the results, and calculating the relative frequency of E. We consider this relative frequency an *estimate* of $P(E)$. The classical approach to this same task would involve no actual flipping of coins. Instead, we would *assume* at a *theoretical level* that we have two perfectly balanced coins, each of whose basic outcomes, h and t, are equally likely. That assumption permits us to conclude that the sample space $S = \{hh, ht, th, tt\}$ has four equally likely outcomes, which in turn allows us to conclude that $P(E) = 1/4$.

In this text we shall use the classical approach as a basis for the development of a theory of probability. We shall concentrate on an understanding of the basic properties of probability, which shall take the form of various theorems in the succeeding sections. Thus we shall be operating at a theoretical level. When we have established a sufficient body of theoretical knowledge about probability, we shall, in the later chapters, apply the theory to practical problems in statistical inference.

Exercises

1. If a chance experiment has a total of m possible outcomes, then how many different events exist which are associated with that chance experiment?

2. Given that the sample space for a particular chance experiment is $S = \{1, 2, 3, 4, 5\}$, indicate which of the following sets represent events associated with that chance experiment.
   (a) ϕ (b) $\{4, 2, 5\}$ (c) $\{1, 2\}$
   (d) $\{4, 5, 6\}$ (e) $\{1, 2, 3, 4, 5\}$

3. A chance experiment involves first flipping a coin and then tossing a die. Find the sample space.

4. Find the sample space for the chance experiment in which three coins are flipped.

5. For the chance experiment in the previous problem, express the following events in set notation and find their probabilities.

   (a) A = event that exactly one head occurs.

   (b) B = event that at least two heads occur.

   (c) C = event that a tail occurs on the first coin.

6.   How many different equally likely outcomes are possible when two balanced dice are tossed? Express the following events in set notation and find their probabilities.

   (a) W = event that the sum on the dice equals 10.

   (b) X = event that at least one of the dice comes up "2."

   (c) Y = event that the first die comes up "5."

   (d) Z = event that the sum on the dice is less than six.

7.   Three marbles are numbered 1, 2, and 3. We put them in a hat, shake it, and then randomly select two marbles. Find the sample space, assuming selection

   (a) without replacement,   (b) with replacement.

8.   In the previous exercise let

        K = event that the numbers on both marbles are odd

        L = event that the sum of the numbers is less than five

        G = event that the same marble is selected twice

   Express these events in set notation and find their probabilities, first (a) assuming replacement, and then (b) assuming no replacement.

9.   Saying that "the odds *in favor* of event E are x *to* y" is equivalent to saying "$P(E) = x/(x + y)$." For example, if the odds in favor of rolling a sum of seven with two dice are *1 to* 5, then P (rolling a seven with two dice) = 1/6. Using this same example, the odds *against* rolling a seven are 5 to 1. With this new concept in mind,

   (a) find probabilities for events A, B, and C, given that the odds in favor of each are 4 to 7, 3 to 2, and 6 to 4, respectively;

   (b) find the odds in *favor* of events D, E, and F, given that $P(D) = 3/8$, $P(E) = 1/2$, and $P(F) = .17$; and

   (c) find $P(E)$, given that the odds *against* E are 15 to 11.

6.4 Counting Principles

From the classical definition of probability (Definition 6.10) we note that in order to calculate $P(E)$ we must be able to *count* the number of outcomes in event E and in sample space S. Up to this point we have listed the outcomes in set notation form and then counted them. The process of listing, however, can become extremely tedious and in fact prohibitive in cases where the number of possible outcomes in a chance experiment runs into the thousands. Since we actually need the *count* rather than the listing, we must learn some useful counting techniques which will eliminate the necessity for listing.

     We have already noted that the number of outcomes in the chance experiment involving the flip of one coin is two ($S = \{h,t\}$). If two coins are flipped, we have four possible outcomes ($S = \{hh, ht, th, tt\}$). If three coins are flipped, we have eight possible outcomes ($S = \{hhh, hht, hth,$

thh, tth, tht, htt, ttt}). The alert reader will note that there appears to be a predictable relationship between the number of coins flipped and the number of outcomes in S. When we replace the two, four, and eight by 2^1, 2^2, and 2^3, the relationship becomes quite clear; *if a coin is flipped k times, the number of possible outcomes is 2^k.* This fact is only a special case of a more general counting principle, which we shall use as a basic principle in this section.

Principle 1:   If k chance experiments have n_1, n_2, n_3, . . . , n_k possible outcomes, respectively, the joint occurrence of the k chance experiments has $(n_1)(n_2)$ . . . (n_k) possible outcomes.

To apply Principle 1 to the k-coin chance experiment we treat the flip of each coin as an individual chance experiment having two possible outcomes. Therefore, according to Principle 1, the joint occurrence of the k flips has

$(2)(2)(2)$ . . . $(2) = 2^k$ possible outcomes.
   (k factors)

Similarly, if a die and a coin were tossed, the number of possible outcomes would be $(6)(2) = 12$. If two dice and three coins were tossed, the number of possible outcomes would be $(6)(6)(2)(2)(2) = 288$. If one die were tossed, one coin flipped, and one card selected from a deck of playing cards, then the number of possible outcomes would be $(6)(2)(52) = 624$.

Consider a chance experiment in which three random selections are made, one at a time, from the letters a, b, and c *with replacement.* That is, the first letter chosen is replaced and is eligible for the next selection, and so on. According to Principle 1, the sample space for this experiment contains $(3)(3)(3) = 27$ possible outcomes. If, on the other hand, we do not allow replacement, the number of outcomes in S according to Principle 1 would be $(3)(2)(1) = 6$ ($S = \{abc, acb, bac, bca, cab, cba\}$). Note that this set is, in fact, an exhaustive list of ways in which the letters a, b, and c can be ordered. Thus, with respect to the set of possible outcomes, we can say that ordering n objects is equivalent to selecting n objects from n objects without replacement. Thus an ordering of n objects can be accomplished in $(n)(n-1)(n-2)$ . . . (1) different ways.

The notation $(n)(n-1)(n-2)$ . . . (1), which represents *the product of the first n positive integers*, is somewhat awkward, so we have a special symbol for it.

Definition 6.11:   If n is a positive integer, the symbol $n!$, pronounced n factorial, represents the product of all positive integers less than or equal to n. If $n = 0$, then $n! = 1$.

Definition 6.11 gives us a short way of writing some long products. For example, 7! is a short way of writing 5040, since $7! = (7)(6)(5)(4)(3)(2)(1) = 5040$. Note also from the definition that $0! = 1$. Referring back to the previous paragraph we can now state a second counting principle.

Principle 2: The number of different ways in which n objects can be ordered is $n!$ [which is equal to the number of different ways in which n objects can be selected from n objects without replacement].

Thus, the number of different ways in which four objects can be selected from themselves without replacement is equal to $4! = (4)(3)(2)(1) = 24$. But how many different ways can two objects be selected from four without replacement? According to Principle 1, the answer is simply $(4)(3) = 12$, since on the first selection there are four possible outcomes and on the second there are only three possible outcomes. This leads us to a third principle.

Principle 3: The number of different ways in which r objects can be selected from n objects without replacement $(r \leq n)$ is symbolized by $_nP_r$ and called *the number of PERMUTATIONS of n objects selected r at a time.* By definition

$$_nP_r = \frac{n!}{(n-r)!}$$

Thus, if we are selecting two objects from four without replacement, the number of permutations, $_4P_2$, can be found by using Principle 3, as follows:

$$_4P_2 = \frac{4!}{(4-2)!} = \frac{4!}{2!} = \frac{(4)(3)(2)(1)}{(2)(1)} = (4)(3) = 12$$

Note that Principle 2 is really a special case of Principle 3 when $r = n$, because in that case

$$_nP_v = {_nP_n} = \frac{n!}{(n-n)!} = \frac{n!}{0!} = \frac{n!}{1} = n!$$

If we take the objects a, b, c, and d and list all possible ways of selecting two of them without replacement, we get $S = \{ab, ac, ad, ba, bc, bd, ca, cb, cd, da, db, dc\}$. Note that in enumerating permutations, not only do we take into account *selection* but also the *order* in which the selections can be made. There are many situations, however, in which the number of different selections of r objects from n objects is important, but the *order* of selection is of no concern. For example, suppose that a, b, c, and d represent four different people who are eligible for election to a two-member committee, and we want to know how many different two-member committees are possible as a result of the election. In this context, outcomes *"ab"* and *"ba"* in the above sample space would represent the same outcome (i.e., the same committee). In situations such as this where the order of selection is unimportant, we count *combinations* instead of permutations, according to Principle 4.

Principle 4: The number of different ways in which r objects can be selected from n objects $(r \leq n)$ without replacement, *and without regard to order of selection,* is symbolized by $\binom{n}{r}$ and called *the number of COMBINATIONS of n objects selected r at a time.* By definition

$$\binom{n}{r} = \frac{n!}{(n-r)!r!}$$

In the preceding example, the number of combinations of four people selected two at a time is

$$\binom{4}{2} = \frac{4!}{2!2!} = \frac{(4)(3)(2)(1)}{(2)(1)(2)(1)} = 6$$

Note the similarity between the formula for $_nP_r$ and $\binom{n}{r}$. They differ only by a factor of $r!$. This is not surprising, since every single combination of r objects can be ordered in $r!$ ways (by Principle 2). The collection of all such orderings is equal to $_nP_r$. That is, mathematically speaking,

$$\binom{n}{r} \cdot r! = \frac{n!}{(n-r)!r!} \cdot r! = \frac{n!}{(n-r)!} = {_nP_r}$$

When confronted with a problem involving repetition of an action, such as repeated flips of a coin, or tosses of a die, or selections from a group of objects, we must first ask ourselves whether replacement is assumed or not? We shall use the following rule of thumb: *if there is no mention in the problem regarding the question of replacement, we shall assume that there is no replacement.* Of course in coin-flipping and die-tossing type problems we have, in essence, implicit replacement; however, in problems where we are making selections from a group of objects, we must depend upon our rule of thumb.

If we can assume replacement, Principle 1 provides a simple counting method. If a coin is flipped six times (or six coins flipped once), the sample space has $(2)(2)(2)(2)(2)(2) = 64$ possible outcomes. If a die is tossed three times (or three dice tossed once), the sample space has $(6)(6)(6) = 216$ possible outcomes. If four selections with replacement are made from a group of five objects, the sample space has $(5)(5)(5)(5) = 625$ possible outcomes.

If we assume no replacement, we must ask ourselves another question, "Is *order* important (permutations) or not important (combinations)?" The answer to this question must come from the wording in the problem.

Problem: Eight runners are entered in a State Championship 100 yard dash semifinal race. The first three finishers will go on to the finals. For this semifinal race, assuming no ties,
(a)   how many different results are possible,
(b)   how many different results are possible with respect to first, second, and third places only, and
(c)   how many different groups of three runners are possible finalists?

Solutions: Clearly the problem involves no replacement (since a single runner cannot occupy more than one finishing position).
(a)   The result of a race obviously involves *order*. We want the number of ways that we can order eight objects, which is $8! = 40,320$. (Principle 2)
(b)   We are interested in order, but only in the first three positions. We want to know how many ways we can fill the first three positions, selecting from eight objects and taking order into account. Thus, we want to find the number of *permutations* of eight objects selected three at a time, which is $_8P_3 = 336$. (Principle 3)
(c)   A given set of three runners, finishing in the first three positions, will go to the finals regardless of who places first, second, or third. Therefore, to count the number of possible groups of three finalists, we need to find how many subsets of three objects can be selected from eight objects without regard to order. That is, we want the number of *combinations*: $\binom{8}{3} = 56$. (Principle 4)

Other applications of the counting principles are represented in the following examples:

Problem: A die is tossed twice.
(a)   How many of the possible 36 outcomes involve no 2s?
(b)   How many of the 36 involve an even number on the first toss and a number greater than four on the second?

Solutions: (a)   If neither toss yields a 2, this means that there are only five possible outcomes for each toss. Thus, there are $(5)(5) = 25$ outcomes involving no 2s. (Principle 1)
(b)   There are three possible even-number outcomes for the first toss and two possible outcomes greater than four for the second toss. Thus, the answer is $(3)(2) = 6$ outcomes. (Principle 1)

Problem: A box of 10 light bulbs contains 7 good bulbs and 3 defective bulbs. If we select 4 bulbs from the box of 10 (without replacement):
(a)   how many outcomes are there in the sample space?
(b)   how many of the outcomes in (a) involve a defective bulb on the first selection and good bulbs on the other three?
(c)   how many of the outcomes in (a) involve one defective bulb and three good bulbs?

Solutions: (a) $(10)(9)(8)(7) = 5040$ outcomes (Principle 1), or simply $_{10}P_4 = 5040$ outcomes (Principle 3)

(b)   Since the first selection is a defective bulb, there are only three possible outcomes. And since the last three selections must be good bulbs, there are seven, six, and five possible outcomes, respectively. Thus, there are $(3)(7)(6)(5) = 630$ outcomes (Principle 1), or simply $(_3P_1)(_7P_3) = 630$ outcomes (Principles 1 and 3)

(c)   At first glance (b) and (c) seem to ask the same question; however, obtaining one defective bulb and three good bulbs in four selections can occur in four distinct ways, depending upon which selection yields the defective bulb. Part (b) refers to only one of those four ways; namely, defective first, then the three good bulbs (*dggg*). But there are three other ways: *gdgg*, *ggdg*, and *gggd*. Therefore, in (c) we want: $(3)(7)(6)(5) + (7)(3)(6)(5) + (7)(6)(3)(5) + (7)(6)(5)(3) = 2520$ outcomes, which, of course, is just four times the answer in (b).

Exercises

1.   Find the value of each of the following:

   (a) $8!$ (b) $_7P_3$ (c) $\binom{5}{2}$

   (d) $\binom{12}{8}$ (e) $_{50}P_2$ (f) $\binom{5}{5}$

   (g) $_5P_5$ (h) $_5P_0$

2.   Show that for any $n \geq 0$,

   (a) $\binom{n}{n} = 1$ (b) $\binom{n}{1} = n$ (c) $\binom{n}{n-1} = n$

   (d) $\binom{n}{0} = 1$ (e) $_nP_{n-1} = n!$ (f) $_nP_0 = 1$

   (g) $_nP_1 = n$ (h) $\binom{n}{n-r} = \binom{n}{r}$

3.   (a)   How many different ways can six books be arranged side by side on a bookshelf?

   (b)   Suppose that of the six books, three are math texts. How many of the arrangements in 3(a) have the three math texts side by side?

   (c)   Suppose that the three math texts are volumes I, II, and III of the same work. How many of the outcomes in 3(a) have the math texts side by side in I-II-III order?

4.   In a room of 20 people everyone shakes hands with everyone else exactly once. How many handshakes take place?

5.   Suppose that in your state the car license plate has seven positions – the first two for letters (*a* through *z*) and the last five for digits (0–9). Under these conditions,

   (a)   how many license plates can be issued,

   (b)   how many of these plates end in "528," and

   (c)   what is the probability that a license plate chosen randomly ends in "528"?

6. How many integers are there *between* 99 and 1000 in which no digit is repeated?

7. Three dice are tossed.
   (a) How many outcomes are in the sample space?
   (b) How many of these outcomes have no 4s?
   (c) How many have at least one 4?
   (d) How many have all even numbers?
   (e) How many have exactly two 4s?

8. A large urn contains five black marbles and three white ones. Four selections are to be made from this urn *with replacement*.
   (a) How many outcomes are there in the sample space?
   °(b) How many outcomes in S involve three black marbles on the first three selections, and a white marble on the fourth selection?
   (c) What is the probability that the first three selections are black and the fourth is white?
   °(d) How many of the outcomes in S involve three black marbles and one white?
   (e) What is the probability that three of the four selected are black?
   (f) What is the probability that the four selections will yield two blacks and two whites?

9. Do Exercise 8 again, assuming *no replacement*.

10. A club contains 10 boys and 8 girls. Three officers are to be elected: President, Vice-President, and Secretary (who form the executive committee). No member can hold more than one office.
    (a) How many different outcomes are possible for this election?
        (i) How many of these outcomes involve all boys?
        (ii) How many involve all girls?
        (iii) How many involve two girls and one boy?
    (b) How many different groups of three people are eligible for election to the executive committee?
        (i) How many of these outcomes involve all boys?
        (ii) How many involve all girls?
        (iii) How many involve two girls and one boy?

11. Assuming that Ann, Bill, and Dave are all in the club mentioned in Exercise 10, and assuming that the election is a random selection process, find the probability that
    (a) Ann is elected President, Bill, Vice-President, and Dave, Secretary.
    (b) Ann, Bill, and Dave are on the executive committee.
    (c) the executive committee includes two girls and one boy.
    (d) the President is a boy and the Vice-President and Secretary are girls.

12. A group contains eight labor and four management representatives.
    (a) How many different committees of three could be selected from this group?

°Note the distinction between questions 8(b) and 8(d).

(b)   What is the probability that a random selection of three from the group would include two from labor and one from management?

(c)   What is the probability that a random selection of four from the group would include at least one representative from labor and at least one from management?

13.   If a box of 25 machine parts contains three defective parts, what is the probability that a random selection of five parts from the box will include the three defectives?

14.   Show that $_nP_r = n(n-1)(n-2) \ldots (n-r+1)$.

6.5 Probability of Compound Events*

In Section 6.1.2 probability notation was first introduced. The symbols $P(E)$, $P(\text{not-}E)$, $P(E \text{ and } F)$, $P(E \text{ or } F)$, and $P(E|F)$ may be used whenever we wish to informally express certain probabilities in a succinct manner. In Section 6.3.2 we formally defined $P(E)$ in the classical context, where the event E is a subset of the sample space S of a chance experiment. Since all possible events for a particular chance experiment must be subsets of sample space S, it is convenient for us to think of S (for a given chance experiment) as the *universal set*. Since not-E represents "the event that E does not occur," it follows that not-E in the classical context is none other than E', the *complement* of E. Therefore, to find the probability that E does not occur, we simply find E' and then use the classical definition of probability (Definition 6.10) to find $P(E')$.

Similarly, we can find set language equivalents to replace "E and F" and "E or F." If E and F are events defined on a particular sample space, then the event that "E and F both occur" is represented by the set $E \cap F$. The event that "at least one of the events E and F occurs" is represented by the set $E \cup F$. Therefore, $P(E \text{ and } F) = P(E \cap F)$ and $P(E \text{ or } F) = P(E \cup F)$ in the classical context. We can, therefore, apply Definition 6.10 to find $P(E \cap F)$ and $P(E \cup F)$.

The following example demonstrates the use of Definition 6.10 to calculate $P(E')$, $P(E \cap F)$, and $P(E \cup F)$. Consider a chance experiment in which three coins are tossed. Let A represent the event that exactly two heads occur, and let B represent the event that a head occurs on the second coin. It follows that (see Figure 6.1)

$$S = \{hhh, hht, hth, thh, tth, tht, htt, ttt\}$$
$$A = \{hht, hth, thh\}$$
$$B = \{hhh, hht, thh, tht\}$$

Let us find the probability that (a) A does not occur, (b) B does not occur, (c) A and B both occur, and (d) at least one of the two events occurs. First

---

*A review of set notation (Section 1.3) is suggested prior to the reading of this section.

we find A', B', $A \cap B$, and $A \cup B$, and then we apply Definition 6.10.

(a) $A' = \{hhh, tth, tht, htt, ttt\}$; $P(A') = 5/8$
(b) $B' = \{hth, tth, htt, ttt\}$; $P(B') = 4/8 = 1/2$
(c) $A \cap B = \{hht, thh\}$; $P(A \cap B) = 2/8 = 1/4$
(d) $A \cup B = \{hhh, hht, hth, thh, tht\}$; $P(A \cup B) = 5/8$

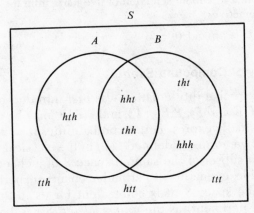

Figure 6.1.   Venn diagram showing outcomes in S, A, and B.

To develop a method for calculating conditional probability, we must carefully consider the meaning of $P(E|F)$ relative to the meaning of $P(E)$. In both cases we are talking about the probability that E occurs. When we ask for $P(E)$, we want to know what proportion of the outcomes in the entire sample space S are in E. When we ask for $P(E|F)$, we want to know what proportion of the outcomes *in* F are in E. In other words, to calculate $P(E|F)$ by Definition 6.10 we must treat F as the sample space. That is, we assume that the outcomes in F are the only possible outcomes. Within that restricted sample space, we then determine what proportion of the outcomes are in set (event) E. Of course, the set of outcomes in F which are in E is $E \cap F$. Therefore, to find $P(E|F)$ we simply divide the number of outcomes in $E \cap F$ by the number of outcomes in F.

Although we could stretch a point and consider the procedure described above as a special application of Definition 6.10, for clarity let us state separately a classical definition for conditional probability.

Definition 6.12: The Classical Definition for Conditional Probability.
Given a finite sample space, S, whose outcomes are equally likely and events E and F defined on S, then

$$P(E|F) = \frac{n(E \cap F)}{n(F)} \quad \text{(assuming } F \neq \phi)$$

In the three-coin example, $P(A|B) = 2/4 = 1/2$. Note also that $P(B|A) = 2/3$, and therefore, in general we must say that $P(E|F) \neq P(F|E)$. Is $P(E \cap F) = P(F \cap E)$? Is $P(E \cup F) = P(F \cup E)$?

Consider the following example. We plan to randomly select one marble from a box of 20 marbles numbered 1, 2, 3, . . . , 20. Let G represent the event that an even-numbered marble is selected. Let H represent the event that the number on the selected marble is a multiple of three. Thus, the pertinent events are

$$S = \{1, 2, 3, \ldots, 20\}$$
$$G = \{2, 4, 6, \ldots, 20\}$$
$$H = \{3, 6, 9, 12, 15, 18\}$$
$$G \cap H = \{6, 12, 18\}$$
$$G \cup H = \{2, 3, 4, 6, 8, 9, 10, 12, 14, 15, 16, 18, 20\}$$
$$G' = \{1, 3, 5, \ldots, 19\}$$
$$H' = \{1, 2, 4, 5, 7, 8, 10, 11, 13, 14, 16, 17, 19, 20\}$$

And Definitions 6.10 and 6.12 give us the following probabilities.

$P(G) = 1/2$ $P(G \cap H) = P(H \cap G) = 3/20$
$P(H) = 3/10$ $P(G \cup H) = P(H \cup G) = 13/20$
$P(G') = 1/2$ $P(G|H) = 1/2$
$P(H') = 7/10$ $P(H|G) = 3/10$

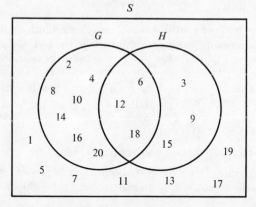

Figure 6.2. Venn diagram showing outcomes in S, G, and H.

Exercises

1. Two cards are to be selected randomly and without replacement from an ordinary deck of playing cards. Let X be the event that a spade is selected on the first draw, and Y be the event that a heart is selected on the second draw. Symbolize each of the following events by
    (i)   writing the statement in set notation form, and
    (ii) shading in the appropriate area on a Venn diagram.

    (a)   The card selected first is not a spade.

    (b)   The first card selected is a spade and the second is a heart.

    (c)   The first card selected is a spade and the second is not a heart.

    (d)   Either the first card selected is a spade or the second is a heart.

    (e)   Either the first card selected is not a spade or the second is a heart.

    (f)   Either the first card selected is not a spade or the second is not a heart.

    (g)   The first card selected is not a spade and the the second is not a heart.

2.   In Exercise 5, page 105, events A, B, and C are defined on the sample space for the three-coin chance experiment. For that same experiment (i) draw a Venn diagram which shows each of the eight outcomes in its proper place, and then (ii) find:

    (a) $P(A \cap B)$ (b) $P(A \cap C)$ (c) $P(B \cap C)$

    (d) $P(A \cup B)$ (e) $P(B \cup C)$ (f) $P(A \cup C)$

    (g) $P(A|B)$ (h) $P(B|A)$ (i) $P(A|C)$

    (j) $P(C|A)$ (k) $P(B|C)$ (l) $P(C|B)$

    (m) $P(C' \cap B)$ (n) $P(A' \cup C)$

3.   In Exercise 6, page 106, events W, X, Y, and Z are defined on the sample space for the two-dice chance experiment. For that experiment find the indicated probabilities.

    (a) $P(W \cap X)$ (b) $P(Y|W)$

    (c) $P(X \cap Z)$ (d) $P(X \cup Z')$

    (e) $P(Y \cap X'|W')$ (f) $P(X \cap Y \cap Z')$

4.   A box contains two red marbles (r_1, r_2) and three white marbles (w_1, w_2, w_3). We plan to select two marbles randomly from the box without replacement. Let R = the event that at least one of the marbles selected is red, and let W = the event that exactly one of the marbles selected is white. List the outcomes in S, R, and W and find the following probabilities.

    (a) $P(R \cap W)$ (b) $P(R \cup W)$

    (c) $P(R|W)$ (d) $P(W|R)$

    (e) $P(R' \cap W')$ (f) $P(R'|W')$

5.   Do Exercise 4 again — this time assuming replacement.

6.   The following table shows the distribution of males and females across the four undergraduate classes in a small college of 1369 undergraduate students.

	Class				
	Freshman	Sophomore	Junior	Senior	Total
Male:	207	183	161	148	699
Female:	201	157	177	135	670
Total:	408	340	338	283	1369

If a single student were randomly selected from the 1369, what is the probability that the student is

(a)   male,

(b)   junior,

(c) male junior,
(d) male, given that he is a junior,
(e) junior, given that he is a male,
(f) freshman or a sophomore,
(g) male freshman or a female sophomore,
(h) female or a senior.

6.6 Properties and Laws of Classical Probability

We are now ready to state and prove some useful theorems regarding certain properties and laws which govern classical probability.

Theorem 6.1: For every event E, $0 \le P(E) \le 1$.

Proof:

$$P(E) = \frac{n(E)}{n(S)} \qquad \text{(Def. 6.10)}$$

$$E \subseteq S \qquad \text{(Def. 6.9)}$$
$$O \le n(E) \le n(S) \qquad \text{(Def. 1.3)}$$
$$O \le P(E) \le 1 \qquad \text{(Divide through by } n(S))$$

Theorem 6.2: For every event E, $P(E') = 1 - P(E)$.

Proof: $P(E') = \dfrac{n(E')}{n(S)} \qquad \text{(Def. 6.10)}$

$$= \frac{n(S) - n(E)}{n(S)} \qquad \text{(Def. 1.4)}$$

$$= \frac{n(S)}{n(S)} - \frac{n(E)}{n(S)}$$

$$= 1 - P(E) \qquad \text{(Def. 6.10)}$$

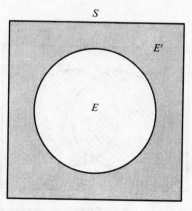

Figure 6.3. Venn diagram showing the relationship between E and E'.

Theorem 6.3: Multiplication Law. For any pair of events E and F ($E \neq \phi$ and $F \neq \phi$) in sample space S,

$$P(E \cap F) = P(E) \cdot P(F|E)$$

Proof:

$$P(F|E) = \frac{n(F \cap E)}{n(E)} \qquad \text{(Def. 6.12)}$$

$$= \frac{n(E \cap F)}{n(E)} \qquad (F \cap E = E \cap F)$$

$$= \frac{n(E \cap F)/n(S)}{n(E)/n(S)} \qquad \text{(Divide numerator and denominator by } n(S))$$

$$= \frac{P(E \cap F)}{P(E)} \qquad \text{(Def. 6.10)}$$

$$\therefore \quad P(E) \cdot P(F|E) = P(E \cap F) \qquad \text{(Mult. both sides by } P(E))$$

Theorem 6.4: Addition Law. For any pair of events E and F in the sample space S,

$$P(E \cup F) = P(E) + P(F) - P(E \cap F)$$

Proof:

$$P(E \cup F) = \frac{n(E \cup F)}{n(S)} \qquad \text{(Def. 6.10)}$$

$$= \frac{n(E) + n(F) - n(E \cap F)}{n(S)} \qquad \text{(By adding } n(E) \text{ and } n(F) \text{ we double count } n(E \cap F). \text{ See Figure 6.4)}$$

$$= \frac{n(E)}{n(S)} + \frac{n(F)}{n(S)} - \frac{n(E \cap F)}{n(S)}$$

$$= P(E) + P(F) - P(E \cap F) \qquad \text{(Def. 6.10)}$$

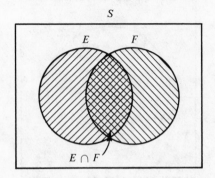

Figure 6.4. Venn diagram showing $E \cup F$ (the total shaded area) and its subsets E, F, and $E \cap F$.

Note that *Theorems 6.2, 6.3, and 6.4 allow us for the first time to calculate probabilities from other probabilities.* For example, if we know $P(E)$, $P(F)$, and $P(E \cap F)$, we can calculate $P(E \cup F)$ directly from Theorem 6.4. In fact, if we know any three of the four probabilities in Theorem 6.4, we can find the fourth. One obvious advantage to having these laws is versatility. For example, to find the probability that a queen or a heart will be selected in a single random draw from an ordinary deck of 52 playing cards, we can either

(a) Let E = the event that a queen or a heart will be selected.
    $P(E)$ is the probability we want.
        $n(E) = 16$ (i.e., 16 cards are either queens, hearts, or *both*)
        $n(S) = 52$
  ∴ $P(E)$ $16/52 = 4/13$

or

(b) Let Q = the event that a queen will be selected.
       H = the event that a heart will be selected.
    $P(Q \cup H)$ is the probability we want.
        $n(Q) = 4$
        $n(H) = 13$
        $n(Q \cap H) = 1$ (the queen of hearts)
        $n(S) = 52$
  ∴ $P(Q \cup H) = P(Q) + P(H) - P(Q \cap H)$
                    $= 4/52 + 13/52 - 1/52$
                    $= 16/52$
                    $= 4/13$

Thus, either approach gives the correct solution.

To demonstrate the use of Theorem 6.3, suppose that two random selections *without replacement* are to be made from a deck of playing cards, and we wish to calculate the probability that both selections will be spades. We can either

(a) Let E = the event that both selections will be spades.
    $P(E)$ is the probability we want.
        $n(E) = (13)(12) = 156$ (Using Principle 1)
        $n(S) = (52)(51) = 2652$
  ∴ $P(E) = 156/2652 = 1/17$

or

(b) Let A = the event that the first selection will be a spade.
       B = the event that the second selection will be a spade.
    $P(A \cap B)$ is the probability we want.
        $P(A) = n(A)/n(S) = 13/52$
        $P(B|A) = n(A \cap B)/n(A) = (13)(12)/(13)(51) = 12/51$
  ∴ $P(A \cap B) = P(A) \cdot P(B|A)$
                $= 13/52 \cdot 12/51 = 1/17$

This second option need not be as complicated as it looks. To find $P(B|A) = 12/51$ in a somewhat more direct manner, we can reason as follows: on the second draw, since we are assuming A (i.e., assuming that a spade is selected on the first draw), therefore there are only 51 cards left, 12 of which are spades. Thus, the probability that a spade occurs on the second draw, *given that a spade has already been removed*, is 12/51.

Before we leave this example, it should be noted that had we designated selection *with replacement*, the two approaches would have given

(a) $P(E) = n(E)/n(S) = (13)(13)/(52)(52) = 1/16$

or

(b) $P(A \cap B) = P(A) \cdot P(B|A) = 13/52 \cdot 13/52 = 1/16$

Note that $P(B|A)$ this time is 13/52, not 12/51, since the first card drawn is replaced before the second draw is made.

We might have occasion to find the probability that three or more events occur jointly. Since $P(A \cap B)$ represents the probability that events A and B occur jointly, it is reasonable that $P(A \cap B \cap C)$ should represent the probability that events A, B, and C occur jointly. Theorem 6.3 can be extended to three events as follows:

$$P(A \cap B \cap C) = P(A) \cdot P(B|A) \cdot P(C|A \cap B)$$

and similarly to four events;

$$P(A \cap B \cap C \cap D) = P(A) \cdot P(B|A) \cdot P(C|A \cap B) \cdot P(D|A \cap B \cap C)$$

The equation governing the joint probability of k events is expressed in Theorem 6.5, which we shall state without proof.

Theorem 6.5:   Generalized Multiplication Law: Given k events $E_1, E_2, E_3, \ldots, E_k$, the probability that they jointly occur is

$$P(E_1 \cap E_2 \cap \ldots \cap E_k)$$
$$= P(E_1) \cdot P(E_2|E_1) \cdot \ldots \cdot P(E_k|E_1 \cap E_2 \cap \ldots \cap E_{k-1})$$

(proof omitted)[*]

Suppose that four selections are to be made without replacement from a deck of playing cards. What is the probability that the first two selections are diamonds and the last two are hearts?

[*]In this theorem and others throughout the remainder of this text for which proofs are omitted without explanation, it may be assumed that the proof requires mathematical competence beyond the level assumed for the reader.

Let D_1 = the event that a diamond is selected on the first draw.
D_2 = the event that a diamond is selected on the second draw.
H_1 = the event that a heart is selected on the third draw.
H_2 = the event that a heart is selected on the fourth draw.

Thus, we are looking for $P(D_1 \cap D_2 \cap H_1 \cap H_2)$. The following probabilities are needed.

$P(D_1) = 13/52$
$P(D_2|D_1) = 12/51$
$P(H_1|D_1 \cap D_2) = 13/50$
$P(H_2|D_1 \cap D_2 \cap H_1) = 12/49$

Therefore, $P(D_1 \cap D_2 \cap H_1 \cap H_2) = 13/52 \cdot 12/51 \cdot 13/50 \cdot 12/49 = 78/20825$. What would be the value of $P(D_1 \cap D_2 \cap H_1 \cap H_2)$ if we were to allow replacement?

In a similar manner, Theorem 6.4 can be generalized. Since $P(A \cup B)$ represents the probability that at least one of the two events occurs, we let $P(A \cup B \cup C)$ represent the probability that at least one of the three events occurs. It can be shown that

$$P(A \cup B \cup C) = P(A) + P(B) + P(C) - P(A \cap B) - P(A \cap C)$$
$$- P(B \cap C) + P(A \cap B \cap C)$$

This extension of Theorem 6.4 follows a definite pattern: first the probabilities of the single events are *added*, then the probabilities of all possible two-way intersections are *subtracted*, and finally the probability of the three-way intersection is *added*. The pattern of adding and subtracting the probabilities of higher and higher order intersections extends to unions of any finite number of events. We shall generalize Theorem 6.4 for k events as follows:

Theorem 6.6: Generalized Addition Law. Given k events E_1, E_2, . . . , E_k, the probability that at least one of them occurs is

$$P(E_1 \cup E_2 \cup \cdots \cup E_k) = P(E_1) + P(E_2) + \cdots + P(E_k)$$

(plus or minus the probabilities of all possible joint occurrences of the events)

This theorem is rather loosely stated because (a) a precise statement for k events would be difficult to write and to apply, and (b) for most practical purposes we can get along without Theorem 6.6 (see Exercise 7(d), page 123). However, some special cases of Theorems 6.5 and 6.6, discussed in the next section, will be invaluable to us.

Exercises

1. Given that $P(X) = 0.3$, $P(Y) = 0.4$, and $P(Y|X) = 0.2$, find:
   (a) $P(X')$　　　　(b) $P(X \cap Y)$　　　(c) $P(X|Y)$
   (d) $P(X \cup Y)$　　(e) $P(Y'|X)$

2. The Venn diagram in Figure 6.5 shows the probabilities associated with certain regions.

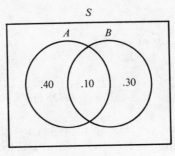

Figure 6.5.

Find
   (a) $P(A)$　　　　　　(b) $P(A \cap B)$　　　(c) $P(A \cup B)$
   (d) $P(A|B)$　　　　　(e) $P(A'|B)$　　　　(f) $P(B|A)$
   (g) $P(B')$　　　　　　(h) $P(A' \cap B')$　　(i) $P(A' \cup B')$
   (j) $P(A \cap B|A \cup B)$

3. Given $P(Q) = 0.46$, $P(T) = 0.58$, $P(R) = 0.32$,
      $P(Q \cap T) = 0.22$, $P(Q \cap R) = 0.15$, $P(T \cap R) = 0.08$,
      and $P(Q \cap T \cap R) = 0.02$.
   (a) Draw a Venn diagram depicting the three events as intersecting circles and showing the probability for each of the eight regions.
   (b) Find $P(Q \cup T \cup R)$ by
      (i) applying Theorem 6.6;
      (ii) adding probabilities for the seven regions inside the circles.

4. From a box of five black marbles and three white marbles, we plan to make four random selections without replacement.
   (a) Find the probability of selecting a black marble, a white marble, and two more black marbles (in that order), using methods discussed prior to this section.
   (b) Let A = event that we get a black marble on the first selection,
         B = event that we get a white marble on the second selection,
         C = event that we get a black marble on the third selection, and
         D = event that we get a black marble on the fourth selection.
      Find $P(A \cap B \cap C \cap D)$, using Theorem 6.5, and compare to the answer to (a).
   (c) Assume replacement and recalculate $P(A \cap B \cap C \cap D)$.

5. Four playing cards are dealt face down.
    (a) Find the probability that the cards are spade, spade, nonspade, spade (in that order).
    (b) Find the probability that the cards are spade, nonspade, spade, spade (in that order).
    (c) How many different orders are there involving three spades and one nonspade?
    (d) Let A = event that the cards are dealt in $sssn$ order (i.e., three spades then a nonspade)
    B = event that the cards are dealt in $ssns$ order,
    C = event that the cards are dealt in $snss$ order, and
    D = event that the cards are dealt in $nsss$ order.
    Use set notation to represent the event that the four cards dealt include three spades and one nonspade.
    (e) Find the probability of that event, using Theorem 6.6.

6. Repeat Exercise 5, assuming replacement.

7. (a) What is the relationship between $A \cup B$ and $A' \cap B'$?
    (b) What is the relationship between $A \cup B \cup C$ and $A' \cap B' \cap C'$?
    (c) Generalize what you have observed in (a) and (b) for events E_1, E_2, . . . , E_k, and explain the relationship intuitively.
    (d) In light of the above, comment on the statement in the text following Theorem 6.6 that "for most practical purposes we can get along without Theorem 6.6."
    (e) For the chance experiment in Exercise 4 find $P(A \cup B \cup C \cup D)$. That is, find the probability that at least one of the events occurs.
    (f) Write an equation involving events E_1, E_2, . . . , E_k which we could use in place of the equation in Theorem 6.6.

8. A balanced coin is flipped 10 times. Find the probability that at least one of the flips will come up heads.

9. For any pair of events E and F ($E \neq \phi$ and $F \neq \phi$) in sample space S, prove that $P(E) \cdot P(F|E) = P(F) \cdot P(E|F)$.

6.7 Special Properties of Events

In this section we shall discuss two special properties of events: *independence* and *mutual-exclusiveness*.

6.7.1 Independent Events

Consider a chance experiment in which two selections are made from a deck of playing cards. Let C_1 be the event that a club is selected on the first draw, and let C_2 be the event that a club is selected on the second draw. If there is *no replacement*, then

$$P(C_1 \cap C_2) = P(C_1) \cdot P(C_2|C_1) = 13/52 \cdot 12/51$$
$$P(C_1' \cap C_2) = P(C_1') \cdot P(C_2|C_1') = 39/52 \cdot 13/51$$

If there *is replacement*, then

$$P(C_1 \cap C_2) = P(C_1) \cdot P(C_2|C_1) = 13/52 \cdot 13/52$$
$$P(C_1' \cap C_2) = P(C_1') \cdot P(C_2|C_1') = 39/52 \cdot 13/52$$

The event C_1' is, of course, the complement of C_1 — that is, C_1' represents the event that a club does not occur on the first draw. Note that when there *is replacement*, the probability that C_2 occurs *in no way depends* upon the occurrence or nonoccurrence of C_1. That is, the probability that C_2 occurs is 13/52, whether C_1 occurs or not! On the other hand, when there is *no replacement*, the probability that C_2 occurs *depends* upon whether or not C_1 occurs (compare 12/51 to 13/51 for $P(C_2|C_1)$ and $P(C_2|C_1')$).

Generally speaking, given two events E and F, if the probability of occurrence of either one is in no way affected by the occurrence or nonoccurrence of the other, we say that E and F are *independent*. On the other hand, if the probability of occurrence of either event is affected by the occurrence or nonoccurrence of the other, we say that E and F are *not independent* events; hence, they are *dependent*. Thus, in our example, events C_1 and C_2 can be labelled independent in the case where the first card is replaced before the second is drawn, but they are dependent in the case where the first card is not replaced before the second is drawn.

From our informal definition of independence, it follows that if events E and F are independent, then

$$P(E|F) = P(E) \qquad \text{and} \qquad P(F|E) = P(F)$$

That is, the probability that E occurs does *not depend* at all upon the occurrence or nonoccurrence of F (and vice versa). Therefore, when E and F are independent, the equation in Theorem 6.3 reduces to

$$P(E \cap F) = P(E) \cdot P(F)$$

In fact, if we have two or more independent events, any expression of conditional probability can be rewritten with the conditional part omitted. For example, if events A, B, and C are independent, we know that

$$P(A|B) = P(A)$$
$$P(C|A) = P(C)$$
$$P(C|A \cap B) = P(C)$$
$$P(B \cup C|A) = P(B \cup C)$$

and so on. We use Definition 6.13 to formally define independence for k events.

Definition 6.13:   Let a set $Q = \{E_1, E_2, \ldots, E_k\}$, where $E_1, E_2, \ldots, E_k$ are events. If for *every* subset of Q containing two or more events, the probability of the intersection of the events equals the product of their individual probabilities, then the k events are said to be *independent*.

Definition 6.13 sounds complicated, but it really is not. To show, for example, that events A, B, and C are independent, we must show that

$$P(A \cap B) = P(A) \cdot P(B)$$
$$P(A \cap C) = P(A) \cdot P(C)$$
$$P(B \cap C) = P(B) \cdot P(C)$$

and

$$P(A \cap B \cap C) = P(A) \cdot P(B) \cdot P(C)$$

Definition 6.13 makes it possible for us to simplify Theorem 6.5 (the Generalized Multiplication Law) for the case when the k events are independent.

Theorem 6.7:   Generalized Multiplication Law for Independent Events. If $E_1, E_2, \ldots, E_k$ are independent events, then

$$P(E_1 \cap E_2 \cap \ldots \cap E_k) = P(E_1) \cdot P(E_2) \cdot \ldots \cdot P(E_k)$$

Proof: Left to the reader.

The reader should carefully compare Theorems 6.5 and 6.7, noting the similarities and differences. Suppose that we want to find the probability of selecting a club (C_1), then a heart (H_1), then another club (C_2) in three random draws from a deck of cards (i.e., we want $P(C_1 \cap H_1 \cap C_2)$). If we do *not allow replacement* (the usual procedure in card games) we need Theorem 6.5.

$$\begin{aligned} P(C_1 \cap H_1 \cap C_2) &= P(C_1) \cdot P(H_1|C_1) \cdot P(C_2|C_1 \cap H_1) \\ &= 13/52 \cdot 13/51 \cdot 12/50 \\ &= 13/850 \end{aligned}$$

On the other hand, if we *do allow replacement*, C_1, H_1, and C_2 are independent events, and therefore, we can use Theorem 6.7.

$$\begin{aligned} P(C_1 \cap H_1 \cap C_2) &= P(C_1) \cdot P(H_1) \cdot P(C_2) \\ &= 13/52 \cdot 13/52 \cdot 13/52 \\ &= 1/64 \end{aligned}$$

It is also important to examine the nature of the complements of independent events. Intuitively it is obvious that if events E and F are independent, then it follows that E and F' are independent, E' and F

are independent, and E' and F' are independent. To generalize this notion, we say that *if k events are independent, any subset of the k events can be replaced by their complements, and the resulting set of k events are still independent.*

6.7.2 *Mutually Exclusive Events*

If two coins are tossed and we let A be the event that exactly one head occurs and B stand for the event that no heads occur, then it is obvious that events A and B *cannot both occur.* That is, $A \cap B = \phi$, because $A = \{ht, th\}$ and $B = \{tt\}$. Any pair of events, such as A and B, which have the property: $A \cap B = \phi$ is said to be *mutually exclusive.* The statement "A and B are mutually exclusive" says, in essence, that the occurrence of either event *precludes* the occurrence of the other. We can easily generalize this property for k events:

> **Definition 6.14:** Given events $E_1, E_2, \ldots, E_k$, if every pair of these events, E_i and E_j, has the property: $E_i \cap E_j = \phi$, then the k events are said to be *mutually exclusive.*

For example, in the two-dice experiment, let

$A =$ the event that a sum of five occurs
$B =$ the event that a sum of 11 occurs
$C =$ the event that the same number comes up on both dice

Thus,

$A = \{(1, 4), (4, 1), (2, 3), (3, 2)\}$
$B = \{(6, 5), (5, 6)\}$
$C = \{(1, 1), (2, 2), (3, 3), (4, 4), (5, 5), (6, 6)\}$

Since $A \cap B = \phi, A \cap C = \phi$, and $B \cap C = \phi$, it follows that events A, B, and C are mutually exclusive by Definition 6.14.

When k events are mutually exclusive, the intersection of every pair of events is ϕ. This in turn implies that *all higher order intersections* (such as $E_1 \cap E_2 \cap E_3, E_2 \cap E_5 \cap E_8 \cap E_9$, etc.) *are also empty.* That is, if no pair of k events can occur jointly, it is impossible for any larger number of the k events to occur jointly. Therefore, when k events are mutually exclusive, *all* conceivable intersections are empty. We can use this fact to prove Theorem 6.8, which is a special case of Theorem 6.6.

> **Theorem 6.8:   Generalized Addition Law for Mutually Exclusive Events.**
> If $E_1, E_2, \ldots, E_k$ are mutually exclusive events, then
>
> $$P(E_1 \cup E_2 \cup \ldots \cup E_k) = P(E_1) + P(E_2) + \ldots + P(E_k)$$
>
> **Proof:**   Left to the reader.

Therefore, in the two-dice experiment, the probability that at least one of the three events occurs is

$$P(A \cup B \cup C) = P(A) + P(B) + P(C)$$
$$= 4/36 + 2/36 + 6/36 = 12/36 = 1/3$$

It is worth noting that the phrase "at least one" (used in the previous sentence) literally means "exactly one" *when the events under consideration are mutually exclusive.* Why?

6.7.3 Independent vs. Mutually Exclusive Events

The properties of independence and mutual-exclusiveness are often confused by new students of probability. There seem to be two "schools" of confusion. First, there are those who think that the two properties are synonymous, probably because "independent" and "exclusive" sound like they should be dictionary synonyms. Of course, this belief cannot be further from the truth. In fact, a pair of events which have nonzero probability cannot possibly be *both* independent and mutually exclusive. That is, if they are independent, then they are definitely not mutually exclusive, and vice versa. This statement is justified by what we already know about these properties. We know that if two events are mutually exclusive, the occurrence of one *precludes* the occurrence of the other. Thus, the probability that one of the events occurs *depends* upon whether or not the other occurs. This means that mutually exclusive events are *dependent* events, and therefore cannot possibly be independent. From a mathematical point of view we can use the following argument. Suppose that we have two events E and F which have nonzero probability (i.e., $P(E) > 0$ and $P(F) > 0$). If the events are independent, it follows that $P(E \cap F) = P(E) \cdot P(F)$, which means that $P(E \cap F) > 0$. This means that E and F cannot possibly be mutually exclusive, because if they were, $P(E \cap F)$ would equal zero!

Those students who understand that *independence* and *mutual-exclusiveness* are not synonymous are sometimes trapped by the second "school" — which takes the position that the properties are opposite in meaning; that is, it is assumed, erroneously, that if two events are not independent, then they must be mutually exclusive, and vice versa. This thinking probably stems from the fact that mutually exclusive events are dependent events. The flaw in the reasoning is the fact that events can be dependent without being mutually exclusive! For example, if we make two random selections *without replacement* from a deck of cards, and let A stand for the event that a spade is drawn first, and B be the event that a heart is drawn second, then A and B are neither mutually exclusive nor independent. Why?

We can easily show the relationship among *mutually exclusive, independent,* and *dependent* sets of events by picturing the entire universe of sets of two or more events and dividing it into two parts—independent and dependent. The mutually exclusive sets of events can then be depicted as a subset of the dependent part. (See Figure 6.6.)

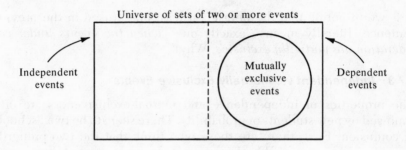

Figure 6.6.   Relationship among mutually exclusive, independent, and dependent events.

It is not uncommon that the notions of *independence* and *mutual-exclusiveness* are present in the same problem. For example, if a die is tossed three times, what is the probability that exactly one "4" occurs? A "4" can occur in three mutually exclusive ways.

(i)   4,   non-4,   non-4
(ii)   non-4,   4,   non-4
(iii) non-4,   non-4,   4

On the other hand, each of these possibilities involves three independent events. Thus, we find the probabilities for (i), (ii), and (iii) separately, according to Theorem 6.7, and then we add the three results, according to Theorem 6.8 to obtain the desired probability.

$$\frac{1}{6}\cdot\frac{5}{6}\cdot\frac{5}{6}+\frac{5}{6}\cdot\frac{1}{6}\cdot\frac{5}{6}+\frac{5}{6}\cdot\frac{5}{6}\cdot\frac{1}{6}=\frac{75}{216}$$

Exercises

1.   In each of the following indicate whether events A and B are
         I.   independent,
        II.   mutually exclusive, or
       III.   dependent but not mutually exclusive.
    (a) $P(A|B) = 0$
    (b) $P(A|B) = 0.30$ and $P(A) = 0.45$
    (c) $P(A \cup B) = 0.85$, $P(A) = 0.30$,   and $P(B) = 0.60$

(d) $P(A \cup B) = 0.70$, $P(A) = 0.50$, and $P(B) = 0.40$

(e) $P(A \cup B) = 0.90$, $P(A|B) = 0.80$, and $P(B) = 0.50$

(f) Two dice are tossed;

   (i) $A =$ first die comes up "4," $B =$ second die comes up "6."

   (ii) $A =$ sum on dice is 4, $B =$ sum on dice is 6.

(g) A coin is flipped twice;

   (i) $A =$ head occurs on first flip, $B =$ tail occurs on first flip.

   (ii) $A =$ head occurs on first flip, $B =$ tail occurs on second flip.

   (iii) $A =$ head occurs on first flip, $B =$ head occurs on second flip.

(h) A coin is flipped five times;

   (i) $A =$ the outcome *hhhtt*, $B =$ the outcome *hthht*.

   (ii) $A =$ the occurrence of three heads and two tails.

   $B =$ the occurrence of one head and four tails.

   (iii) $A =$ the occurrence of four heads on the first four flips.

   $B =$ the occurrence of a head on the fifth flip.

(i) Two objects are randomly selected without replacement from a group of 10 objects (four green and six yellow);

   (i) $A =$ a green object is selected first.

   $B =$ a yellow object is selected second.

   (ii) same — but with replacement.

2. Given that A, B, and C are mutually exclusive events, $P(A) = 0.62$, $P(B) = 0.20$, and $P(C) = 0.15$, find:

   (a) $P(A \cap B)$ (b) $P(A|C)$

   (c) $P(A \cup C)$ (d) $P(A \cup B \cup C)$

   (e) $P(A \cap B \cap C)$ (f) $P(A' \cap B' \cap C')$

   (g) $P(A' \cup B)$ (h) $P(A' \cap B)$

   (It sometimes helps to draw the appropriate Venn diagram, filling in the probabilities associated with each region.)

3. Given that A, B, and C are independent events, $P(A) = 0.60$, $P(B) = 0.50$, and $P(C) = 0.90$, find:

   (a) $P(A \cap B)$ (b) $P(B|C)$

   (c) $P(A \cup C)$ (d) $P(A' \cap B)$

   (e) $P(B' \cap C')$ (f) $P(A \cup B \cup C)$

   (g) $P(A \cap B \cap C)$ (h) $P(A \cup B')$

4. Why are events A and B, defined on page 127, neither mutually exclusive nor independent?

6.8 Bayes' Theorem

In this section we shall extend our knowledge of conditional probability by considering a particular type of situation in which the occurrence of a certain event (E) depends upon the occurrence of one of k mutually exclusive events ($F_1, F_2, \ldots, F_k$). For example, the probability that the local tax rate will increase within the next year may depend upon which of three mayoral candidates is elected this year. The event that a new washing machine will need repair within the first two years of use depends upon which one of several brands is purchased. The probability that a particular car manufacturer will get this year's models on the market by November 1st depends upon whether or not the union strikes before that date. We shall develop two theorems which are useful in this type of situation. The first tells us how to find $P(E)$, and the second how to find $P(F_i|E)$ for any $i = 1, 2, \ldots, k$. The latter theorem is a famous one called *Bayes' Theorem.*

We begin by defining what we mean by a *partition* of the sample space.

> **Definition 6.15:** A *partition* of the sample space S is a set of events $\{F_1, F_2, \ldots, F_k\}$, such that
>
> (i)   each event is a nonempty subset of S
> (ii) the events are mutually exclusive
> (iii) $F_1 \cup F_2 \cup \ldots \cup F_k = S$

Figure 6.7 shows the basic structure of a partition of S into four events. For example, suppose we plan to select at random one member of a college student body. Then S would be the set of all students in the college. We can think of S as being partitioned into four mutually exclusive sets: freshmen (F_1), sophomores (F_2), juniors (F_3), and seniors (F_4).

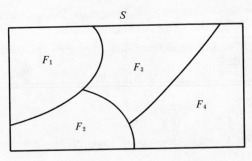

Figure 6.7. A partition of S into four subsets.

> **Theorem 6.9:** If the set of events $\{F_1, F_2, \ldots, F_k\}$ is a partition of sample space S and E is any event in S, then
>
> $$P(E) = P(F_1) \cdot P(E|F_1) + P(F_2) \cdot P(E|F_2) + \ldots + P(F_k) \cdot P(E|F_k)$$

We can visualize E as an event in S which may possibly (but not necessarily) intersect each event of the partition (see Figure 6.8). Note that E can be thought of as the union of k intersections: $F_1 \cap E, F_2 \cap E, \ldots, F_k \cap E$. Note also that the k intersections are mutually exclusive and their union equals E. Thus, the proof of Theorem 6.9 is rather straightforward.

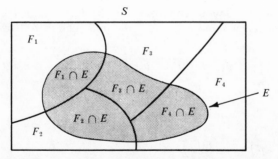

Figure 6.8. Partition of S showing E intersecting each F_i.

Proof of Theorem 6.9:

$$E = (F_1 \cap E) \cup (F_2 \cap E) \cup \ldots \cup (F_k \cap E)$$
$$P(E) = P(F_1 \cap E) + P(F_2 \cap E) + \ldots + P(F_k \cap E) \quad \text{(Thm. 6.8)}$$
$$\therefore \quad P(E) = P(F_1) \cdot P(E|F_1) + P(F_2) \cdot P(E|F_2) + \ldots + P(F_k) \cdot P(E|F_k)$$
$$\text{(Thm. 6.3)}$$

In the college student body example, we could let E represent the event that a student chosen at random is an English major. Table 6.1 shows an enrollment breakdown by class and major. If we want to calculate $P(E)$, we can do it rather quickly by dividing 240 by 2900, since by Definition 6.10 $P(E) = n(E)/n(S)$. To show that Theorem 6.9 is also applicable, we get the necessary information from Table 6.1, as follows:

$P(F_1) = 1000/2900$ $P(E|F_1) = 100/1000$
$P(F_2) = 800/2900$ $P(E|F_2) = 50/800$
$P(F_3) = 600/2900$ $P(E|F_3) = 40/600$
$P(F_4) = 500/2900$ $P(E|F_4) = 50/500$

The reader should verify each of the above probabilities from Table 6.1. Applying these values to Theorem 6.9 we get

$$P(E) = \left(\frac{1000}{2900} \cdot \frac{100}{1000}\right) + \left(\frac{800}{2900} \cdot \frac{50}{800}\right) + \left(\frac{600}{2900} \cdot \frac{40}{600}\right) + \left(\frac{500}{2900} \cdot \frac{50}{500}\right)$$

$$= \frac{240}{2900}$$

$$= \frac{24}{290}$$

Table 6.1. Breakdown of a College Student Body by Class and Major

		Class					
		Frosh	Soph	Junior	Senior		
Major {	English:	100	50	40	50	240	E
	Non-English:	900	750	560	450	2660	E'
		1000	800	600	500	2900	
		F_1	F_2	F_3	F_4		S

Theorem 6.10: Bayes' Theorem. If $\{F_1, F_2, \ldots , F_k\}$ is a partition of the sample space S and E is any event in S such that $P(E) \neq 0$, then

$$P(F_i|E) = \frac{P(F_i) \cdot P(E|F_i)}{P(F_1) \cdot P(E|F_1) + P(F_2) \cdot P(E|F_2) + \ldots + P(F_k) \cdot P(E|F_k)}$$

for all $i = 1, 2, 3, \ldots k$

Proof:

$$P(F_i|E) = \frac{n(F_i \cap E)}{n(E)} \qquad \text{(Def. 6.12)}$$

$$P(F_i|E) = \frac{P(F_i \cap E)}{P(E)} \qquad \text{(Divide numerator and denominator by } n(S))$$

$$P(F_i|E) = \frac{P(F_i) \cdot P(E|F_i)}{P(E)} \qquad \text{(Thm. 6.3)}$$

$$\therefore \quad P(F_i|E) = \frac{P(F_i) \cdot P(E|F_i)}{P(F_1) \cdot P(E|F_1) + P(F_2) \cdot P(E|F_2) + \ldots + P(F_k) \cdot P(E|F_k)}$$
$$\text{(Thm. 6.9)}$$

Suppose we want to find $P(F_2|E)$ in the college student body example. We can find $P(F_2|E)$ directly from Table 6.1 by dividing 50 by 240, since by Definition 6.12 $P(F_2|E) = n(F_2 \cap E)/n(E)$. Or we could use Bayes' Theorem,

$$P(F_2|E) = \frac{P(F_2) \cdot P(E|F_2)}{P(F_1) \cdot P(E|F_1) + P(F_2) \cdot P(E|F_2) + P(F_3) \cdot P(E|F_3) + P(F_4) \cdot P(E|F_4)}$$

$$= \frac{\dfrac{800}{2900} \cdot \dfrac{50}{800}}{\left(\dfrac{1000}{2900} \cdot \dfrac{100}{1000}\right) + \left(\dfrac{800}{2900} \cdot \dfrac{50}{800}\right) + \left(\dfrac{600}{2900} \cdot \dfrac{40}{600}\right) + \left(\dfrac{500}{2900} \cdot \dfrac{50}{500}\right)}$$

$$= \frac{\dfrac{50}{2900}}{\dfrac{240}{2900}} = \frac{50}{240} = \frac{5}{24}$$

At this point the reader is no doubt questioning, "Why do we need such involved methods for finding $P(E)$ and $P(F_i|E)$ when the answers can be found so quickly from Table 6.1 by applying Definitions 6.10 and 6.12"? As a matter of fact, neither Theorem 6.9 nor Theorem 6.10 is needed when a complete breakdown of the sample space is available. As we have just witnessed, Definitions 6.10 and 6.12 are quite sufficient and much easier to apply. The reason for the rather tedious demonstration above is simply to show that Theorems 6.9 and 6.10 do in fact give results identical to those obtained by application of Definitions 6.10 and 6.12. In addition, we must recognize the fact that in many practical situations of this general type, a complete breakdown of the sample space is *not* available. When this happens, Theorems 6.9 and 6.10 provide the *only* means for finding $P(E)$ and $P(F_i|E)$. Let us consider such an example.

Suppose that a particular test for diabetes can detect the disease in 95 percent of the people who actually have diabetes. Among people who do not have diabetes, the test will claim detection of the disease in two percent of the cases. If we assume that three percent of the U.S. population has diabetes, then (a) what is the probability that a person selected randomly off the street would be declared a diabetic by the test?, and (b) if a person is declared a diabetic by the test, what is the probability that he really is one? Let the sample space S be the U.S. population, let F_1 equal the event that a person randomly selected is a diabetic and F_2 the event that he is not a diabetic. Thus, $\{F_1, F_2\}$ is a partition of S. Let E represent the event that the test detects diabetes. It is given, therefore, that $P(F_1) = 0.03$, $P(F_2) = 0.97$, $P(E|F_1) = 0.95$, and $P(E|F_2) = 0.02$. In question (a) we want to find $P(E)$. Therefore, by Theorem 6.9,

$$
\begin{aligned}
P(E) &= P(F_1) \cdot P(E|F_1) + P(F_2) \cdot P(E|F_2) \\
&= (0.03)(0.95) + (0.97)(0.02) \\
&= 0.0479
\end{aligned}
$$

Thus, although only three percent of the population actually have diabetes, this test detects diabetes in about five percent of the people tested (in the long run).

Question (b) requires that we find $P(F_1|E)$, and therefore we need to apply Bayes' Theorem.

$$
\begin{aligned}
P(F_1|E) &= \frac{P(F_1) \cdot P(E|F_1)}{P(F_1) \cdot P(E|F_1) + P(F_2) \cdot P(E|F_2)} \\
&= \frac{(0.03)(0.95)}{(0.03)(0.95) + (0.97)(0.02)} = \frac{0.0285}{0.0479} = 0.59
\end{aligned}
$$

Thus, without the test the probability that a particular person has diabetes is 0.03 (i.e., $P(F_1) = 0.03$); but, the probability that a person has diabetes, *given that his test says he does*, is 0.59 (i.e., $P(F_1|E) = 0.59$).

Exercises

1. Let $\{X, Y, Z\}$ be a partition of sample space S and W be a subset of S. We shall assume that $P(X) = 0.20$, $P(Y) = 0.50$, $P(W|X) = 0.30$, $P(W|Y) = 0.60$, and $P(W|Z) = 0.80$. Find:
    (a) $P(Z)$ (b) $P(W)$ (c) $P(X|W)$
    (d) $P(Y|W)$ (e) $P(Z|W)$

2. In a college fraternity 20 percent of the membership are seniors, 25 percent are juniors, 30 percent are sophomores, and 25 percent are freshmen. None of the seniors are math majors. However, 10 percent of the juniors, five percent of the sophomores, and eight percent of the freshmen are math majors. If a person were selected randomly from the total membership of this fraternity,
    (a) what is the probability that he would be a math major?
    (b) what is the probability that he would be a sophomore, given that he is a math major?

3. Container A holds five pennies and three nickels, B holds two pennies and four nickels, and C holds three pennies and three nickels. The task is to choose one of the containers and then randomly select a coin from it. The container is chosen by the following method: toss a die — if a one appears, choose A, if a two or three appears, choose B, and if a four, five, or six appears, choose C.
    (a) What is the probability that the coin we select will be a nickel?
    (b) If someone else performed this experiment and told you, after he was done, that he selected a nickel, what is the probability that he selected a nickel from container B?

4. An old machine produces three defective parts in every 100 parts produced. A newer machine produces only one defective part in every 1000 parts produced. Furthermore, the new machine yields three times the output of the old machine. During a quality control check, a part is selected from the joint output of the two machines and found to be defective. What is the probability that the defective part was produced by the old machine?

5. The probability that John's office door is locked is 0.80. The key to the office is one of four identical-looking keys in John's pocket. If he selects a key at random to try in the door, what is the probability that he can open the door (i.e., that either the key works, or the door was already unlocked)?

6.9   General Application of Classical Probability Theory

In this chapter we have been developing from the *classical* approach a *theory of probability*. It is important to remember one significant limitation of the classical approach, namely, that it assumes that the sample space involved is a set of *equally likely* outcomes. Unless this condition is satisfied, our two basic definitions (Definitions 6.10 and 6.12) cannot be used to find probabilities. The theorems which describe the basic laws governing probability (Theorems 6.1–6.10) are based on the same condition. It is true, however, that *these laws apply generally to all kinds of events—even those which cannot reasonably be expressed as subsets of a sample space which has equally likely outcomes.*[*] For example, if we let A be the event that Bob will be late to work tomorrow, and B the event that a National League team will win the next World Series, then we know, for instance, that $P(A \cup B) = P(A) + P(B) - P(A \cap B)$, even though we do not know the exact value of any one of the four probabilities. Thus, if we wish to hypothesize values for any three of the four probabilities, the fourth one can be calculated. If, for example, we assume that the probability of rain on any particular day is 0.50, and that the probability of rain on a particular day, given that it rained the day before, is 0.60, then the probability of rain on any two successive days is $(0.50)(0.60) = 0.30$ (an application of Theorem 6.3).

Therefore, from this point on we shall assume that Theorems 6.1–6.10 apply to all conceptual events, and we shall use these theorems to calculate probabilities under the assumption that other probabilities are known. The following exercises should make this important point clear, and also serve as a useful review of the theorems.

Exercises

1. On Friday night Bill is free to choose one (but not more than one) of the following activities: (a) go to the movies, (b) go to a dance, (c) stay home and watch TV, and (d) stay home and study. If the probabilities that he elects these alternatives are 0.45, 0.14, 0.30, and 0.04, respectively, what is the probability that
    (a) Bill will not go to the movies?
    (b) Bill either goes to a dance or stays home and studies?
    (c) Bill does none of the four activities?

2. Assume that the probability of rain next Saturday is 0.70 and the probability that State U. will win its Saturday football game is 0.60. If it rains on Saturday, the probability that State U. wins the game is 0.50. What is the probability that
    (a) either State U. wins or it rains next Saturday?
    (b) State U. wins and it rains next Saturday?
    (c) it rains next Saturday, given that State U. wins?

3. Three women independently attempt to swim the English Channel. If we assume that their probabilities of succeeding are 0.60, 0.30, and 0.20, what is the probability that
    (a) none of them succeed?
    (b) only one of them is successful?
    (c) at least one is successful?

[*]We are saving some time here due to the fact that the reader is not assumed to be an accomplished mathematician. To develop a rigorous theory of probability which applies to all types of events, the mathematical statistician would assume only two or three of these basic laws and then prove the rest, using an axiomatic approach (for example, see Meyer, 1965, page 13).

4. Five people are selected at random. What is the probability that at least two of them were born on the same day of the week? (Assume that the seven days are equally likely.)

5. The probability that a bank burglar alarm system fails (i.e., alarm fails to go off) is 0.005. If the system fails, a backup system is automatically activated. The probability that the alternative system fails is 0.08. What is the probability that in the next burglary at the bank
   (a) the alternate system is not activated?
   (b) the alternate system calls the alarm?
   (c) both systems fail?
   (d) an alarm goes off?

6. In the long run a certain bowler gets a strike (knocks down 10 pins with one ball) once out of every five attempts on the average. If the bowler attempts to get a strike 10 times during a game, what is the probability that
   (a) he does not get a strike until the 10th attempt?
   (b) he gets exactly one strike?
   (c) he gets at least one strike?

7. When the temperature is below zero, Mr. Green has experienced trouble starting his two cars. If the probabilities that they will start are 0.40 (first car) and 0.90 (second car), and the probability that they both start is 0.30, what is the probability that
   (a) at least one of them starts?
   (b) neither car starts?
   (c) the second car starts, given that the first car starts?
   (d) the first car starts, given that the second car does not start?

8. Three men, working independently, attempt to decode a secret message. If their individual probabilities of successfully decoding messages are 0.2, 0.4, and 0.5, what is the probability that the message is successfully decoded?

9. A rifleman shoots at a target until he hits the bull's-eye. If on each shot the probability of hitting the bull's-eye is 1/4, what is the probability that
   (a) he will take exactly four shots?
   (b) he will take at least four shots?

10. In a biology class the instructor has noted that students who complete their lab work do much better on the final exam than those who do not. Assume that in the long run, (i) 50 percent of those who complete the lab get As, (ii) five percent of those who do not complete the lab get As, and (iii) 40 percent of the class completes the lab.
   (a) If it is known that Bob got an A on the final exam, what is the probability that he did not complete the lab?
   (b) If it is known that Bob did not get an A on the final exam, what is the probability that he did not complete the lab?

Summary

In everyday situations we make decisions in the face of uncertainty. We develop the concept of *probability* in order that we may express in a consistent and meaningful way our feelings of uncertainty with regard to the occurrence of events. The task of defining probability can be approached in two different, but related, ways: (a) the *empirical* approach, in which the probability of an event is conceived as the *relative frequency* of that event in the *long run*, and (b) the *classical* approach, in which the probability of an event is defined as a proportion of the outcomes in a *sample space*, assuming that all the outcomes in the sample space are *equally likely*. Certain counting principles, which include the notions of *permutations* and *combinations*, help us to find the number of outcomes in a sample space or an event, thus avoiding the necessity for listing outcomes. A group of two or more events can always be described as either *independent* or *dependent*. If the events are dependent, they may or may not be *mutually exclusive*. Several properties and laws governing classical probability give us means for calculating probabilities from other probabilities. *Bayes' Theorem* is one of the more important of the probability laws. Although we have based our development of probability on the assumption that the sample space has equally likely outcomes, in fact these same properties and laws are applicable outside the restrictions of the classical model. Therefore, we shall from now on assume general application for the properties and laws we have developed.

7

Discrete Probability Functions

7.1 Random Variables

When three coins are flipped, the sample space contains eight equally likely outcomes, $S = \{hhh, hht, hth, thh, htt, tht, tth, ttt\}$. If we let x represent the number of heads obtained in this experiment, then the possible values of x are 0, 1, 2, and 3. We could let y stand for the number of heads minus the number of tails occurring in the three flips, in which case y could have values $-3, -1, 1,$ or 3. Or we might let z equal zero when the second coin is a head and one when the second coin is a tail. The letters $x, y,$ and z are variables in the same sense that variables are used in algebra. The letter represents any one of a *set* of numbers called the *domain* of the variable. When we define a variable on a sample space, we are essentially assigning a number to each outcome of the chance experiment. Table 7.1 shows the assignments made when we defined $x, y,$ and z above. The domains of $x, y,$ and z are $\{0, 1, 2, 3\}$, $\{-3, -1, 1, 3\}$, and $\{0, 1\}$, respectively. If the three coins are tossed, and the result is hht, we can describe that result by saying that $x = 2$, or $y = 1$, or $z = 0$. Since these variables take on particular numerical values as a direct result of the outcome of a *chance* experiment, we say that x, y, and z are *random variables*. Moreover, since the domain of each variable is a finite set, we say that x, y, and z are *discrete* random variables.

Table 7.1.   Variables Defined on Three-Coin Sample Space

Outcomes	x	y	z
hhh	3	3	0
hht	2	1	0
hth	2	1	1
thh	2	1	0
htt	1	−1	1
tht	1	−1	0
tth	1	−1	1
ttt	0	−3	1

Definition 7.1:   A variable x which (i) assigns a numerical value to each outcome in a sample space, and (ii) has either a finite domain or a domain with as many values as there are whole numbers, is called a *discrete random variable.*

Note that a random variable need not have a finite domain to qualify as a discrete random variable. For example, if we were to flip a coin until a head appears and we let x equal the number of flips necessary before the first head occurs, then $S = \{h, th, tth, ttth, \ldots\}$, and therefore the domain of x is $\{0, 1, 2, 3, \ldots\}$, the set of all whole numbers.

If a random variable is not discrete, it is said to be *continuous*. Continuous random variables can take any real number values on a continuous scale (i.e., within a given interval on the real number line). For convenience we can think of most *discrete* random variables as *counts* (how many heads, children, spades, or accidents?) and most *continuous* random variables as *measures* (how tall, long, heavy, or intelligent?). We shall discuss continuous random variables in the next chapter.

7.2   The Probability Function

One of the most fundamental concepts in mathematics is the concept of *function*. The equation: $y = x^2 + 2x - 5$ implies that y is a function of x, usually expressed as $y = f(x)$ ($f(x)$ is read *f of x*). Thus, we might find the equation written: $f(x) = x^2 + 2x - 5$. It is customary to indicate what set of numbers are possible values for x. This set is called the *domain* of the function. Suppose that the domain of the function above is $\{0, 1, 2, 3\}$.

If we let $x = 0$, then it follows that $f(0) = -5$; that is, the value of the function at $x = 0$ is -5. Similarly, $f(1) = -2$, $f(2) = 3$, and $f(3) = 10$. In other words, the function itself determines a matching between the values in the domain of x and a new set of unique values called *the values of the function*. Such a matching can be demonstrated in tabular form as follows:

x:	0	1	2	3
	↓	↓	↓	↓
$f(x)$:	-5	-2	3	10

Consider the equation $y = 3x + 4$, where the domain of the function (the possible values of x) is still the set $\{0,1,2,3\}$. Here again y is a function of x, but it is a *different* function of x from that in the previous example. To distinguish this function from the function $f(x)$ above, we can use a different letter* in place of f. For example, we could call this function $g(x)$. Thus, the value of this function at $x = 0$ is $g(0)$, which equals four. Similarly, $g(1) = 7$, $g(2) = 10$ and $g(3) = 13$.

x:	0	1	2	3
$g(x)$:	4	7	10	13

Now let us apply the notion of function to random variables and probability. If random variable x is the number of heads occurring in the three-coin chance experiment, then the domain of x (the set of possible values of x) is $\{0, 1, 2, 3\}$. We know from Definition 6.10 that

$$P(x = 0) = 1/8$$
$$P(x = 1) = 3/8$$
$$P(x = 2) = 3/8$$
$$P(x = 3) = 1/8$$

What we have actually done here is match each value of x with a probability. This defines a function $f(x)$ which tells us the probability that the random variable has particular values.

x:	0	1	2	3
	↓	↓	↓	↓
$f(x)$:	1/8	3/8	3/8	1/8

Since this function matches values of x to *probabilities*, we call $f(x)$ a *probability function*. The tabular form above, which demonstrates the matching of each x value with a probability value, is really quite sufficient to define this particular probability function (although the arrows are not necessary). In addition it is sometimes possible to define the

---

*In mathematics the letters f, g, and h are commonly used to denote functions.

probability function $f(x)$ by an equation. In this case we could write

$$f(x) = \frac{\binom{3}{x}}{8}, \qquad \text{for } x = 0, 1, 2, 3$$

The reader should verify that the *tabular* form of $f(x)$ and the *equation* form give identical information.*

Definition 7.2:   If x is a discrete random variable, then any function $f(x)$ whose value at $x = a$ is *the probability that $x = a$*, is called a *discrete probability function*. Such a function has the following properties:

(i) $f(x) \geq 0$ for all x (i.e., $f(x)$ is never negative)

(ii) $\sum\limits_{\text{all } x} f(x) = 1$ (i.e., the sum of all values of $f(x)$ equals one)

Note that the function $f(x)$ in the three-coin experiment satisfies the conditions of Definition 7.2. That is, $f(x)$ is never negative and

$$\sum\limits_{\text{all } x} f(x) = 1/8 + 3/8 + 3/8 + 1/8 = 1$$

The reader should find the probability functions for y and z (which we could call $g(y)$ and $h(z)$, respectively), defined on page 138, and make sure that they satisfy the conditions in Definition 7.2 (see Exercise 1, page 142).

The graph of the discrete probability function for the three-coin experiment is shown in Figure 7.1. Since x has only four values which have nonzero probability, the graph of $f(x)$ runs along the x axis, except

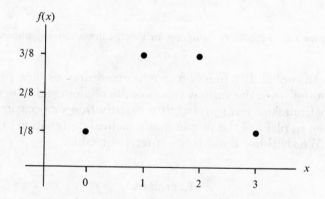

Figure 7.1.   Graph of $f(x)$ in the three-coin experiment.

*Note that subscripts have been omitted for the sake of simplicity. It is understood that each x is multiplied by its corresponding $f(x)$, and then all the resulting products are added to give μ_x.

at points $x = 0, 1, 2,$ and 3. For convenience we simply graph the nonzero values of the function.

Another way to picture graphically the distribution of probability for the random variable x in the three-coin experiment is to draw a *probability histogram*, which is similar to a frequency histogram (Section 3.3) except that values of x replace class marks and probabilities replace frequencies.

Note in Figure 7.2 that the height of the first rectangle is 1/8, the height of the second is 3/8, and so on, and the width of each rectangle is one unit. Therefore, the area of the first rectangle equals $P(x = 0)$, the area of the second is $P(x = 1)$, etc., which means that the total area covered by this probability histogram is one. This property holds only when the possible values of x are integers. It is under this condition that we find the probability histogram especially useful for pictorial description of discrete probability functions.

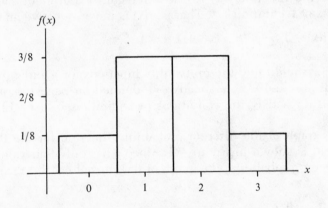

Figure 7.2. Probability histogram for x in the three-coin experiment.

Since a probability function in essence describes how probability is "distributed" over the various values of the random variable, we sometimes substitute the term *probability distribution* or occasionally just *distribution* in place of the proper mathematical designation *probability function*. We shall use these terms interchangeably.

Exercises

1. Find the probability functions $g(y)$ and $h(z)$ referred to in the text on page 141, and show that they satisfy the conditions in Definition 7.2.

2. Discrete random variable x has the following probability distribution $f(x)$.

x:	1	2	3	4
$f(x)$:	2/14	3/14	4/14	5/14

    (a)   Verify that $f(x)$ satisfies the two properties in Definition 7.2.
    (b)   What is the probability that $x = 4$?
    (c)   What is the probability that x is greater than 1?
    (d)   What is the probability that x is less than 3?
    (e)   Define the function $f(x)$ as an equation.

3.   Two dice are tossed. Let x represent the sum showing on the two dice. Find the probability function for x either in tabular form or equation form.

4.   For the same chance experiment in Exercise 3, let $y =$ the number of 4s occurring in a toss of two dice. Find the probability function for y in tabular form.

5.   Draw probability histograms for each of the functions in Exercises 2, 3, and 4.

6.   The discrete probability distribution for random variable z is defined by the following equation.

$$g(z) = \frac{z^2 + 1}{18}, \qquad z = 0, 1, 2, 3$$

Express this function in tabular form and verify that the function satisfies the two properties in Definition 7.2.

7.   From a box of 10 marbles, four of which are red, two random selections are made without replacement. Let x be the number of red marbles obtained in this chance experiment. Find the probability function for x in tabular form.

8.   Repeat Exercise 7, assuming replacement.

9.   One hundred tickets are sold in a raffle at two dollars each. Prizes are awarded by randomly selecting ticket stubs from a hat. The first prize is 10 dollars, there are five second prizes of 5 dollars, and 10 third prizes of one dollar each (i.e., there are 16 prizes awarded). Bill buys one ticket. Let x represent the amount he will win. Find the probability function for x.

10.   In the previous problem, let y equal Bill's *net* winnings (i.e., winnings minus cost of the ticket) and find the probability distribution for y.

11.   Player A plays a gambling game against player B. Player A tosses a single die. If the die comes up "6," he wins 5 dollars from B; if the die comes up either a "2" or a "5," he wins 2 dollars from B; and for any other outcome he loses and must pay 4 dollars to B. Let $y =$ the amount A wins in the game (when A loses, the value of y is negative). Find the probability function for y.

7.3 Parameters of a Discrete Probability Function

7.3.1 *Mean of a Random Variable*

Suppose that we were to take three coins and flip them 10 times, each time recording the number of heads (x). We might get the following results: 3, 1, 1, 0, 3, 2, 1, 2, 0, 1. Thus, the mean number of heads obtained in the 10 flips would be

$$\bar{x} = \frac{\sum_{i=1}^{10} x_i}{10} = \frac{14}{10} = 1.4$$

Then we might ask ourselves, what would be a reasonable value for the mean number of heads if we were to flip three perfectly balanced coins an infinite number of times? In other words, what value would $\bar{x}$ approach in the *long run*? This is, of course, a theoretical question, since we are talking about "perfectly balanced coins" and the "long run." However, we now have sufficient groundwork in probability theory to answer this question. We know from the theory that x (the number of heads obtained when three coins are flipped) has the probability function

x:	0	1	2	3
$f(x)$:	1/8	3/8	3/8	1/8

which tells us that in the long run we should expect no heads to occur 1/8 of the time, exactly one head to occur 3/8 of the time, and so on. A careful look at the table above tells us intuitively that the mean value of x in the long run should be 1.5 (note also that 1.5 is the midpoint in the probability histogram for this function—Figure 7.2). We do not need to rely on intuition to find the value 1.5. An application of Definition 7.3 is sufficient.

Definition 7.3:   Given a discrete random variable x which has probability function $f(x)$, the mean of x, denoted by μ_x, is[*]

$$\mu_x = \sum_{\text{all } x} xf(x)$$

Thus, for the three-coin experiment

$$\mu_x = (0)(1/8) + (1)(3/8) + (2)(3/8) + (3)(1/8) = 1.5$$

as we expected intuitively.

It is important at this point to clarify the difference between $\bar{x}$ and μ_x, both of which are means. The difference is analogous to the difference between "relative frequency" and "probability." The relative frequency of an event is calculated from actual data, and it can be used to estimate the probability of that event, which is a long run concept, and therefore theoretical in nature. Similarly, $\bar{x}$, the sample mean, is calculated from actual data, and it can be used to estimate the unknown population mean. But μ_x, the mean of random variable x, is strictly a

---

[*]It is understood in both the tabular and equation forms that $f(x) = 0$ for all x other than 0, 1, 2, and 3.

theoretical value based on probability. The mean of random variable x (μ_x) is considered a *parameter* of the probability function $f(x)$, and it measures central location in $f(x)$, just as $\bar{x}$ does in a sample. Other commonly used names for μ_x are *expected value of x*, *theoretical mean*, and *population mean*, depending upon the context and the audience.

7.3.2   *Variance and Standard Deviation of a Random Variable*

The sample variance of x (s_x^2) and the sample standard deviation (s_x) also have theoretical counterparts which are parameters of the probability function of random variable x. The theoretical values measure dispersion in $f(x)$, just as s_x^2 and s_x measure dispersion in a sample.

Definition 7.4:   Given a discrete random variable x which has probability function $f(x)$,

$$\text{variance of } x = \sigma_x^2 = \sum_{\text{all } x} [(x - \mu_x)^2 \cdot f(x)]$$

$$\text{standard deviation of } x = \sigma_x = \sqrt{\sigma_x^2}$$

Before we waste time using Definition 7.4 to calculate σ_x^2, we shall prove a "calculation formula" for σ_x^2, similar to the calculation formula for s_x^2 introduced in Chapter 2 (Theorem 2.3).

Theorem 7.1:

$$\sigma_x^2 = \sum_{\text{all } x} x^2 \cdot f(x) - \mu_x^2$$

Proof:

$$\sigma_x^2 = \sum_{\text{all } x} [(x - \mu_x)^2 \cdot f(x)] \qquad \text{(Def. 7.4)}$$

$$= \sum_{\text{all } x} [(x^2 - 2\mu_x x + \mu_x^2) \cdot f(x)] \qquad \text{(Expand } (x - \mu_x)^2)$$

$$= \sum_{\text{all } x} [x^2 \cdot f(x) - 2\mu_x x \cdot f(x) + \mu_x^2 \cdot f(x)] \qquad \text{(Mult. through by } f(x))$$

$$= \sum_{\text{all } x} x^2 \cdot f(x) - \sum_{\text{all } x} 2\mu_x x \cdot f(x) + \sum_{\text{all } x} \mu_x^2 f(x) \qquad \text{(Dist. property for } \textstyle\sum)$$

$$= \sum_{\text{all } x} x^2 \cdot f(x) - 2\mu_x \sum_{\text{all } x} x \cdot f(x) + \mu_x^2 \sum_{\text{all } x} f(x) \qquad \text{(Thm. 1.2)}$$

$$= \sum_{\text{all } x} x^2 \cdot f(x) - 2\mu_x \cdot \mu_x + \mu_x^2 \cdot 1 \qquad \text{(Defs. 7.3 and 7.2)}$$

$$= \sum_{\text{all } x} x^2 \cdot f(x) - \mu_x^2$$

Using the same sample data for 10 flips of three coins: 3, 1, 1, 0, 3, 2, 1, 2, 0, 1, we can apply definitions from Chapter 2 to find that $s_x^2 = 1.16$ and $s_x = 1.08$. To compare these statistics to the theoretical, long-run variance and standard deviation, we first apply Theorem 7.1 to obtain the variance, and then we find the positive square root of the variance to give us the standard deviation. Thus, from Table 7.2 we get $\sigma_x^2 = 3 - (1.5)^2 = 0.75$. And therefore $\sigma_x = \sqrt{0.75} = 0.87$.

Table 7.2. Sums Needed for the Calculation of σ_x^2

x	$f(x)$	$x \cdot f(x)$	$x^2 \cdot f(x)$
0	1/8	0	0
1	3/8	3/8	3/8
2	3/8	6/8	12/8
3	1/8	3/8	9/8
Σ		12/8 = 1.5	24/8 = 3

The reader should very carefully compare the formulas for s_x^2 and s_x with those for σ_x^2 and σ_x, taking note of the similarities—and especially the differences.

7.3.3 Transformed Random Variables

Theorem 2.4 expresses the effect of linear transformations of sample data on $\bar{x}$, s_x^2, and s_x. If we were to transform a random variable in the same way (i.e., multiply each value of a random variable x by a and then add b to each result), we would affect the mean (μ_x), variance (σ_x^2), and standard deviation (σ_x) in ways identical to those expressed in Theorem 2.4. That is, if x is a random variable, then the linear transformation $ax + b$ produces a new random variable called, let us say, w. The parameters of random variable w relate to the parameters of x in the following way.

> **Theorem 7.2:** Given a discrete random variable x with mean μ_x, variance σ_x^2 and standard deviation σ_x, if $w = ax + b$, where a and b are any real numbers, then $\mu_w = a\mu_x + b$, $\sigma_w^2 = a^2\sigma_x^2$ and $\sigma_w = |a|\sigma_x$.

> **Proof:** Similar to the proof of Theorem 2.4.

For example, if $\mu_x = 50$, $\sigma_x^2 = 16$, $\sigma_x = 4$, and $w = 3x + 2.5$, then $\mu_w = 3(50) + 2.5 = 152.5$, $\sigma_w^2 = 3^2(16) = 144$, and $\sigma_w = |3| \cdot 4 = 12$.

The following transformation from random variable x to random variable x' is of special importance to us.

$$x' = \frac{x - \mu_x}{\sigma_x}$$

This transformation may not at first appear to be linear, but it is linear because it can be expressed in the $x' = ax + b$ form as follows:

$$x' = \frac{1}{\sigma_x}x + \frac{-\mu_x}{\sigma_x}$$

This special transformation converts the values of any random variable to *standard units*. The expression "standard units" is short for *standard deviation units. Each value of x' tells how many standard deviations a value of x is from the mean.* If a particular value of x is a distance of 2.5 standard deviations *above* the mean, then the corresponding value of x' is 2.5. If an x value is a distance of 1.2 standard deviations *below* the mean, then the corresponding x' value is -1.2. For example, if $\mu_x = 50$ and $\sigma_x = 5$, then for $x = 40$, x' would be -2, since 40 is 2 standard deviations below μ_x. That is,

$$x' = \frac{x - \mu_x}{\sigma_x} = \frac{40 - 50}{5} = \frac{-10}{5} = -2$$

Similarly, if $x = 59$, then $x' = 1.8$, because

$$x' = \frac{59 - 50}{5} = \frac{9}{5} = 1.8$$

The nature of this special transformation gives the random variable x' some very special properties, namely, that the mean of x' is always zero and the variance and standard deviation are both equal to one.

Theorem 7.3: Given a discrete random variable x with mean μ_x and standard deviation σ_x, if $x' = (x - \mu_x)/\sigma_x$, then $\mu_{x'} = 0$ and $\sigma_{x'}^2 = \sigma_{x'} = 1$.

Proof:

$$x' = \frac{x - \mu_x}{\sigma_x} = \frac{1}{\sigma_x}x + \frac{-\mu_x}{\sigma_x} \qquad \text{(Given)}$$

therefore,

$$\mu_{x'} = \frac{1}{\sigma_x}\mu_x + \frac{-\mu_x}{\sigma_x} = \frac{\mu_x}{\sigma_x} - \frac{\mu_x}{\sigma_x} = 0,$$

$$\sigma_{x'}^2 = \left(\frac{1}{\sigma_x}\right)^2 \sigma_x^2 = 1,$$

and

$$\sigma_{x'} = \left|\frac{1}{\sigma_x}\right|\sigma_x = 1 \qquad \text{(Thm. 7.2)}$$

We shall be using this transformation often in later chapters.

Exercises

1. Find μ_x, σ_x^2, and σ_x for the following probability function.

x:	0	2	5	10
$f(x)$:	2/10	4/10	3/10	1/10

2. Find the mean, variance, and standard deviation for the random variable involved in
   - (a) Exercise 3, page 143.
   - (b) Exercise 4, page 143.
   - (c) Exercise 7, page 143.
   - (d) Exercise 8, page 143.

3. In Exercise 11, page 143:
   - (a) find the mean of y, which in this context is sometimes called the *expected value* of y, or player As *expectation*.
   - (b) give an interpretation of your answer in (a).
   - (c) what is player Bs expectation?
   - (d) it is intuitively clear that for the game to be considered "fair," As expectation and Bs expectation should be equal. Therefore, if this game were "fair," what would μ_y equal?

4. The probability that a given car in a drive-in theatre contains x persons is given by the probability distribution below.

x:	1	2	3	4	5	6	7
$f(x)$:	.05	.40	.15	.30	.05	.04	.01

   - (a) Find the mean number of persons per car.
   - (b) Find the standard deviation of x.

5. In Exercise 9, page 143, what is Bill's expectation in the raffle? (In other words, what is the ticket really worth? Since raffles are designed to make money for the organization sponsoring them, we can expect Bill's expectation to be less than the amount he paid for the ticket.)

6. Two players are gambling with a pair of dice. The dice are to be tossed. Player A must pay B five dollars if the sum on the dice is either 7 or 11. Player B must pay A 10 dollars if the sum on the dice is either 2 or 10. No money changes hands on the other possible outcomes. Is this game "fair"? (See Exercise 3, above.)

7. Random variable x has $\mu_x = 50$ and $\sigma_x = 5$. If $w = (x - 20)/2$, then find μ_w, σ_w^2, and σ_w.

8. Random variable y is transformed into a new random variable w, according to the equation $w = (y + 10)/5$. Given that $\mu_w = 20$ and $\sigma_w^2 = 9$, find μ_y, σ_y^2, and σ_y.

9. Let x have probability function

$$f(x) = x/10, \qquad x = 1, 2, 3, 4$$

and let random variable y equal $3x - 1$.

(a) Express $f(x)$ in tabular form.

(b) Express $g(y)$, the probability function for random variable y, in tabular form.

(c) Calculate: μ_x, μ_y, σ_x, σ_y.

(d) Verify that Theorem 7.2 holds for the transformation $y = 3x - 1$.

10. If $\mu_y = 3$ and $\sigma_y^2 = 1.96$, then what are the values of random variable y corresponding to $y = -1, 2, 3,$ and 5? (I.e., convert the values $-1, 2, 3,$ and 5 to standard units.)

11. Given the following probability function $f(x)$ for random variable x,

x:	1	2	3
$f(x)$:	2/5	1/5	2/5

(a) find μ_x;

(b) find σ_x;

(c) how many standard deviations greater than the mean is the value $x = 3$?

12. Let random variable x be the sum of the numerical outcomes when two dice are thrown (refer back to Exercise 3, page 143). For the values of x tell how many standard deviations above or below the mean each is.

13. Prove Theorem 7.2.

7.4   The Binomial Probability Function

In Section 7.2 we introduced the general notion of the discrete probability function. There are three discrete probability functions which deserve our special attention because of their high frequency of use in statistical application. In this section we shall discuss the *binomial probability function*, which we shall compare in later sections to the *hypergeometric* and *Poisson* probability functions.

7.4.1   The Binomial Expansion

High school students usually learn rather early how to expand a binomial; that is, to take an expression of the form $(a + b)^n$, n a positive integer, and express it as the sum of $n + 1$ terms. For example,

$$(a + b)^2 = a^2 + 2ab + b^2$$
$$(a + b)^3 = a^3 + 3a^2b + 3ab^2 + b^3$$
$$(a + b)^4 = a^4 + 4a^3b + 6a^2b^2 + 4ab^3 + b^4$$

Careful examination of these expressions reveals certain patterns.

(i) When $a + b$ is raised to the nth power, there are always $n + 1$ *terms* in the expression.

(ii) Each of the $n + 1$ terms has an $a^{n-r}b^r$ *component*, where $r = 0$

for the first term, $r = 1$ for the second, and so forth, until finally $r = n$ for the last term. (Note that the sum of the exponents $n - r$ and r equals n.)

(iii)   Each of the $n + 1$ terms has a *coefficient*.

To demonstrate these points, the above expansions can be rewritten as follows:

$$(a + b)^2 = 1 \cdot a^2 b^0 + 2 \cdot a^1 b^1 + 1 \cdot a^0 b^2$$
$$(a + b)^3 = 1 \cdot a^3 b^0 + 3 \cdot a^2 b^1 + 3 \cdot a^1 b^2 + 1 \cdot a^0 b^3$$
$$(a + b)^4 = 1 \cdot a^4 b^0 + 4 \cdot a^3 b^1 + 6 \cdot a^2 b^2 + 4 \cdot a^1 b^3 + 1 \cdot a^0 b^4$$

We can observe from the examples that each term of the expansion of $(a + b)^n$ is of the form $ca^x b^y$, where c is the appropriate *coefficient*, $x + y = n$, and the exponent of b is always one less than the number of the term. The fifth term of the expansion of $(a + b)^{10}$, for example, is $ca^6 b^4$. The only remaining question is, "How do we find the coefficient c for each term"? It turns out that the coefficients for the expansion of $(a+b)^n$ are in fact numbers of the form

$$\binom{n}{r}, \quad \text{where } r = 0, 1, 2, \ldots, n$$

For example, the coefficients for the expansion of $(a + b)^3$ are

$$\binom{3}{0}, \binom{3}{1}, \binom{3}{2}, \text{ and } \binom{3}{3},$$

which are equal to 1, 3, 3, and 1, respectively. Therefore, we can write

$$(a + b)^2 = \binom{2}{0}a^2 + \binom{2}{1}ab + \binom{2}{2}b^2$$

$$(a + b)^3 = \binom{3}{0}a^3 + \binom{3}{1}a^2 b + \binom{3}{2}ab^2 + \binom{3}{3}b^3$$

$$(a + b)^4 = \binom{4}{0}a^4 + \binom{4}{1}a^3 b + \binom{4}{2}a^2 b^2 + \binom{4}{3}ab^3 + \binom{4}{4}b^4$$

So, the fifth term of the expansion of

$$(a + b)^{10} \text{ is } \binom{10}{4}a^6 b^4, \text{ or } 210a^6 b^4.$$

A generalization of the examples above is expressed in Theorem 7.4, which we shall accept without formal proof.

Theorem 7.4:   Binomial Expansion.   If n is a positive integer and a and b are any real numbers, then

$$(a + b)^n = \binom{n}{0}a^n + \binom{n}{1}a^{n-1}b + \binom{n}{2}a^{n-2}b^2 + \cdots + \binom{n}{n}b^n$$

Problem: Expand the following binomial expression.

$(2x + 3)^5$

Solution: According to Theorem 7.4,

$$(2x + 3)^5 = \binom{5}{0}(2x)^5 + \binom{5}{1}(2x)^4(3)^1 + \binom{5}{2}(2x)^3(3)^2 +$$

$$\binom{5}{3}(2x)^2(3)^3 + \binom{5}{4}(2x)^1(3)^4 + \binom{5}{5}(3)^5$$

$$(2x + 3)^5 = 32x^5 + 240x^4 + 720x^3 + 1080x^2 + 810x + 243$$

Problem: Find the third term of the expansion of

$$\left(\frac{1}{3} + \frac{2}{3}\right)^6$$

Solution: According to Theorem 7.4, the third term of the expansion of $(a + b)^n$ is

$$\binom{n}{2}a^{n-2}b^2.$$

Therefore, the third term of the expansion of $(1/3 + 2/3)^6$ is

$$\binom{6}{2}\left(\frac{1}{3}\right)^4\left(\frac{2}{3}\right)^2 = 15\left(\frac{1}{81}\right)\left(\frac{4}{9}\right) = \frac{20}{243}$$

If we were to let $a = 1$ and $b = 1$ in Theorem 7.4, we would get the following interesting result.

$$2^n = \binom{n}{0} + \binom{n}{1} + \binom{n}{2} + \cdots + \binom{n}{n}$$

This special case of Theorem 7.4 can be thought of as a restatement of a fact that we noted earlier (page 7); namely, that for a set of n elements there are 2^n different subsets.

7.4.2   Distinguishable Permutations

We learned earlier that the number of different ways in which n objects can be ordered is $n!$, (i.e., $_nP_n$). For example, the letters a, b, c, and d can be ordered in $4! = 24$ different ways, including $abcd$, $abdc$, $acbd$, etc. Since the objects are different letters of the alphabet, each of the 24 permutations (orders) is *distinguishable*; that is, each has a different appearance. However, if the four objects are a, b, b, b, then not all of the permutations are distinguishable. There are still 24 permutations of these four objects, but the only distinguishable permutations (by appearance) are $abbb$, $babb$, $bbab$, and $bbba$. On the other hand, if the four objects are a, a, b, b, there are six distinguishable permutations. The reader should verify this statement by listing the six permutations.

Generally speaking, in situations where we are considering n objects, r of which are indistinguishable among themselves (for example, all as), and the remaining $n - r$ objects are indistinguishable among themselves (for example, all bs), the number of *distinguishable* permutations of these n objects is

$$\frac{n!}{(n - r)!\, r!}$$

which, of course, is the number of *combinations* of n objects selected r at a time, or $\binom{n}{r}$. To intuitively justify this statement examine the objects a, b, b, b. Finding a distinguishable permutation is equivalent to selecting one position (out of the four positions) in which to place the a. The number of different selections of one position from four positions is $\binom{4}{1}$. Similarly, finding a distinguishable permutation of the objects a, a, b, b is equivalent to selecting a pair of positions in which to place the as. The number of different pairs of positions, selected from four positions, is $\binom{4}{2}$. Thus, we have what appears on the surface to be a contradiction in terms; namely, a special situation in which the number of distinguishable permutations equals a number of combinations! We shall add this counting principle to the four discussed in Section 6.4.

Principle 5: Given n objects, such that r of them are indistinguishable among themselves, and the remaining $n - r$ are distinguishable from the first r, but are indistinguishable among themselves; then *the number of distinguishable permutations* of these n objects is $\binom{n}{r}$.*

Problem: How many distinguishable permutations are there of the following objects?

x,y,y,x,y,x,x,x

Solution: Here we have $n = 8$ objects, five of which are indistinguishable among themselves (the xs) and the remaining three indistinguishable among themselves, but distinguishable from the xs (the ys). Thus, by Principle 5 the number of distinguishable permutations of these eight objects is $\binom{8}{5} = 56$

---

*Principle 5 is a special case of a more general principle, though it is sufficient for our immediate purposes. (See Exercise 17, page 159.)

Principle 5 is of special importance to us in certain types of problems. For example, if 10 coins are flipped, what is the probability that exactly six of them are heads? The sample space for this experiment has 2^{10}, or 1024 outcomes. The outcomes involving exactly six heads are represented by a series of six *h*s and four *t*s in various orders. For example, *hthttthhhh* is one way to get exactly six heads. To find the total number of outcomes which involve exactly six heads we simply answer the question, "How many distinguishable permutations are there for the letters: $h, h, h, h, h, h, t, t, t, t$?" The answer, according to Principle 5, is $\binom{10}{6} = 210$. Each of the 210 outcomes has probability 1/1024. Therefore, the probability of getting exactly six heads in 10 flips of a coin is $(210)(1/1024) = 105/512$.

Suppose that we plan to randomly select five cards *with replacement* from an ordinary deck; what is the probability that exactly three spades are selected? We can think of any card as being either a spade or a nonspade. Therefore, every occurrence of exactly three spades will involve three spades and two nonspades in some order. The probability of getting a spade on any draw is 1/4, and a nonspade, 3/4. Principle 5 tells us how many distinguishable permutations there are of three spades and two nonspades; namely, $\binom{5}{3} = 10$. Thus, the probability of getting exactly three spades is $(10)(1/4)^3(3/4)^2 = (10)(9/1024) = 45/512$.

7.4.3   Binomial Probability

The two chance experiments just discussed are typical examples of a certain class of experiments in which the random variable involved is said to have a *binomial probability function*. This type of experiment has several specific characteristics:

(a) *The experiment involves a series of repeated trials.* A coin may be repeatedly flipped, a die repeatedly tossed, or objects repeatedly selected from a group of objects.

(b) *The possible outcomes for each trial can be expressed in terms of a dichotomy.* Examples include head or tail, spade or nonspade, yes or no, and male or female. We generalize the notion of dichotomy by referring to the outcomes as *success* or *failure*.

(c) *The random variable involved is the number of successes in n trials.* Random variable *x* may be the number of heads occurring in 10 flips of a coin, the number of spades selected in five random draws from a deck of cards, and so on.

(d) *The probability of success on each trial is constant from trial to trial.* On every flip of a coin the probability of getting a head is 1/2. On every random draw *with replacement* from a deck of cards the probability of getting a spade is 1/4.

(e) *The repeated trials are independent.* That is, from trial to trial the probability of success (or failure) is in no way affected by the occurrence or nonoccurrence of success (or failure) on previous trials.

Theorem 7.5: The probability of getting exactly x successes in n independent trials, where the probability of success on each trial is p, is defined by the following probability function.

$$b(x) = \binom{n}{x} p^x (1-p)^{n-x}, \qquad x = 0, 1, 2, \ldots, n$$

This function is called the *binomial probability function.**

The proof of Theorem 7.5 is somewhat implicit in the discussion and examples already presented, and therefore, no formal proof is presented.

If the probabilities of the various values of random variable x are determined by a binomial probability function, then we say that x is *binomially distributed*, or that x has a *binomial distribution*. With Theorem 7.5 we can easily find the probability that exactly six heads occur when 10 coins are flipped;

$$
\begin{aligned}
&n = 10 \\
&p = 1/2 \qquad b(6) = \binom{10}{6}\left(\frac{1}{2}\right)^6 \left(\frac{1}{2}\right)^4 = \frac{105}{512} \\
&x = 6
\end{aligned}
$$

or the probability that exactly three spades are drawn in five selections *with replacement* from an ordinary deck;

$$
\begin{aligned}
&n = 5 \\
&p = 1/4 \qquad b(3) = \binom{5}{3}\left(\frac{1}{4}\right)^3 \left(\frac{3}{4}\right)^2 = \frac{45}{512} \\
&x = 3
\end{aligned}
$$

Situations involving the selection of objects *with replacement,* or ones involving coin-flipping, die-tossing, and other such operations (which are really automatic-replacement situations) present relatively straightforward applications of the binomial probability function. These two types of situations present models which readily conform to the five binomial characteristics listed earlier. It is true, however, that the binomial function can be used in certain everyday situations which do not completely conform to either of the replacement models. The decision to apply or not to apply the binomial function in a particular situation depends upon how closely the five binomial characteristics (or assumptions) are satisfied.

---

*This is our first *special* probability function. Thus, we assign the letter b to represent it. As we introduce other special probability functions, we shall assign other letters to represent them exclusively.

For example, suppose that we want to find the probability that of the next eight babies born at a given hospital, exactly three of them will be males. If we want to use the binomial function, we assume that the five binomial characteristics are reasonably satisfied. We can think of births as repeated trials, where the possible outcomes for each trial can be expressed in terms of the dichotomy, male vs. female. The random variable (x) is the number of males (successes) in the eight trials. We must assume that on each trial the probability of a male being born is *constant*. From empirical evidence we would probably use $p = 1/2$. Finally, we must assume that the trials are *independent*. That is, the determination of sex on a particular trial is in no way affected by sex determination on any of the other trials. The last two binomial characteristics are clearly the key assumptions which differentiate binomial from nonbinomial applications. In this example, since $n = 8$ and $p = 1/2$,

$$P(x = 3) = b(3) = \binom{8}{3}\left(\frac{1}{2}\right)^3\left(\frac{1}{2}\right)^5 = \frac{7}{32}$$

Problem: Suppose it is known from past experience that approximately 70 percent of humans contracting sleeping sickness die from the disease. Given that five people in a certain town contract the disease, use the binomial probability function to find the probability that exactly two of them will die.

Solution: Assuming that we are dealing with five independent trials with probability 0.7 for success (death) on each trial, it follows that $n = 5$ and $p = 0.7$. Therefore,

$$P(x = 2) = b(2) = \binom{5}{2}(0.7)^2(0.3)^3 = 0.1323$$

The really interesting thing about Theorem 7.5 is its intimate relationship to the binomial expansion, Theorem 7.4. Let x represent the number of heads occurring in three flips of a balanced coin. Thus x could equal 0, 1, 2, or 3. The probabilities for each coin can be found by using Theorem 7.5 to find $b(0)$, $b(1)$, $b(2)$, and $b(3)$.

$$b(0) = \binom{3}{0}\left(\frac{1}{2}\right)^0\left(\frac{1}{2}\right)^3 = \frac{1}{8}$$

$$b(1) = \binom{3}{1}\left(\frac{1}{2}\right)^1\left(\frac{1}{2}\right)^2 = \frac{3}{8}$$

$$b(2) = \binom{3}{2}\left(\frac{1}{2}\right)^2\left(\frac{1}{2}\right)^1 = \frac{3}{8}$$

$$b(3) = \binom{3}{3}\left(\frac{1}{2}\right)^3\left(\frac{1}{2}\right)^0 = \frac{1}{8}$$

Compare these figures to the expansion of $(a+b)^n$, where $a = 1/2, b = 1/2$, and $n = 3$.

$$\left(\frac{1}{2}+\frac{1}{2}\right)^3 = \binom{3}{0}\left(\frac{1}{2}\right)^0\left(\frac{1}{2}\right)^3 + \binom{3}{1}\left(\frac{1}{2}\right)^1\left(\frac{1}{2}\right)^2 + \binom{3}{2}\left(\frac{1}{2}\right)^2\left(\frac{1}{2}\right)^1 + \binom{3}{3}\left(\frac{1}{2}\right)^3\left(\frac{1}{2}\right)^0$$

$$1 \quad = \quad \frac{1}{8} \quad + \quad \frac{3}{8} \quad + \quad \frac{3}{8} \quad + \quad \frac{1}{8}$$

This example points out that when n is the number of independent trials, and p the probability of success on each trial, then

$$[(1-p)+p]^n = b(0) + b(1) + b(2) + \ldots + b(n) = 1$$

Problem: In three tosses of a die, what is the probability that a "5" results *at least* once?

Solution: In this problem we can assume that the three trials are independent, with probability 1/6 of getting a "5" on each trial. If we let x be the number of 5s (successes) occurring in the three tosses, then the binomial probability function which applies in this case is

$$b(x) = \binom{3}{x}\left(\frac{1}{6}\right)^x\left(\frac{5}{6}\right)^{3-x}$$

This function gives us the probability of *exactly* x successes. However, the problem calls for *at least* one success, which means "exactly 1 *or* exactly 2, or exactly 3 successes." Since we want the probability that *at least* one of these three *mutually exclusive* events occurs, we want $b(1) + b(2) + b(3)$, which is three applications of Theorem 7.5 with the results added;

$$\binom{3}{1}\left(\frac{1}{6}\right)^1\left(\frac{5}{6}\right)^2 + \binom{3}{2}\left(\frac{1}{6}\right)^2\left(\frac{5}{6}\right)^1 + \binom{3}{3}\left(\frac{1}{6}\right)^3\left(\frac{5}{6}\right)^0 = \frac{91}{216}$$

Alternate Solution: Since we know that $b(0) + b(1) + b(2) + b(3) = 1$ (see paragraph above), we note that

$$b(1) + b(2) + b(3) = 1 - b(0)$$

In other words, the probability of *at least one success* equals one minus the probability of no successes. This approach is much faster, since it requires only *one* application of Theorem 7.5, as follows:

$$1 - b(0) = 1 - \binom{3}{0}\left(\frac{1}{6}\right)^0\left(\frac{5}{6}\right)^3 = 1 - \frac{125}{216} = \frac{91}{216}$$

(Note that we have applied Theorem 6.2 in this alternate approach.)

7.4.4 *Binomial Parameters*

In Section 7.2 the general notion of the discrete probability function was first introduced. The binomial probability function (defined by Theorem 7.5) is our first *special* discrete probability function. In fact, Theorem

7.5 defines not just one, but an entire family of binomial probability functions; the members of this family differ only in the value of n or the value of p (or both). For example,

$$b(x) = \binom{5}{x}\left(\frac{1}{4}\right)^x\left(\frac{3}{4}\right)^{5-x}, \qquad x = 0, 1, \ldots, 5$$

and

$$b(y) = \binom{8}{y}\left(\frac{2}{7}\right)^y\left(\frac{5}{7}\right)^{8-y}, \qquad y = 0, 1, \ldots, 8$$

are different probability functions, although both are binomial.* In the former case $n = 5$ and $p = 1/4$; in the latter case $n = 8$ and $p = 2/7$.

A binomially distributed random variable x has a mean μ_x, a variance σ_x^2, and a standard deviation σ_x, just like any other random variable. Definitions 7.3 and 7.4 and Theorem 7.1 give the appropriate formulas for calculating these parameters. As a matter of fact, the example used in Section 7.3 (the three-coin experiment) to demonstrate the calculation of these parameters involves a binomially distributed random variable (number of heads in three flips of a coin). The values obtained were $\mu_x = 1.5$, $\sigma_x^2 = 0.75$, and $\sigma_x = 0.87$. Because of the special nature of binomial probability functions, μ_x, σ_x^2, and σ_x for binomially distributed random variables can be calculated directly from n and p.

Theorem 7.6: If random variable x is the number of successes in n independent trials, where the probability of success on each trial is p (i.e., *x is binomially distributed*), then

$$\mu_x = np \qquad \text{and} \qquad \sigma_x^2 = np(1 - p)$$

In the three-coin experiment, for example,

$$\mu_x = (3)(1/2) = 1.5$$
$$\sigma_x^2 = (3)(1/2)(1/2) = 0.75$$
$$\sigma_x = \sqrt{0.75} = 0.87$$

Thus, for binomially distributed random variables *only*, μ_x, σ_x^2 and σ_x can be calculated from Theorem 7.6.

Exercises†

1. Expand the following binomial expressions and reduce them to simplest form:

  (a) $(a + b)^5$ (b) $(2x - 1)^4$ (c) $(y/2 + 5)^3$

---

*Mathematicians would emphasize the fact that these two binomial functions are different by using the notation: $b(x;5,1/4)$ and $b(y;8,2/7)$, since the different values of n and p make the difference.

†In the interest of time-saving in this set of exercises the reader may wish to use Table C in the Appendix to find binomial coefficients corresponding to given values of n and r.

2. Find the
   (a) coefficient of the fourth term of the expansion of $(x + y)^{12}$.
   (b) coefficient of a^3b^2 in the expansion of $(a + b)^5$.
   (c) coefficient of a^3b^2 in the expansion of $(2a + 3b)^5$.
   (d) sixth term of the expansion $(2/5 + 3/5)^8$.
   (e) 150th term of the expansion $(0.25 + 0.75)^{265}$.

3. In how many distinguishable ways can four tanks and three jeeps be parked side-by-side? (Assume that all tanks look alike and all jeeps look alike.)

4. Six boys are lining up to get tickets to a movie.
   (a) Assuming that an observer can distinguish the boys from one another, how many distinguishable ways can they line up?
   (b) Suppose that four boys are wearing red hats and the other two blue hats. If an observer is standing so far away from the line that hat color is the only distinguishing feature among the boys, then how many distinguishable ways can the boys line up?

5. How many different five-letter code words can be formed from the letters b, b, e, e, e?

6. In eight consecutive tosses of a coin, how many of the possible 2^8 outcomes contain exactly five heads?

7. In 12 independent trials, each resulting in either "success" or "failure," how many of the possible outcomes involve exactly four successes?

8. Let random variable x be the number of successes in n independent trials, where the probability of success on each trial is p. Find the probability that
   (a) $x = 2$, given that $n = 6$ and $p = 1/4$;
   (b) $x > 2$, given that $n = 4$ and $p = 1/3$;
   (c) $x < 3$, given that $n = 8$ and $p = 1/2$;
   (d) $x > 1$, given that $n = 200$ and $p = 2/7$.

9. Three dice are to be tossed. What is the probability that
   (a) exactly one "5" occurs?
   (b) at least two "5s" occur?
   (c) no "5s" occur?

10. If we assume that human births occur independently and that the probability of getting a boy is always 1/2, then of the next 10 babies born in the U.S., what is the probability that
    (a) exactly half will be boys?
    (b) at least half will be boys?

11. On an eight-question multiple-choice test (four choices for each question of which only one is correct), a score of five correct answers is necessary for a passing grade. If a student were to answer all eight randomly, without even looking at the questions,
    (a) what is the probability that he would get a passing grade?
    (b) what is the probability that the third correct answer occurs on the sixth question?

12. John plays a game in which a pair of dice are tossed. If a "2" comes up on

either die, he wins. If the dice are tossed six times, what is the probability that John wins exactly three times?

13. Given that four people are selected randomly from the population, what is the probability that exactly three of them were born on either a Monday or a Wednesday? (Assume that each day of the week is equally likely to be one's day of birth.)

14. Do Exercise 2, page 148 again, parts (b) and (d), using Theorem 7.6, and compare your answers to your previous answers.

15. In Exercises 10, 11, 12, and 13 above,
   (a) express in equation form the binomial probability function used in the exercise;
   (b) calculate the mean, variance, and standard deviation for the random variable defined in the exercise.

16. If a random variable y is a binomially distributed random variable with probability function

$$b(y) = \binom{14}{y}(2/9)^y(7/9)^{14-y}, \qquad y = 0, 1, 2, \ldots, 14,$$

find the mean, variance, and standard deviation of y.

17. Principle 5 can be generalized to more than two groups. Given n objects of k distinguishable types such that n_1 are indistinguishable among themselves, n_2 are indistinguishable among themselves, and so on up to n_k, and $n_1 + n_2 + \ldots + n_k = n$, then the number of distinguishable permutations of the n objects is

$$\frac{n!}{n_1! \, n_2! \, \ldots \, n_k!}$$

   (a) Twelve marbles are identical except for color; three are red, two are black, four are blue, and three are yellow. Find the number of distinguishable permutations of these 12 marbles.
   (b) How many nine-letter code words can be formed by rearranging the letters in the word STATISTIC?

7.5 The Hypergeometric Probability Function

In Section 7.4.3 we noted that the binomial probability function readily applies in situations involving selection of objects *with replacement*. However, when we select objects *without replacement*, the two most important assumptions underlying the binomial distribution are not satisfied. That is, probability of success from trial to trial is *not constant* and the trials are *not independent*.

Take, for example, the chance experiment in which three random selections are made from a box containing eight marbles, two of which are red. Let x equal the number of red marbles selected. If replacement were allowed, then the domain of x would be $\{0, 1, 2, 3\}$ and the probability of getting exactly x red marbles in the three trials would be determined by a binomial probability function; namely,

$$b(x) = \binom{3}{x}\left(\frac{1}{4}\right)^x\left(\frac{3}{4}\right)^{3-x}, \qquad x = 0, 1, 2, 3$$

In tabular form we would get

x:	0	1	2	3
$b(x)$:	$\dfrac{27}{64}$	$\dfrac{27}{64}$	$\dfrac{9}{64}$	$\dfrac{1}{64}$

If replacement were *not* allowed, however, we would approach the problem quite differently. With no replacement the trials are no longer independent, and the probability of success on each trial is no longer a constant p. With no replacement the probability of obtaining exactly x successes is found by a "combinations" approach. For example, to find the probability that $x = 1$ (in the same marble problem) we reason that after the three selections have been made, if the event $x = 1$ occurs we will have before us three marbles exactly one of which is red. That set of three marbles is one of a possible $\binom{8}{3}$, or 56, different sets of three marbles which have equal probability of selection. But how many of those 56 possibilities would involve exactly one red marble? Since there are only two red marbles in the box, there are $\binom{2}{1}$, or 2, different sets of one red marble that could possibly be selected. Similarly, there are $\binom{6}{2}$, or 15, different sets of two non-red marbles that could possibly be selected. Therefore, there are $\binom{5}{1} \times \binom{6}{2}$, or 30, equally likely possibilities out of the 56 which contain one red marble and two non-red marbles. In other words, if we let $g(x)$ represent the probability function for x (the number of red marbles) when there is no replacement,

$$g(1) = \frac{\binom{2}{1}\binom{6}{2}}{\binom{8}{3}} = \frac{(2)(15)}{56} = \frac{15}{28}$$

Since there is no replacement, x can have only the values 0, 1, or 2; therefore, using similar reasoning,

$$g(0) = \frac{\binom{2}{0}\binom{6}{3}}{\binom{8}{3}} = \frac{(1)(20)}{56} = \frac{10}{28}$$

$$g(2) = \frac{\binom{2}{2}\binom{6}{1}}{\binom{8}{3}} = \frac{(1)(6)}{56} = \frac{3}{28}$$

We can reduce the above results to tabular form,

x:	0	1	2
$g(x)$:	$\dfrac{10}{28}$	$\dfrac{15}{28}$	$\dfrac{3}{28}$

and the probability function for random variable x can be expressed as the equation below.

$$g(x) = \frac{\binom{2}{x}\binom{6}{3-x}}{\binom{8}{3}}, \qquad x = 0, 1, 2$$

This function is one example of a family of functions called *hypergeometric* probability functions. The general form of this function and the conditions under which it applies are spelled out in Theorem 7.7.

Theorem 7.7:   Given a set of N objects, including M of a particular type called "successes," *the probability of getting exactly x successes in n selections without replacement* is defined by the following probability function.

$$h(x) = \frac{\binom{M}{x}\binom{N-M}{n-x}}{\binom{N}{n}}, \qquad x = 0, 1, 2, \ldots, M$$

This function is called the *hypergeometric probability function.* Again, the proof of this theorem is implicit in the preceding discussion and examples, and therefore, no formal proof is presented.

Note in Theorem 7.7 that as in the case of binomial probability (a) we are dealing with a dichotomy, success or nonsuccess (failure), (b) n is the number of trials (selections), and (c) the random variable is the number of successes occurring in n trials. The major difference between the two situations is the "no replacement" condition in the hypergeometric case. Without replacement the n trials are not independent, and the probability of success on each trial is not constant.

Sometimes the binomial function is used as an approximation to the hypergeometric function. The best approximations occur where N is large and n is small. For example, if we make three random selections without replacement from a box of 100 marbles, 40 of which are red, the probability of getting exactly one red marble is

$$h(1) = \frac{\binom{40}{1}\binom{60}{2}}{\binom{100}{3}} = 0.438$$

If we were to use the binomial probability function to approximate this answer, we would get

$$b(1) = \binom{3}{1}\left(\frac{40}{100}\right)^1\left(\frac{60}{100}\right)^2 = 0.432$$

The reader should check out the two calculations above and, in doing so, note that the binomial calculation is somewhat easier. There are, however, occasions when calculating binomial probability can be extremely tedious. For example, the probability of getting a sum of seven at least 20 times in 100 tosses of a pair of dice is

$$b(20) + b(21) + \cdots + b(100) = \sum_{x=20}^{100} b(x) = \sum_{x=20}^{100} \binom{100}{x}\left(\frac{1}{6}\right)^x\left(\frac{5}{6}\right)^{100-x}$$

With our present methods we cannot think of attempting such a calculation. We shall, however, discuss in Chapter 8 a method for approximating binomial probability. In fact, in the next section we shall discuss the Poisson probability function, which can be used as an approximation to the binomial function under certain special conditions.

Before we proceed to Poisson probability we should note that just as in the binomial case there are special shortcut formulas for calculating the mean, variance, and standard deviation for hypergeometrically distributed random variables.

Theorem 7.8: If a random variable x is hypergeometrically distributed (as defined in Theorem 7.7), then

$$\mu_x = \frac{nM}{N} \quad \text{and} \quad \sigma_x^2 = \frac{M(N-M)(n)(N-n)}{(N^2)(N-1)}$$

Exercises

1. Five girls and seven boys are members of a club. They randomly select three of their membership to represent them at a state-level meeting. Let x represent the number of girls on the committee. Express the probability function of x in both equation and tabular form.

2. In a factory, bolts are packed in boxes of 50. Suppose that a particular box contains five defective bolts. If a quality control investigator were to select four bolts at random from the box, what is the probability that

(a) exactly two would be defective?

(b) all would be defective?

(c) at least one would be defective?

3. A bridge player is dealt 13 cards from an ordinary deck. Let y equal the number of hearts he is dealt.

(a) Express in equation form the probability function for y.

(b) Express (do not calculate) the probability that he gets exactly two hearts.

(c) Express the probability that he gets at least five hearts.

(d) Express the probability that the fifth card dealt is the third heart dealt.

4. Four objects are randomly selected without replacement from 20 objects, of which six are green.

(a) Find the probability that exactly two of the objects selected are green.

(b) Find the probability that more than two of the objects selected are green.

(c) Use the binomial probability function to approximate your answers to (a) and (b). That is, do (a) and (b) again, assuming a binomial distribution for the number of green objects selected.

5. In a political convention suppose that 300 of the 500 delegates are supporters of candidate A. Five of the delegates are randomly selected.

(a) Express the probability that exactly four of those selected are for candidate A.

(b) Approximate the answer to (a) by using the binomial probability function.

6. R objects include Q red objects. We plan to select k objects from the R. Let x be the number of red objects selected. Express the probability function for x, assuming that the selections were made

(a) without replacement;

(b) with replacement.

7. In Exercises 1, 2, and 3 above, find the mean, variance, and standard deviation of the random variable involved.

8. If x is a hypergeometrically distributed random variable with probability function

$$h(x) = \frac{\binom{15}{x}\binom{25}{10-x}}{\binom{40}{10}}, \qquad x = 0, 1, 2, \ldots, 15$$

then find μ_x, σ_x^2, and σ_x.

7.6 The Poisson Probability Function

The Poisson probability function is used in many practical problems to find the probability that x "successes" occur over a given interval of time or region of space—especially when few such successes are likely to occur. For example, the random variable might be the number of defective light bulbs coming off a factory assembly line in an hour, the number of telephone calls received at a certain residence in a day, or the number of honeymoon couples registered in a particular hotel during the second week in June.

Definition 7.5: If m is a positive real number, then the function

$$p(x) = \frac{e^{-m}m^x}{x!}, \qquad x = 0, 1, 2, \ldots$$

is called the *Poisson probability function.*°

The symbol e is standard notation for a real number used by mathematicians, which is approximately equal to 2.7183. (See Table D in the Appendix for values of e^{-m}.)

Knowing when to use the Poisson probability function is not as clear cut in practice as is the application of the binomial and hypergeometric functions. Assumptions underlying the correct use of the Poisson function are

(i)   each "success" occurs independently of the others;
(ii) successes are uniformly distributed throughout the interval of time or region of space;
(iii) the probability of success in the interval (region) is very small.

The first assumption is almost identical to the binomial assumption of independence; the second says essentially that the probability of success is proportional to the length of the time interval or region of space; and the last assumption means that most of the probability is concentrated at the low end of the domain of x.

In order to use Definition 7.5 we need to know the value of m. It turns out that m equals the mean of random variable x *and also the variance of* x. That is, when x is Poisson-distributed, $m = \mu_x = \sigma_x^2$.

Theorem 7.9: If random variable x is Poisson-distributed with probability function

$$p(x) = \frac{e^{-m}m^x}{x!}, \qquad x = 0, 1, 2, \ldots$$

then $\mu_x = m$ and $\sigma_x^2 = m$.

For example, consider the following situation. An insurance company knows from experience that about 0.004 percent of the population die

°Named after its creator, Simeon Poisson (1781–1840).

from a specific kind of accident. What is the probability that exactly two of 40,000 insured people die in such an accident in a given year? Let x represent the number of people who die from the accident in a given year. The mean of $x = (0.00004)(40,000) = 1.6 = m$. Thus, the probability that $x = 2$ is $p(2)$, where

$$p(x) = \frac{e^{-1.6}(1.6)^x}{x!}, \qquad x = 0, 1, 2, \ldots$$

Therefore,

$$p(2) = \frac{e^{-1.6}(1.6)^2}{2!} = \frac{(0.2019)(2.56)}{2} = 0.26$$

Note that, from Table D in the Appendix, $e^{-1.6} = 0.2019$.

Table 7.3 represents the function in tabular form. Note that although the Poisson function theoretically has as many values as there are whole numbers (i.e., $x = 0, 1, 2, \ldots$), in this particular application $x = 0, 1, 2, \ldots, 40,000$. More significant than this is the fact that the probability that an insured person will die from this type of accident is so low that there are no values of $p(x)$ of significant size past $p(4)$.

Table 7.3. Poisson Probabilities Where x = Number of Deaths Out of 40,000 People

x	$p(x)$
0	.202
1	.323
2	.258
3	.138
4	.055
5	.018
6	.005
.	.
.	.
.	.

In addition to using the Poisson probability function in practical applications where the three assumptions seem reasonable, we use this probability function to approximate binomial probability *when the number of independent trials (n) is large and the probability of success (p) on each trial is small.* For example, let x be the number of successes in 24 independent trials where the probability of success on each trial is 0.05. The distribution of probability for random variable x is binomial. However, the actual calculations are quite tedious. These binomial probabilities can be approximated very closely by Poisson probabilities, where $m = np = 24(0.05) = 1.2$ (i.e., letting $m =$ the binomial mean). A comparison of the actual binomial probabilities and the Poisson approximations appear in Table 7.4.

Table 7.4. Comparison of Binomial and Poisson Probabilities for $n = 24$ and $p = .05$

x	(Binomial) $b(x)$	(Poisson) $p(x)$
0	.292	.301
1	.369	.361
2	.223	.217
3	.086	.087
4	.024	.026
5	.005	.006
6	.001	.001
.	.	.
.	.	.
.	.	.
24	.000	.000

Although the Poisson probability function can be useful under certain conditions (n large and p small) as an approximation to the binomial probability function, the Poisson approximations are somewhat tedious to calculate themselves, especially when we wish to find, for example, the probability of obtaining *at least* 15 successes in 25 independent trials. In Chapter 8 we shall learn to approximate probabilities of this type with the *normal probability* function. It should be noted also that there are books available which contain extensive tables of binomial, hypergeometric, and Poisson probabilities—see Owen (1962) and Pearson and Hartley (1967).

Exercises

1. Mr. Brown receives on the average 3.2 phone calls per day. On a particular day what is the probability that he will receive
    (a) at most one call?
    (b) at least one call?
    (c) exactly one call?

2. Use the Poisson probability function to approximate the probability that at least two successes occur in 120 independent trials, where the probability of success on each trial is 0.02.

3. Given that a random variable x is Poisson-distributed with probability function

$$p(x) = \frac{e^{-2.56}(2.56)^x}{x!}, \qquad x = 0, 1, 2, \ldots ,$$

find the mean, variance, and standard deviation of x.

4. A typist averages 2.8 errors per page. What is the probability that on a particular page she exceeds her average number of errors?

5. Assume that the number of major snow storms in Oswego, New York, is Poisson-distributed. If Oswego averages 2.2 major snow storms per year and the city has enough money to provide for heavy snow-removal equipment to be used a maximum of three times per year without the city going into debt, then what is the probability that the city will go into debt next year due to an excess of major snow storms?

6. Rained-out games are always somewhat of a financial loss to baseball clubs, since the games are usually made up later in the season as part of a double-header—with no increase in the price of admission. Assume that the average number of rained-out games per season is four, and that the club owners have enough money budgeted to cover the expected losses from five rained-out games. Next season, what is the probability that
   (a) only two games are rained out?
   (b) no games are rained out?
   (c) the "rain-out" budget is used up?

7. On the average 5,000 cars pass through a certain downtown intersection between 4:30 and 6:00 P.M. on weekdays. Each car has probability 0.0003 of having an accident at the intersection during that period. What is the probability that exactly one accident occurs at that intersection next Monday between 4:30 and 6:00 P.M.?

7.7 Joint Probability Functions

7.7.1 Introduction

We have been concentrating on discrete probability functions for individual variables. If "5" is a possible value of the discrete random variable x, and $f(x)$ is the probability function for x, then the value of the function at 5 is the probability that $x = 5$; that is, $f(5) = P(x = 5)$. In general, $f(a) = P(x = a)$. It is possible to define probability functions for two (or more) variables. Such functions are called *joint probability functions*. For two discrete random variables, say x and y, the joint probability function for x and y can be represented by $f(x,y)$. If "5" is a possible value of x, and "7" a possible value of y, then $f(5,7)$ is the probability that $x = 5$ *and* $y = 7$. The subject of joint probability functions is rather extensive, and we shall not attempt a thorough study of it. However, it is necessary that we become familiar with the basic notion of the joint probability function in two variables, if for no other reason than to enable us to define the *correlation* between two random variables.

7.7.2 Basic Notions

We shall begin with an example. Suppose that two dice are to be tossed. Let x equal the number of "5s" that result, and let y be the number of odd numbers that result. Thus, the domain for both variables is $\{0, 1, 2\}$. The probability function for each variable is shown in Table 7.5.*

Table 7.5. Probability Functions for x and y in the Two-Dice Experiment

x	$f(x)$	y	$g(y)$
0	25/36	0	9/36
1	10/36	1	18/36
2	1/36	2	9/36

Now let us consider the possible joint probabilities. For example, what is the probability that exactly one "5" occurs *and* both dice yield odd numbers? That is, what is the probability that "$x = 1$" *and* "$y = 2$"? If we let $j(x,y)$ represent the joint probability function for x and y, then we can rephrase the question, "What is $j(1,2)$?" Since there are only four outcomes which satisfy this joint condition (namely: (5, 1), (5, 3), (1, 5), and (3, 5)), therefore, $j(1, 2) = 4/36$. This probability and the other values of $j(x,y)$ are displayed in Table 7.6. Note that the nine joint probabilities in the body of the table add up to one. Note also that the

Table 7.6. Joint Probability Function for x and y in the Two-Dice Experiment

x \ y	$j(x,y)$ 0	1	2	$f(x)$
0	9/36	12/36	4/36	25/36
1	0	6/36	4/36	10/36
2	0	0	1/36	1/36
$g(y)$	9/36	18/36	9/36	1

*Even though these are both binomial functions, we shall use f and g to represent them so that we can easily distinguish between them.

marginal sums in Table 7.6 are the probabilities in Table 7.5. The reader should verify each of the entries in Table 7.6.

Definition 7.6:   If x and y are discrete random variables, then any function $j(x,y)$ whose value at $x = a$ and $y = b$ is *the probability that $x = a$ and $y = b$*, is called a *discrete joint probability function*. Such a function has the following properties;

(i) $j(x,y) \geq 0$, for all x and y

(ii) $\displaystyle\sum_{\text{all } x,y} j(x,y) = 1$

Some joint probability functions can be defined in equation form, for example,

$$j(x,y) = \frac{1}{32}(x^2 + y^2), \qquad \text{for } x = 0, 1 \qquad \text{and } y = 0, 1, 2, 3$$

defines a joint probability function. The tabular form of this function can be constructed easily by substituting all possible pairs (x,y) into the equation to obtain the individual joint probabilities. For example, $P(x = 1 \text{ and } y = 2) = j(1,2) = 1/32 \, (1^2 + 2^2) = 5/32$, and so on.

7.7.3   Linear Correlation Between Random Variables

In Chapter 5 we discussed correlation with respect to *sample* data. Just as we have theoretical parameters μ_x, σ_x^2, and σ_x as the counterparts to sample statistics $\bar{x}$, s_x^2, and s_x, we also define a theoretical correlation between random variables x and y, denoted by ρ_{xy}, which is the counterpart to the sample statistic r_{xy}.

Definition 7.7:   Let x and y be random variables with joint probability function $j(x,y)$, means μ_x and μ_y, and standard deviations $\sigma_x \neq 0$ and $\sigma_y \neq 0$. The *linear correlation between random variables x and y*, denoted by ρ_{xy}, is

$$\rho_{xy} = \frac{\displaystyle\sum_{\text{all } x,y} [(x - \mu_x)(y - \mu_y) \cdot j(x,y)]}{(\sigma_x)(\sigma_y)}$$

The equation in Definition 7.7 can be written in a somewhat simpler form, according to Theorem 7.10.

Theorem 7.10:

$$\rho_{xy} = \frac{\displaystyle\sum_{\text{all } x,y} xy \cdot j(x,y) - (\mu_x)(\mu_y)}{(\sigma_x)(\sigma_y)}$$

Using Theorem 7.10 and the information from Table 7.6, we calculate ρ_{xy} as follows:

$$\sum_{\text{all } x,y} xy \cdot j(x,y) = (0)(0)\frac{9}{36} + (0)(1)\frac{12}{36} + (0)(2)\frac{4}{36} + (1)(0)(0)$$

$$+ (1)(1)\frac{6}{36} + (1)(2)\frac{4}{36} + (2)(0)(0)$$

$$+ (2)(1)(0) + (2)(2)\frac{1}{36}$$

$$= \frac{1}{2}$$

$$\mu_x = \sum_{\text{all } x} xf(x) = (0)\frac{25}{36} + (1)\frac{10}{36} + (2)\frac{1}{36} = \frac{1}{3}$$

$$\mu_y = \sum_{\text{all } y} yg(y) = (0)\frac{9}{36} + (1)\frac{18}{36} + (2)\frac{9}{36} = 1$$

$$\sigma_x^2 = \sum_{\text{all } x} x^2 f(x) - \mu_x^2 = (0)^2\frac{25}{36} + (1)^2\frac{10}{36} + (2)^2\frac{1}{36} - \left(\frac{1}{3}\right)^2 = \frac{5}{18}$$

$$\sigma_y^2 = \sum_{\text{all } y} y^2 g(y) - \mu_y^2 = (0)^2\frac{9}{36} + (1)^2\frac{18}{36} + (2)^2\frac{9}{36} - (1)^2 = \frac{1}{2}$$

Therefore,

$$\rho_{xy} = \frac{\frac{1}{2} - \left(\frac{1}{3}\right)(1)}{\sqrt{\frac{5}{18}} \cdot \sqrt{\frac{1}{2}}} = 0.45$$

The interpretation of ρ_{xy} is the same as that for r_{xy} (review Section 5.4), with the exception that ρ_{xy} represents the true linear correlation between x and y, not an estimate (see footnote at the bottom of page 83).

7.7.4 Independence of Random Variables

Definition 7.8: Random variables x and y, with probability functions $f(x)$ and $g(y)$ and joint probability function $j(x,y)$, are said to be *independent* if and only if

$$j(x,y) = f(x) \cdot g(y)$$

for every ordered pair of values (x,y).

Definition 7.8 is a logical extension of the concept of independent events. If x and y are independent random variables, then for every pair of events $x = a$ and $y = b$, *the probability that $x = a$ and $y = b$ equals the probability that $x = a$ multiplied by the probability that $y = b$* (i.e., $j(a,b) = f(a) \cdot g(b)$). We can easily verify in Table 7.6 that random variables x and y are *not independent* by noting, for example, that $j(1,2) \neq f(1) \cdot g(2)$.

Let us consider, then, two random variables which *are* independent. Suppose that three coins are flipped and two dice are tossed. Let x represent the number of heads that occur on the coins, and let y stand for the number of odd numbers that occur on the dice. Thus, the domains for x and y are $\{0, 1, 2, 3\}$ and $\{0, 1, 2\}$, respectively. Therefore,

x:	0	1	2	3
$f(x)$:	1/8	3/8	3/8	1/8

and

y:	0	1	2
$g(y)$:	1/4	1/2	1/4

are the probability functions for x and y, and the joint probability function $j(x,y)$ is shown in Table 7.7. The reader should verify that all of the probability entries in Table 7.7 are correct and note that Definition 7.8 is satisfied.

Table 7.7. Joint Probability Function for the Three-Coin, Two-Dice Experiment

	$j(x,y)$			
x \ y	0	1	2	$f(x)$
0	1/32	1/16	1/32	1/8
1	3/32	3/16	3/32	3/8
2	3/32	3/16	3/32	3/8
3	1/32	1/16	1/32	1/8
$g(y)$	1/4	1/2	1/4	1

Theorem 7.11: If random variables x and y are *independent*, then $\rho_{xy} = 0$.

Theorem 7.11 formalizes the connection between the notions of *independence* and *linear correlation*. Oddly enough, the converse of Theorem 7.11 is not valid. That is, it is not always true that when $\rho_{xy} = 0$, the random variables are independent. It is possible to have no *linear* relationship, but at the same time have a nonlinear dependence. Such a case is shown in Exercise 5, page 170.

Exercises

1. For the two-dice experiment in the text (page 168) let x remain as defined, but let

$$y = \begin{cases} 0, \text{ if the sum on the dice is 2, 3, 4, or 5} \\ 1, \text{ if the sum on the dice is 6, 7, or 8} \\ 2, \text{ if the sum on the dice is 9, 10, 11, or 12} \end{cases}$$

   (a) Construct in tabular form the joint probability function for x and y, including the marginal totals.
   (b) Are x and y independent?

2. (a) Write in tabular form the function $j(x,y) = (x^2 + y^2)/32$, for $x = 0, 1$ and $y = 0, 1, 2, 3$ mentioned in the text (page 169), including the marginal totals.
   (b) Find $P(x = 1)$, $P(y = 2)$, $P(x = 0$ and $y = 3)$, $P(x = 1$ or $y = 2)$, and $P(x = 0 | y = 1)$.

3. Given the following joint probability distribution:

x \ y	0	2	4
1	.15	.22	.03
2	.25	.08	.27

   (a) find μ_x, μ_y, σ_x, σ_y, and ρ_{xy}.
   (b) find $P(x = 2)$, $P(y = 4)$, $P(x = 2$ and $y = 4)$, and $P(x = 2 | y = 4)$.
   (c) are x and y independent?

4. Verify that for independent random variables x and y, whose joint probability function appears in Table 7.7, $\rho_{xy} = 0$.

5. Show that the following joint probability function supports the fact that the converse of Theorem 7.11 is not true.

x \ y	1	2	3
0	.1	.3	.1
1	.2	.1	.2

Summary

A *discrete random variable* is a variable (a) whose values are associated with outcomes of a chance experiment, and (b) whose set of possible outcomes (domain) is either finite or contains as many elements as there are whole numbers.

A *discrete probability function* for discrete random variable x, denoted by $f(x)$, matches each value a of x with the probability that $x = a$; that is, $f(a) = P(x = a)$, for all values of x. The value of the function at $x = a$ is, therefore, always nonnegative, and $\sum_{\text{all } x} f(x) = 1$. The mean ($\mu_x$), variance (σ_x^2), and standard deviation (σ_x) are defined for random variable x. The terms "probability function" and "probability distribution" are used interchangeably.

The *binomial, hypergeometric,* and *Poisson* probability functions are widely used in statistical work. The mean, variance, and standard deviation for these special functions can be found by the usual definition or by special shortcut formulas.

A *discrete joint probability function* for random variables x and y, denoted by $j(x,y)$, matches every pair of values a and b (where a is a value of x, and b is a value of y), with the probability that $x = a$ and $y = b$, that is, $j(x,y) = P(x = a$ and $y = b)$, for all possible pairs of values – one for each variable. The *linear correlation* between random variables x and y, denoted by ρ_{xy}, is defined. The property of *independence* between random variables is also defined. If two variables are independent, then $\rho_{xy} = 0$.

8 | Continuous Probability Functions

8.1   Continuous vs. Discrete Probability Functions

8.1.1   Introduction

In Chapter 7 we confined ourselves to a discussion of probability functions (distributions) for *discrete* random variables. Customarily such variables take on a finite set of values, usually a subset of the integers. We know that discrete random variables typically deal with counting (how many successes, etc.), and we use discrete probability functions to describe the manner in which probability is distributed among the possible values of random variables. The discrete probability function allows us to find such things as: $P(x = a)$, $P(x < b)$, and so forth. For example, if a discrete random variable x, which can take on values 0, 1, 2, 3, and 4, has a probability function $f(x)$, then $P(x = 2) = f(2)$ and $P(x < 3) = f(0) + f(1) + f(2)$.

By contrast, in this chapter we shall concentrate on *continuous* random variables and the methods which have been developed to describe the distribution of probability for such variables. Continuous random variables deal with *measurement* on a continuous scale, such as measurement of weight, speed, or distance. It is true that when such variables are measured for practical purposes the results are rounded to the nearest unit or tenth of a unit; however, the exact weight, speed, or distance being measured can theoretically take on any one of an infinite number of real number values within a certain interval on the real number line.

8.1.2 Continuous Probability

Definition 8.1:   A variable x which (i) assigns a numerical value to each outcome in a sample space, and (ii) has as a domain the set of all real numbers in an interval on the real number line, is called a *continuous random variable*.

Probability functions (distributions) associated with *continuous* random variables have similarities to discrete probability functions *and* some very significant differences. We shall use the same notation "$f(x)$" to denote a continuous probability function, and such a function shall assign a nonnegative number to each possible value of the continuous random variable x, just as a discrete probability function does.

Definition 8.2:   Given a continuous random variable x, such that $a < x < b$ (where a and b are real numbers), a function $f(x)$ which satisfies the following conditions is called a *continuous probability function** for random variable x.
   (i) $f(x) \geq 0$, for all x.
   (ii)   The area under the graph of $f(x)$ equals one.†
   (iii)   If c and d are any two values of x such that $c \leq d$, then $P(c \leq x \leq d) =$ the area under the graph of $f(x)$ from c to d.†

Definition 8.2 should be compared to Definition 7.2. Since continuous probability functions assign a value to every one of an *infinite* number of values in an interval on the real number line, the graph of $f(x)$ is a solid line (not necessarily straight) above the x axis. Figure 8.1 shows schematically how different types of continuous probability functions appear when graphed; in each case the shaded area equals one.

In Chapter 7 we discussed briefly the probability histogram. This graphing method for discrete random variables (which take on integer values) is in a sense analogous to the shaded graphs in Figure 8.1; the

---

*Also called a *probability density function*, or simply a *density function*.

†For students who have had calculus, conditions (ii) and (iii) can be written:

(ii) $\int_a^b f(x)dx = 1$,   and

(iii) $P(c < x < d) = \int_c^d f(x)dx$.

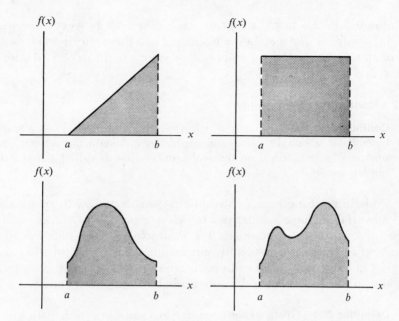

Figure 8.1. Some examples of graphs for continuous probability functions.

reader should recall that the area under such a probability histogram is

$$\sum_{\text{all } x} f(x) = 1$$

To find $P(2 \leq x \leq 5)$ for a *discrete* random variable we would calculate $f(2) + f(3) + f(4) + f(5)$, each term of which is equal to a rectangular area in the histogram. Similarly, to find $P(2 \leq x \leq 5)$ for a *continuous* random variable we would find the area under the graph of $f(x)$ from two to five. Without reference to specific functions, Figure 8.2 shows for the discrete and continuous cases the areas representing $P(2 \leq x \leq 5)$.

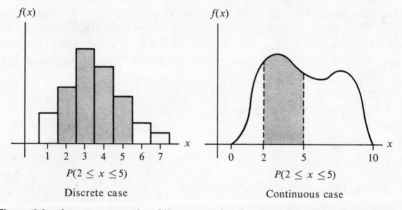

Figure 8.2. Areas representing $P(2 \leq x \leq 5)$ for the discrete and continuous cases.

8.1.3 P(x = k) = 0

We must emphasize one very important difference between discrete and continuous probability functions. *In the discrete case $P(x = k) = f(k)$, but in the continuous case $P(x = k) = 0$.* That is, the probability that a *continuous* random variable equals any specific value k is 0. This fact is always difficult for beginning students to understand. It seems paradoxical that a "possible" value of x should have probability equal to zero, since we have come to understand that $P(E) = 0$ means that "the occurrence of E is *impossible*"! Some intuitive justification is in order.

First, we can see from Definition 8.2 that in the continuous case probability is always expressed as *area* under the graph of $f(x)$. Thus, since a point k alone produces no area, $P(x = k) = 0$. That is, $P(c \le x \le d) = P(k \le x \le k) = 0$, under condition (iii) of Definition 8.2. Thus an assignment of zero probability to each individual value of x is consistent with Definition 8.2. To justify such an assignment, the following analogy is sometimes helpful. Suppose that we have 10 marbles, one of which is red, in a hat. In one random draw from the hat the probability of selecting that red marble is 1/10. If we had the same situation with 1000 marbles, the probability of selecting the red one would be 1/1000. If $n = 1,000,000$, then $P(\text{red}) = 1/1,000,000$. Thus, as n approaches infinite size, $P(\text{red})$ approaches zero.* We shall agree, therefore, that $P(x = k) = 0$, for every value k when x is a continuous random variable, since k is only one of an *infinite* number of possible x values. What then does $f(k)$ represent in the continuous case? The answer is simple: $f(k)$ is the vertical distance from the x axis to the graph at the point $x = k$. Actually $f(k)$ has the same graphical meaning in both the discrete and continuous cases. However, in the *discrete* case $f(k) = P(x = k)$, but in the *continuous case* $f(k)$ has no meaning with respect to probability.

Since $P(x = c) = 0$ and since $P(c \le x \le d) = P(x = c) + P(c < x \le d)$, it follows that in the continuous case

$$P(c \le x \le d) = P(c < x \le d) = P(c \le x < d) = P(c < x < d)$$

Does this equation hold for discrete probability? Also, we should note that continuous probability functions are expressed in equation form. Since x has an infinite number of values, and since the probability that x equals any one of those values is zero, it is impossible to express $f(x)$ in the type of tabular form used in Chapter 7.

8.1.4 Examples

To find any probabilities associated with a continuous random variable it is necessary to find the *area* under the graph of the probability function

---

*Students who have studied *limits* know that $\lim_{n \to \infty}(1/n) = 0$.

$f(x)$. Finding area under a graph is done by a mathematical process called *integration*. This process falls under the general heading of *calculus*, which is not an assumed prerequisite for this text. Therefore, we are somewhat restricted in what we can actually do in this chapter without help from some standard statistical tables which do the integration (area-finding) for us. Before we consider the more popular probability functions (for which we need tables to find areas under the graph of $f(x)$), let us consider two examples for which we can find probabilities (areas) with some simple geometry.

Let the probability function for continuous random variable x be

$$f(x) = \frac{1}{7}, \quad \text{for } 2 \le x \le 9$$

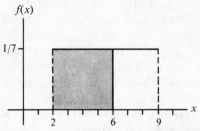

Figure 8.3.   Graph of $f(x) = 1/7$, showing $P(x < 6)$, the shaded area.

Therefore, the graph of $f(x)$ is a straight line parallel to the x axis at a height of 1/7 for all x in the interval two to nine (see Figure 8.3). We can easily verify that $f(x)$ satisfies conditions (i) and (ii) of Definition 8.2; that is, for every x in the interval, $f(x) \ge 0$ (and $f(x) = 0$ outside the interval), and the total area under the graph is base $\times$ height $= (7)(1/7)$ $= 1$. To find $P(x < 6)$, for example, we must find the area under the graph from two to six (shaded in Figure 8.3). Since this area is the area of a rectangle, it is easy to show that $P(x < 6) = (4)(1/7) = 4/7$. The reader should verify the following statements.

$$
\begin{aligned}
P(x > 4) &= 5/7 \\
P(x \ge 4) &= 5/7 \\
P(3 < x < 6) &= 3/7 \\
P(3 \le x \le 6) &= 3/7 \\
P(x = 5) &= 0
\end{aligned}
$$

This continuous probability distribution is a special case of what statisticians call a *uniform probability function.*

Consider another example. Suppose that continuous random variable x has probability function,

$$g(x) = \frac{x - 3}{8}, \quad 3 < x < 7$$

Between $x = 3$ and $x = 7$ the graph of this function is a straight line with slope $= 1/8$ and y-intercept $= -3/8$. The function equals zero outside the interval. The reader should verify that $g(x)$ satisfies conditions (i) and (ii) of Definition 8.2. To find, for example, $P(x \le 5)$ we must find the area under the graph from three to five (shaded in Figure 8.4). Since this area is the area of a triangle, it follows that $P(x \le 5) = 1/2 \times$ base $\times$ height $= (1/2)(2)(1/4) = 1/4$. (We know that the height of the triangle is $1/4$ because $g(5) = 1/4$.) This continuous probability distribution is called a *triangular probability function*, for obvious reasons.

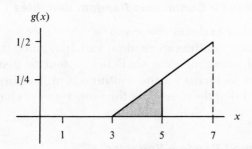

Figure 8.4. Graph of $g(x) = (x - 3)/8$, showing $P(x \le 5)$, the shaded area.

8.1.5 Percentile

Definition 8.3: Given a continuous random variable x with probability function $f(x)$, if the area under the graph of $f(x)$ to the left of $x = k$ is p, then we say that $x_p = k$.

Definition 8.3 simply introduces the new notation "x_p" which denotes a value of random variable x below which the area under the graph of $f(x)$ is p (see Figure 8.5).

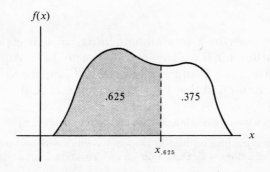

Figure 8.5. Area under the graph of $f(x)$ to the left of $x_{.625}$ is .625.

Definition 8.4: Given a continuous random variable x with probability function $f(x)$. If $100p$ is a whole number, then x_p is called the $(100p)$th *percentile* in $f(x)$.

Definition 8.4 defines the counterpart to the rth percentile, P_r, defined in Definition 4.6, page 62. If, for example, the area under the graph of $f(x)$ to the left of $x = 28$ is 0.35, then we say that 28 is the 35th percentile in $f(x)$. This means, of course, that 35 percent of the area under the graph of $f(x)$ lies below $x = 28$, or, equivalently, $P(x < 28) = 0.35$.

8.1.6 Parameters for Continuous Random Variables

We shall not bother to define the mean (μ_x), variance (σ_x^2), and standard deviation (σ_x) for a continuous random variable, since these definitions involve integral notation.* We shall be content to assume that these parameters do in fact exist for the continuous probability functions that we consider, and that they measure the same types of characteristics that they do for discrete random variables.

8.1.7 Transformed Random Variables

As in the discrete case, we can define the notion of linear transformation for continuous random variables (i.e., transformations of the form $w = ax + b$, where a and b are real numbers). Such transformations have the same effect as in the discrete case (see Theorem 7.2) on the parameters μ_x, σ_x^2, and σ_x; namely,

$$\mu_w = a\mu_x + b$$
$$\sigma_w^2 = a^2\sigma_x^2$$
$$\sigma_w = |a|\,\sigma_x$$

In addition, if x is a continuous random variable, then the transformation

$$x' = \frac{x - \mu_x}{\sigma_x}$$

converts the values of x to *standard units*, an expression which we introduced earlier for the discrete case (page 147). And again, as in the discrete case, $\mu_{x'} = 0$ and $\sigma_{x'}^2 = \sigma_{x'} = 1$. (Compare to Theorem 7.3.) The reader is reminded that for a given value of x, the corresponding

---

*For those readers who have had integral calculus; $\mu_x = \int_{-\infty}^{\infty} xf(x)dx$, $\sigma_x^2 = \int_{-\infty}^{\infty} x^2f(x)dx - (\mu_x)^2$, and, as usual, $\sigma_x = \sqrt{\sigma_x^2}$ for a continuous random variable x with probability function $f(x)$. (Compare these definitions with Definition 7.3 and Theorem 7.1.)

value of x' tells how many standard deviations x is from μ_x and in which direction. For example, if $\mu_x = 10$ and $\sigma_x = 8$, then $x = 20$ converts to $x' = 1.25$, which means that 20 is 1.25 standard deviations *above* the mean in the probability distribution $f(x)$. On the other hand, the value $x = 6$ converts to $x' = -0.5$, which means that six is 0.5 standard deviations *below* the mean. We shall be using this transformation in various contexts throughout the remainder of this text.

Exercises

1. For the triangular distribution defined on page 179, find:
   (a) $P(x < 5)$.
   (b) $P(4 < x < 6)$.
   (c) $P(x = 6.5)$.
   (d) the 25th percentile.
   (e) $x_{0.64}$.
   (f) the *area* to the right of $x_{0.427}$.

2. Consider the following probability function:

   $$f(x) = \frac{1}{10}, \quad 50 < x < 60$$

   (a) Graph $f(x)$ and shade the area under the graph equal to $P(x > 52.6)$.
   (b) Find $P(x > 52.6)$.
   (c) Find $P(51.8 < x < 56.2)$.
   (d) Find the area under the graph between $x_{0.140}$ and $x_{0.456}$.

3. Given the following probability function:

   $$g(x) = \frac{x + 1}{6}, \ 1 < x < 3,$$

   (a) graph $g(x)$,
   (b) show that $g(x)$ satisfies (i) and (ii) in Definition 8.2,
   (c) find $P(x < 2)$,
   (d) find $P(1.5 < x < 5)$.

4. If random variable y has $\mu_y = 24$ and $\sigma_y = 8$ and $w = (y - 10)/2$, then find μ_w and σ_w^2.

5. Random variable x has $\mu_x = 65$ and $\sigma_x = 5$.
   (a) What are the values of x' which correspond to $x = 50, 64$, and 71?
   (b) What are $\mu_{x'}$ and $\sigma_{x'}^2$?

6. The *median* of random variable x is defined as the 50th percentile $(x_{0.50})$. (*Note:* this relationship parallels the relationship between the *sample median* and P_{50}.) Find this parameter for
   (a) the uniform probability distribution defined on page 178.
   (b) the triangular probability distribution defined on page 179.
   (c) the probability distribution defined in Exercise 3.

7. The general form for the "family" of uniform probability functions is

$$f(x) = \frac{1}{b-a}, \, a < x < b$$

For such a function

$$\mu_x = \frac{a+b}{2} \quad \text{and} \quad \sigma_x^2 = \frac{(b-a)^2}{12}$$

For the uniform distribution in Exercise 2 find:
(a) μ_x and σ_x.
(b)   convert $x = 58$ to standard units.
(c)   What is the relationship between μ_x and the median (which is true for all uniform probability functions)?

8.2   The Normal Probability Function

The most important probability functions for the statistician are not as simply defined as the uniform and triangular distributions. The most useful and most famous continuous probability distribution of them all is the *normal probability function*. Many physical variables common in nature have frequency distributions which closely resemble the graph of this function.

> **Definition 8.5:**   A continuous random variable x is said to have a *normal probability function* $n(x)$ (i.e., is *normally distributed*), if
>
> $$n(x) = \frac{1}{d\sqrt{2\pi}} e^{-\frac{1}{2}\left(\frac{x-m}{d}\right)^2}, \quad \text{for} -\infty < x < \infty$$
>
> where
>
> m = any real number
> d = any positive real number
> e = approximately 2.7183
> π = approximately 3.1416

The graph of a normal probability function is bell-shaped and symmetric along a line perpendicular to the x axis at $x = m$, and there is, of course, a total area of one between the graph and x axis. Oddly enough, however, the graph never actually touches the x axis (although it gets closer and closer to the axis as x approaches $-\infty$ and as x approaches $+\infty$ (see Figure 8.6).

Since m can be any real number and d any positive real number, it is clear that the function $n(x)$ in Definition 8.5 describes an infinitely large family of functions with bell-shaped graphs, each of which differs only in the specific values for m and d. The values m and d determine central location and spread in the distribution. As a matter of fact, it can be proved that

$$m = \mu_x \quad \text{and} \quad d = \sigma_x$$

Therefore, we could rewrite $f(x)$ in Definition 8.5 as follows:

$$n(x) = \frac{1}{\sigma_x \sqrt{2\pi}} e^{-\frac{1}{2}\left(\frac{x - \mu_x}{\sigma_x}\right)^2}, \quad -\infty < x < \infty$$

Suppose, for example, that random variable x has a normal distribution with $\mu_x = 50$ and $\sigma_x^2 = 16$; it follows that the probability function for x is

$$n(x) = \frac{1}{4\sqrt{2\pi}} e^{-\frac{1}{2}\left(\frac{x - 50}{4}\right)^2}, \quad -\infty < x < \infty$$

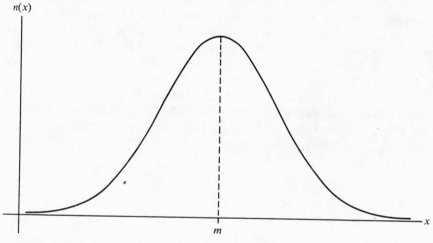

Figure 8.6. Graph of a typical normal probability function.

Theorem 8.1: If random variable x has a normal probability function, then random variable z, where $z = (x - \mu_x)/\sigma_x$, has a normal probability function with a mean of zero and standard deviation one; that is,

$$n(z) = \frac{1}{\sqrt{2\pi}} e^{-\frac{1}{2}z^2}, \quad -\infty < z < \infty$$

Note that according to our previous notation $z = x'$; that is, the transformation $z = (x - \mu_x)/\sigma_x$ converts values of x to standard units. And since z is normally distributed, we shall call z *the standard normal random variable*. Theorem 8.1 says in essence that (a) any random variable x which is normally distributed can be transformed into a new random variable z which is *also normally distributed*, and (b) that $\mu_z = 0$ and

$\sigma_z = 1$, which comes as no surprise (refer to Section 8.1.7). Thus, the one specific normal probability function with $m = 0$ and $d = 1$ we shall call the *standard normal probability function*. And we shall reserve the letter z to represent the standard normal random variable.

The graph of the standard normal probability function is, of course, bell-shaped and symmetric along the vertical axis at $z = 0$, and the total area between the graph and the z axis is one. We shall need this important function to help us find probabilities associated with normally distributed random variables. But first we must learn how to find probabilities associated with random variable z. For example, to find the probability that z has a value less than 1.62, we must find the area under the graph of $n(z)$ to the left of 1.62; that is, $z_p = 1.62$, and we want to find p. Earlier we discussed the fact that finding areas under graphs of continuous functions involves the application of calculus. Since the reader is not expected to have a calculus background, a table is provided (Table E, Appendix) which gives areas to the left of z for the standard normal probability function $n(z)$.* When we look up $z = 1.62$ in the table, we get $p = 0.9474$. That is, $z_{0.9474} = 1.62$. Therefore, $P(z < 1.62) = 0.9474$ (see Figure 8.7). To find $P(z > -0.88)$ we look for $1 - P(z \le -0.88) = 1 - 0.1894 = 0.8106$ (i.e., $z_{0.8106} = -0.88$). And to find $P(-0.26 < z < 1.07)$ we would calculate the area *between* $z = -0.26$ and $z = 1.07$, which is

$$P(z < 1.07) - P(z < -0.26) = 0.8577 - 0.3974 = 0.4603$$

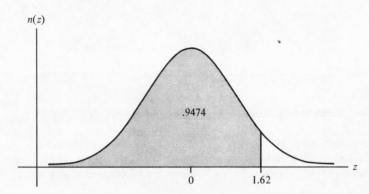

Figure 8.7. Graph of the standard normal probability function with $P(z < 1.62)$ shaded.

It is always helpful to draw a sketch of the graph when using Table E. It is important to remember that the area given in the table for a particular z value is the area to the left of z under the graph of $n(z)$. Note from Table E that areas are given for z values from -3.49 to 3.49 only, even though theoretically z varies from $-\infty$ to ∞. Note, however, that the area between $z = -3.49$ and $z = 3.49$ is 0.9996. This means that only 0.04 percent of the area under the graph is unaccounted for by the table.

*Actually, even if the reader were familiar with integral calculus, we would still use Table E for convenience.

Suppose that continuous random variable x has a normal distribution with a mean of 28 and standard deviation of 4, and we want to find $P(x < 30)$. This probability is equal to the area to the left of 30 under the graph of

$$n(x) = \frac{1}{4\sqrt{2\pi}} e^{-\frac{1}{2}\left(\frac{x-28}{4}\right)^2}, \quad -\infty < x < \infty$$

We do not have areas under this particular graph tabled for us. The only areas we have tabled are those for the *standard* normal probability function. However, the following theorem will let us use Table E to find $P(x < 30)$.

Theorem 8.2: If x has a normal distribution with mean $= \mu_x$ and standard deviation $= \sigma_x$, then $P(x < k) = P(z < (k - \mu_x)\sigma_x)$ where z has a standard normal distribution.

Theorem 8.2 says that the area under the graph of $n(x)$ to the left of k for a normally distributed random variable x with mean μ_x and standard deviation σ_x is exactly equal to the area under the graph of $n(z)$ to the left of $(k - \mu_x)/\sigma_x$ for the standard normal variable z. Therefore, in the example posed above

$$P(x < 30) = P\left(z < \frac{30 - 28}{4}\right) = P(z < 0.5) = 0.6915$$

Theorem 8.2 is an extremely powerful tool, since it allows us to find *any* area we want under the graph of *any* normal probability function $n(x)$ simply by "standardizing" the appropriate value of x (which gives us a value of z to look up in Table E). See Figure 8.8 for the graphical representation of the above example.

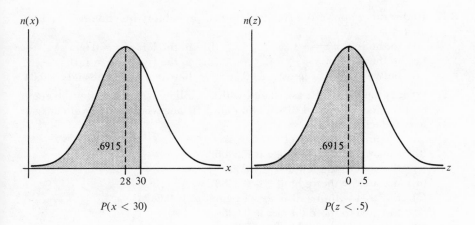

Figure 8.8. Transformation from x to z, showing $P(x < 30) = P(z < .5) = .6915$.

Let us consider another example. Suppose that the weight of 10-year-old boys is a normally distributed random variable x with $\mu_x = 60.6$ lb and standard deviation = 6.5 lb. What is the probability that a boy selected randomly from a large group of 10-year-old boys weighs less than 70 lb? The problem asks us to find $P(x < 70)$; thus, according to Theorem 8.2,

$$P(x < 70) = P\left(z < \frac{70 - 60.6}{6.5}\right) = P(z < 1.45) = 0.9265$$

Another way to ask the same question is, "What percentage of a large group of 10-year-old boys weigh less than 70 lb?" The answer is 92.65 percent.

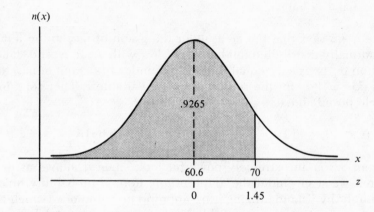

Figure 8.9. Distribution of weights of ten-year-old boys.

Exercises

1. Under the graph of the standard normal probability function $n(z)$ find the area
   (a) to the left of $z = -1.28$.
   (b) to the left of $z = 0.69$.
   (c) to the right of $z = -2.74$.
   (d) to the right of $z = 1.61$.
   (e) between $z = 0.52$ and $z = 2.06$.
   (f) between $z = -1.88$ and $z = 0.04$.

2. What is the value of z associated with the following areas under the graph of the standard normal distribution $n(z)$? (If necessary, round your answer to two decimal places.)
   (a) the area to the left of z is 0.9732.
   (b) the area to the right of z is 0.6406.
   (c) the area to the left of z is 0.0409.
   (d) the area to the right of z is 0.2389.
   (e) the area within a distance of z from μ_z is 0.6424.
   (f) the area to the right of z is 0.025.
   (g) the area to the right of z is 0.01.
   (h) the area between z and $-z$ is 0.99.

3. Find: (a) $z_{0.7852}$.   (b) $z_{0.1335}$.   (c) $z_{0.025}$.   (d) $z_{0.01}$.   (e) $z_{0.05}$.

4. Random variable x is normally distributed with $\mu_x = 85$ and $\sigma_x^2 = 25$.
   (a) How many standard deviations *above* the mean is $x = 92$? Convert 92 to standard units.
   (b) How many standard deviations *below* the mean is $x = 81$? Convert 81 to standard units.
   (c) Find $P(x < 81)$, $P(x < 92)$, and $P(81 < x < 92)$.
   (d) Find $P(76 < x < 84)$.
   (e) Find the 20th percentile (i.e., find $x_{0.20}$).

5. In any normal distribution what is the area within
   (a) one standard deviation from the mean?
   (b) two standard deviations from the mean?
   (c) three standard deviations from the mean?

6. Random variable y is normally distributed with $\mu_y = 48$ and $\sigma_y = 10$. Find:
   (a) $P(y = 48)$.   (b) $P(y > 55)$.
   (c) $P(40 > y > 50)$.   (d) $P(40 \le y \le 50)$.
   (e) the median of y (defined in Exercise 6, page 181).
   (f) In any normal distribution what is the relationship between the mean and the median?

7. A normally distributed random variable x has $\mu_x = 50$. Find its standard deviation given that 87.7 percent of the area under the graph lies to the left of $x = 55$.

8. A certain type of transistor has a mean life span of 100 hours, with a standard deviation of 20 hours. Let random variable x represent transistor life in hours, and assume that x is normally distributed. Find the probability that a new transistor selected at random will
   (a) live less than 75.6 hours.   (b) lives at least 112 hours.
   (c) lives exactly 105 hours.   (d) lives between 90 and 110 hours.

9. A new strain of corn under ordinary conditions will in three weeks grow from seed to a mean height of 5.5 inches, with a standard deviation of 1.2 inches. If a large field is planted with this corn, what percentage of the seeds can be expected to grow to a height of at least six inches?

10. The weight of a certain machine is normally distributed with mean = 12.8 lb and variance = 2.25 lb. In a large warehouse full of these parts, what percentage of them weigh between 13 and 14 pounds?

11. The life span (in years) for a particular brand of hot water heater is normally distributed with mean equal to 7.5 and a standard deviation of 1.4. If the heater is guaranteed against breakdown for five years, what proportion of the heaters sold need to be replaced by the company under the guarantee?

12. Bill's time (x) for running 100 yards is assumed to be normally distributed with $\mu_x = 9.8$ seconds and $\sigma_x = 0.15$ seconds. John's time (y) for 100 yards is assumed to be normally distributed with $\mu_y = 10.0$ and $\sigma_y = 0.12$. When Bill and John compete against each other, what is the probability that John wins if
   (a) Bill runs his "average" race?
   (b) John runs his "average" race?

8.3   Normal Approximation to Discrete Probability Functions

8.3.1   Approximation to the Binomial

We noted in Chapter 7 that some binomial probability calculations can be very laborious. For example, let x be the number of successes in 21 independent trials, where the probability of success on each trial is 0.3. Therefore, the probability function for x is

$$b(x) = \binom{21}{x}(0.3)^x(0.7)^{21-x}, \quad x = 0, 1, 2, \ldots, 21$$

If we want the probability that x is less than 8, we need to find

$$b(0) + b(1) + \cdots + b(7)$$

which is the summation of results from eight fairly long calculations. The following theorem suggests a relatively simple method for approximating this sum.

Theorem 8.3:   If x is the number of successes in n independent trials, where the probability of success on each trial is p, then the probability function for x can be approximated by a normal probability function with mean $= np$ and variance $= np(1 - p)$.

Figure 8.10 shows a probability histogram for a binomially distributed variable x with $n = 4$ and $p = 1/2$, and the graph of a normally distributed random variable with mean $= np$ and standard deviation $= \sqrt{np(1 - p)}$ superimposed upon it.

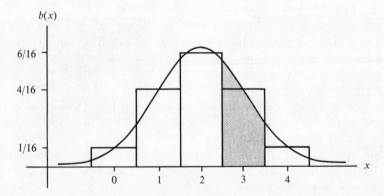

Figure 8.10.   Normal approximation of $P(x = 3)$.

Note that the area of each rectangle in the histogram is approximated closely by the area under the graph between the lower and upper boundaries of the rectangle. For example, the area of the rectangle which represents $P(x = 3)$ in Figure 8.10 is approximately equal to the shaded area from 2.5 to 3.5 under the graph. The probability that $x = 3$ is exactly equal to the area of the fourth rectangle, which is

$$b(3) = \binom{4}{3}\left(\frac{1}{2}\right)^3\left(\frac{1}{2}\right)^1 = 0.25$$

The shaded area under the graph is

$$P(2.5 < x < 3.5) = P(x < 3.5) - P(x < 2.5)$$

$$= P\left(z < \frac{3.5 - 2}{1}\right) - P\left(z < \frac{2.5 - 2}{1}\right)$$

$$= P(z < 1.5) - P(z < 0.5)$$

$$= 0.9332 - 0.6915$$

$$= 0.2417$$

Thus, the difference between the exact probability and the approximation is only 0.0083.

This approximation method can be used to find $P(x < 8)$ in the problem posed earlier, where x equals the number of successes in 21 independent trials, and $p = 0.3$. Since we are approximating the total area in the first eight rectangles of the binomial probability histogram, the corresponding area under the graph of the normal distribution begins at $x = -0.5^*$ and ends at $x = 7.5$ (see Figure 8.11). Since $\mu_x = np = 6.3$ and $\sqrt{np(1-p)} = 2.1$,

$$P(-0.5 < x < 7.5) = P(x < 7.5) - P(x < -0.5)$$

$$= P\left(z < \frac{7.5 - 6.3}{2.1}\right) - P\left(z < \frac{-0.5 - 6.3}{2.1}\right)$$

$$= P(z < 0.57) - P(z < -3.24)$$

$$= 0.7157 - 0.0006$$

$$= 0.7151$$

The difference between this approximation and the exact binomial probability is less than 0.008.

Students sometimes have difficulty determining exactly what limits to use in the normal approximation procedure. For example, why do we use -0.5 rather than zero in the previous problem? Or, why do we use 7.5 rather than 7 or 8? A glance at Figure 8.11 should clear up this

---

*Usually (when n is large) it is not necessary to worry about this lower limit; instead we simply find the area under the graph to the left of $x = 7.5$; i.e., $P(x < 7.5)$. Note that even in this problem, where $n = 21$, the area below the lower limit is only 0.0006.

type of difficulty once and for all. The problem is to estimate the probability that x will take on a value between zero and seven. The exact probability of this event is represented by the total area of the rectangles in the probability histogram whose midpoints are 0, 1, 2, . . . , 7. This total area covers an interval on the x axis from −0.5 to 7.5. Incidentally, the difference between 7 and 7.5 and between 0 and −0.5 is called a *correction for continuity*.

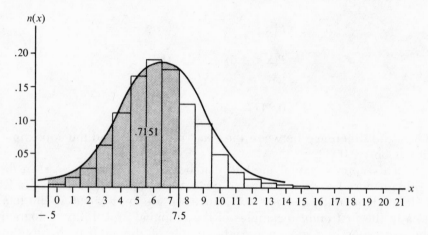

Figure 8.11. Normal approximation of $P(x < 8)$.

The conditions which produce the closest approximations are (a) n large, and (b) $p = 1/2$. As n gets larger, the histogram has more and more rectangles, which in turn get narrower and narrower and this causes the top lines of the histogram to approach a smooth curve. When $p = 1/2$, the histogram is symmetric (as in Figure 8.10). But when $p \neq 1/2$, the histogram is skewed. Since the graph of a normal distribution is always symmetric, it stands to reason that the best approximation (best "fit") occurs when $p = 1/2$. It is interesting to note, however, that when $n = 21$ (not a particularly large n) and $p = 0.3$ (not very near $p = 1/2$) we still get a normal approximation for $P(x < 8)$ which is within 0.008 of the exact binomial probability. This normal approximation procedure is therefore an extremely valuable statistical tool.

8.3.2 Approximation to Other Discrete Distributions

The approximation procedure used in the previous section differs in one very important aspect from the straightforward application of the normal distribution to find probabilities associated with normally distributed random variables. For example, when a random variable x is normally distributed (and therefore *continuous*), $P(x > 6)$ is the area under the graph of a normal probability function to the right of $x = 6$. However, if x is binomially distributed, then $P(x > 6)$ is approximated by the area under the graph of a normal probability function to the right of $x = 6.5$. The half unit difference between 6 and 6.5, called the *correction for continuity*, is necessary whenever a normal distribution is used to approximate binomial probability. In fact, *the correction for continuity should be used whenever a normal distribution is used to approximate "discrete" probability*. Approximating binomial probability is only one specific type of normal approximation.

For example, let discrete random variable x represent the number of babies born in a certain hospital each week. Suppose that we know from experience that the mean number born per week is around 50 with a standard deviation of 16. What is the probability that in a given week the number of babies born is less than 30? If past records show that the distribution of x values from week to week tends toward a bell-shaped, symmetric *frequency distribution*, then we might use the normal distribution to approximate $P(x < 30)$. Since x is clearly discrete, we can visualize a probability histogram which is approximated by the graph of a normal probability function. Therefore, we use the correction for continuity and calculate the area under the graph of a normal distribution (with $\mu_x = 50$ and $\sigma_x = 16$) to the left of 29.5, as follows:

$$P(x < 29.5) = P\left(z < \frac{29.5 - 50}{16}\right) = P(z < -1.28) = 0.1003$$

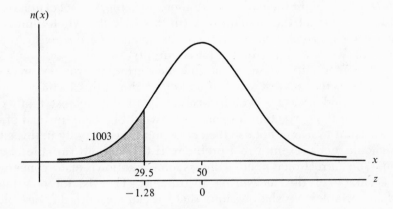

Figure 8.12. Normal approximation to discrete probability in the hospital births example.

The following example is somewhat more subtle, since it involves a continuous random variable which, because of measurement limitations, requires a correction for continuity. Suppose the weights of male college freshmen are normally distributed with $\mu_x = 150\,\text{lb}$ and $\sigma_x = 15\,\text{lb}$.

Then the proportion of male college freshmen who weigh more than 165 lb is $P(x > 165)$, which is

$$1 - P(x \leq 165) = 1 - P\left(z \leq \frac{165 - 150}{15}\right)$$

$$= 1 - P(z \leq 1)$$

$$= 1 - 0.8413$$

$$= 0.1587$$

In the practical situation, however, let us assume that these freshmen are weighed on a scale which is accurate only to the *nearest pound*. Therefore, a particular freshman who "weighs in" at a recorded weight of 165 lb could actually weigh anywhere from 164.5 to 165.5 lb. Therefore, those freshmen whose recorded weights exceed 165 lb actually weigh 165.5 or more. Thus, the proportion of freshmen whose *recorded* weights exceed 165 lb is

$$1 - P(x \leq 165.5) = 1 - P\left(z \leq \frac{165.5 - 150}{15}\right)$$

$$= 1 - P(z \leq 1.03)$$

$$= 1 - 0.8485$$

$$= 0.1515$$

Even though weight is clearly a continuous variable, the act of measuring weight can restrict the domain of x (in this case the whole numbers), which makes x (recorded weight) strictly speaking a *discrete* variable. Therefore, we need a correction for continuity.

Of course, the accuracy of the measuring instrument may not always be to the nearest whole number. If the scales in the previous example could measure weight accurately to the nearest tenth of a pound, then the correction for continuity would be 0.05, instead of 0.5. The decision to use or not use the correction for continuity in this situation should not present a real problem. If the random variable is continuous, no correction is needed unless some mention is made concerning the accuracy of the measuring instrument. That is, for continuous random variables we shall assume that we can measure the variable to any desired degree of accuracy, *unless* the problem states or implies something to the contrary.

Exercises

1. For a binomially distributed random variable x, with $n = 9$ and $p = 1/2$, find $P(2 \leq x \leq 5)$
    (a) using the binomial probability function;
    (b) using the normal approximation to the binomial.

2. Do Exercise 10, page 158 again using the normal approximation, and compare your answers to the exact binomial probabilities which you found earlier.

3. On a 100-question True-False test the examinee answers each question by flipping a coin and answering "true" if a head occurs and "false" if a tail occurs. If a score of at least 60 is necessary for a passing grade, what is the probability that this examinee passes?

4. Let x = the number of successes in 60 independent trials where the probability of success on each trial is 5/8. Use the normal approximation method to find:
    (a) $P(32 < x < 40)$,
    (b) $P(32 \leq x \leq 40)$.

5. A basketball player makes only 60 percent of his free throws. In his next 64 attempts, what is the approximate probability that he will make less than 50 percent of them?

6. A car dealer put his salesmen on a salary which was based on the expectation that each salesman would sell a minimum of 10 cars per month. Let x represent the number of cars sold per salesman per month and assume that the probability function for x can be approximated by a normal distribution with mean = 12.5 cars and standard deviation = 2.4 cars. In a given month what proportion of salesmen sell less than the expected 10 cars?

7. The heights of a large population of men are normally distributed with $\mu_x = 69.3$ inches and $\sigma_x = 3$ inches. A man is selected randomly from the population.
    (a) What is the probability that he is taller than six feet?
    (b) If this man's height is measured and then recorded to the nearest inch, what is the probability that his recorded height is greater than six feet?

8. Each year a large population of high school students take a special examination in competition for college scholarships. They receive whole-number scores on a scale from 0 to 100. Only students who score in the top 15 percent of those taking the exam are eligible for a scholarship. If we assume that the distribution of scores on such a test is approximately normal with $\mu_x = 70$ and $\sigma_x = 10$, then what is the lowest score which qualifies the examinee for a scholarship?

9. In major league baseball let us assume that the pennant winner averages 95 wins per season with a standard deviation of 4.2. Using a normal approximation, find the probability that a pennant-winning team will win more than 100 games.

10. Several thousand young brook trout are released into a stream. Assume that the lengths of these trout are normally distributed with $\mu_x = 6$ inches and $\sigma_x = 1.2$ inches. The next day a fisherman catches one of these trout. What is the probability that the trout is less than seven inches long?

11. Make the following change in the previous exercise. Replace the last sentence with, "The fisherman measures the length of his catch to the nearest 1/4 of an inch. What is the probability that the measured length is less than seven inches?", and recalculate the probability.

8.4   Other Important Functions

The reader should note at this point that although the equation form for the normal probability function is somewhat complicated (refer back to Definition 8.5), we need not be overly concerned. It has been demonstrated in the previous sections that we can use this function without direct reference to its equation form—thanks to the standard normal table.

In this section we shall discuss three other continuous probability functions which play an important role in statistical work: the *Student-t*, *chi-square*, and *F* probability functions. These functions also have rather complicated-looking equation forms (see Definitions 8.6, 8.7, and 8.8). However, as in the case of the normal probability function, we do not need to deal directly with the equations. For each function we have sufficient information tabled in the Appendix for practical application. Thus, the reader should not be overly concerned about the equation forms for these functions, but instead should develop facility with the tables in preparation for the work in later chapters.

8.4.1   Parameters and "Degrees of Freedom"

We have been using the term *parameter* to refer to μ_x, σ_x^2, and σ_x, which are measures that describe certain important characteristics of a probability function. That is, parameters μ_x, σ_x^2, and σ_x describe a random variable x whose distribution of probability is determined by a probability function $f(x)$. These parameters are *variables* in the sense that they may take on different values from function to function. They are also *constants* in the sense that each has a specific value for a given probability function. Not only does a parameter have a constant value for a given function, but that value helps to determine the shape of the graph for that function.

We have been using the term "parameter" in a much too restrictive sense. In fact, we have been using parameters to define certain special families of probability functions, namely the binomial (with parameters n and p), the hypergeometric (N, M, and n), the Poisson (m), and the normal (m and d). All of these parameters have the parameter qualities described in the previous paragraph; that is, within a given family the parameters are *constant* for a specific function, but *vary* from function to function. As a matter of fact two of them are means (m for the Poisson and normal), one is a variance (m for the Poisson), and one is a standard deviation (d for the normal). In all four families that we have discussed so far, μ_x, σ_x^2, and σ_x can be defined in terms of the family parameters; that is,

$$\text{Binomial:}\quad \mu_x = np, \qquad \sigma_x^2 = np(1-p)$$

Hypergeometric: $\mu_x = \dfrac{nM}{N}$, $\sigma_x^2 = \dfrac{M(N-M)(n)(N-n)}{(N^2)(N-1)}$

Poisson: $\mu_x = m$, $\sigma_x^2 = m$

Normal: $\mu_x = m$, $\sigma_x^2 = d^2$

(Of course, $\sigma_x = \sqrt{\sigma_x^2}$ as always)

The three continuous probability functions to be discussed in this section also have parameters in their functional definitions. The *Student-t probability function* has a parameter r which is called the *number of degrees of freedom*. At this point we shall not attempt to explain why parameter r should have this particular name. The reason will become clearer later on when we use the Student-t distribution for practical purposes. The *chi-square probability function* has a parameter r, also designated as the number of degrees of freedom. The *F probability function* has two parameters, r_1 and r_2, each of which is a number of degrees of freedom, but determined from different sources. Again, the practical explanation for the expression "degrees of freedom" shall be discussed later (see page 228).

8.4.2 The Student-t Probability Function

Definition 8.6: A continuous random variable x is said to have a Student-t probability function* $s(x)$ if

$$s(x) = \frac{\left(\dfrac{r-1}{2}\right)!}{\sqrt{r\pi}\left(\dfrac{r-2}{2}\right)!\left(1+\dfrac{x^2}{r}\right)^{\left(\frac{r+1}{2}\right)}}, \quad -\infty < x < \infty$$

where r = the number of degrees of freedom (any positive integer)
  π = approximately 3.1416

Definition 8.6 defines a family of Student-t functions, each one having a different value of r. Each function in the family has

mean $= 0$ and   variance $= \dfrac{r}{r-2}$

All Student-t distributions are symmetric along the vertical axis through $x = 0$, and, in fact, appear in graph form to be standard normal probability functions. Actually *if we let r approach infinite size, the Student-t probability function becomes the standard normal probability function in the limit!* (Note that the mean is zero, as in the standard normal distribution, and, for very large r, the variance $r/(r-2)$ is virtually equal

---

*The Student-t function was originally derived by W. S. Gosset who published the results secretly under the name of "Student," due to certain publication restrictions which prevented him from using his own name.

to one.) Generally speaking then, the graph of a particular Student-t probability function with r degrees of freedom looks like a slightly flattened-out graph of the standard normal probability function. The smaller the value of r the flatter the graph (see Figure 8.13).

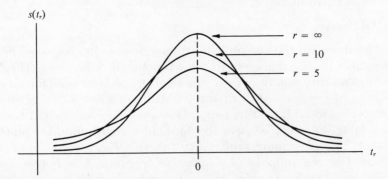

Figure 8.13. Graphs of three Student-t probability functions, one of which is the standard normal probability function ($r = \infty$).

A random variable which has a Student-t distribution with r degrees of freedom shall be denoted by t_r. And if we wish to talk about a particular value of t_r below which the area under the graph equals p (remember that area = probability), then we shall use the notation $t_{r,p}$. For example, consider the graph of the Student-t probability function with 20 degrees of freedom. The value of t_{20} below which the area under the graph is 0.90 is denoted by $t_{20;0.90}$. Remember that the total area under all probability functions is 1.00. (See Figure 8.14.) The value of $t_{20;0.90}$ can be found in Table F (Appendix). Note the difference between this table and Table E: (a) the body of this table contains values of t_r not areas under the graph of the probability function as in Table E, and (b) certain values of t_r are given for selected areas under the graph of $f(t_r)$ for 37 different Student-t probability functions (i.e., for $r = 1, 2, 3, \ldots, 30, 40, 60, 80, 100, 200, 500, \infty$). To find $t_{20;0.90}$ we simply look across from $r = 20$ and down from $p = 0.90$. Therefore, $t_{20;0.90} = 1.33$.

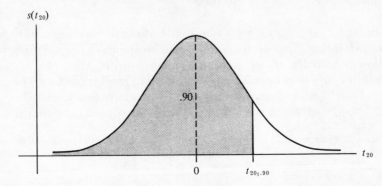

Figure 8.14. Graph of the Student-t probability function with 20 degrees of freedom.

We cannot use Table F to find such things as $P(t_{10} < k)$ unless k just happens to be listed in row $r = 10$. Suppose, for example, that we want $P(t_{10} < 1.67)$. In this case the best we can do is to say that $P(t_{10} < 1.67)$ is somewhere between 0.90 and 0.95 because 1.67 lies between 1.37 $(t_{10;0.90})$ and 1.81 $(t_{10;0.95})$ in row $r = 10$. On the other hand, if we want $P(t_{13} < 2.65)$, the table gives us $p = 0.99$ exactly. This limitation of Table F is not serious because we shall be using this table primarily to find values of t_r, given certain areas under the graph, rather than the other way around.

One last note: row $r = \infty$ in Table F displays t_∞ values which are in fact values of the standard normal random variable z. For example, $z_{0.95} = 1.645$ (from Table E), which is the same value as $t_{\infty;0.95}$ (from Table F).

8.4.3 The Chi-Square (χ^2) Probability Function

Definition 8.7: A continuous random variable x is said to have a *chi-square probability function* $c(x)$ if

$$c(x) = \frac{e^{-\frac{x}{2}} x^{\frac{r-2}{2}}}{\left(\frac{r-2}{2}\right)! \, 2^{\frac{r}{2}}}, \qquad 0 < x < \infty$$

where $r =$ the number of degrees of freedom (any positive integer)
       $e =$ approximately 2.7183

Again, we have in Definition 8.7 the definition for a family of probability functions, one for each positive integer (i.e., one for each value of r). Whereas the normal and Student-t probability functions have symmetric graphs and the domain of the random variable extends from $-\infty$ to ∞, the chi-square probability function has a graph which is in general skewed to the right, and the domain of a chi-square-distributed random variable extends only from 0 to ∞. Each function in the family has

mean $= r$ and   variance $= 2r$

A random variable which has a chi-square distribution with r degrees of freedom shall be denoted by χ_r^2. If we wish to denote a particular value of χ_r^2 below which the area under the graph is p, then we shall use the notation $\chi_{r;p}^2$. For example, consider the graph of the chi-square probability function with 25 degrees of freedom. The value of χ_{25}^2 below which the area under the graph is 0.95 is denoted by $\chi_{25;0.95}^2$. This value can be found in Table G (Appendix). The chi-square table is set up like the Student-t table. Therefore, to find the value $\chi_{25;0.95}^2$ we simply look in the body of the table in row $r = 25$ and column $p = 0.95$ and get the value 37.65. As we shall see later, the chi-square table is used primarily for the purpose of finding an individual value of χ_r^2, given a specific area to the left under the graph.

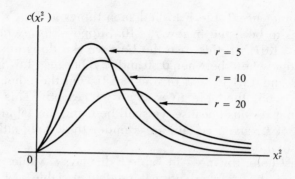

Figure 8.15.   Graphs of three chi-square probability functions.

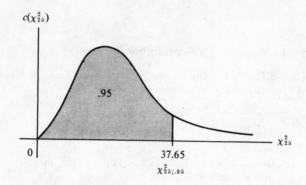

Figure 8.16.   Graph of the chi-square probability function with 25 degrees of freedom.

8.4.4   The F Probability Function

Definition 8.8:   A continuous random variable x is said to have an F *probability function* $f(x)$ if

$$f(x) = \frac{\left(\frac{r_1 + r_2 - 2}{2}\right)!}{\left(\frac{r_1 - 2}{2}\right)!\left(\frac{r_2 - 2}{2}\right)!} \cdot \left(\frac{r_1}{r_2}\right)^{\frac{r_1}{2}} \cdot \frac{x^{\frac{r_1 - 2}{2}}}{\left(1 + \frac{r_1 x}{r_2}\right)^{\frac{r_1 + r_2}{2}}}, \qquad 0 < x < \infty$$

where $r_1 =$ the number of degrees of freedom from one source (any positive integer)

$r_2 =$ the number of degrees of freedom from a second source (any positive integer)

Each member of the family of F probability functions has a domain extending from 0 to ∞, and the graph of each is skewed to the right—

looking very much like the graph of a chi-square probability function. Each F probability function has

$$\text{mean} = \frac{r_2}{r_2 - 2} \quad \text{and} \quad \text{variance} = \frac{2r_2^2 \, (r_1 + r_2 - 2)}{r_1(r_2 - 2)^2(r_2 - 4)}$$

We let random variables which have an F distribution with r_1 and r_2 degrees of freedom be denoted by F_{r_1, r_2}. And a specific value of F_{r_1, r_2} below which the area under the graph is equal to p is denoted as $F_{r_1, r_2; p}$. The major difference between the chi-square and F probability functions is that the F has two parameters (r_1 and r_2). This fact complicates the use of tables for the F function. We need a separate F table for every area p under the graph of F_{r_1, r_2} which may be of some practical use to us. Therefore, we find in the Appendix, Table H, four F tables, pages 326–29 for $p = 0.95, 0.975, 0.99$, and 0.995. For example, $F_{5,8;0.95} = 3.69$. We will need F_{r_1, r_2} values for $p = 0.005, 0.01, 0.025$, and 0.05 also, but we can obtain these values from the given tables by applying the following relationship.

$$F_{r_1, r_2; p} = \frac{1}{F_{r_2, r_1; 1-p}}$$

Therefore, if we want $F_{5,8;0.05}$, we find $F_{8,5;0.95}$ in Table H, page 326 (which is 4.82) and take its reciprocal (which is 0.21). Thus,

$$F_{5,8;0.05} = \frac{1}{F_{8,5;0.95}} = \frac{1}{4.82} = 0.21$$

It follows, therefore, that the area under the graph of $f(F_{5,8})$ *below* 0.21 is 0.05, and the area *above* 3.69 is 0.05 (see Figure 8.17). What is the area *between* 0.21 and 3.69?

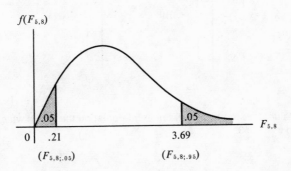

Figure 8.17. Graph of the F probability function with five and eight degrees of freedom.

8.4.5 Linear Interpolation

The Appendix tables are set up so that for most practical purposes the values we want are the tabled values. However, on occasion we may wish to find a value which falls somewhere between two tabled values — either because the table is partially incomplete, or because we want greater accuracy. In either case we rely on a method called *linear interpolation* to do the job.

For example, suppose that we want to find the area under the standard normal probability distribution to the left of $z = 1.053$. As a general rule in this text, we would round this z value to $z = 1.05$ and then look up the corresponding area in Table E. This situation occurs often when we wish to find probabilities for a normally distributed random variable, and we can get adequate accuracy this way for most purposes. However, if for some reason we want the exact area to the left of $z = 1.053$, we first find the areas to the left of $z = 1.05$ and $z = 1.06$, which are 0.8531 and 0.8554, respectively. Then we take the difference between the two areas, multiply it by 0.3, round the result to four decimal places and add it to 0.8531. Therefore, the area to the left of $z = 1.053$ under the graph is

$$p = 0.8531 + \frac{0.003}{0.010}\,(0.0023) = 0.8541$$

Figure 8.18 shows where the numbers used above come from.

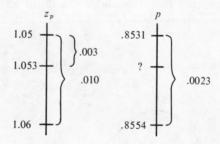

Figure 8.18. Values needed to find by linear interpolation the area p which corresponds to $z = 1.053$.

Reversing the situation, suppose that we want to find the z value which has an area to the left equal to 0.9000. From Table E we find that 0.9750 falls between 0.8997 and 0.9015, which correspond to $z = 1.28$ and $z = 1.29$, respectively. If we want z to two decimal places, then

$z = 1.28$, because 0.9000 is closer to 0.8997 than to 0.9015. However, if we want z to three decimal places, we need to interpolate. Thus,

$$z_{0.9000} = 1.28 + \frac{0.0003}{0.0018} (0.01) = 1.282$$

See Figure 8.19 for the values needed in the above calculation.

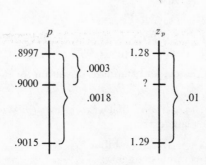

Figure 8.19. Values needed to find $z_{.9000}$ by linear interpolation.

The same procedure can be used to interpolate in the other tables. For tables involving degrees of freedom, such as the chi-square table (Table G), the values are not given for every possible number of degrees of freedom (r). For example, if we want $\chi^2_{47;0.95}$, we need to interpolate between $\chi^2_{40;0.95}$, which is 55.76, and $\chi^2_{50;0.95}$, which is 67.50. Therefore,

$$\chi^2_{47;0.95} = 55.76 + 0.7(67.50 - 55.76) = 63.98$$

8.4.6 *Continuous Joint Probability Functions*

In the previous chapter we discussed joint probability functions for discrete random variables. Joint probability functions exist also for continuous random variables. However, without a solid background in calculus the reader cannot really appreciate a discussion on continuous joint probability functions. Therefore, this subject is considered beyond the scope of this text.

Exercises

1. Find the area under the standard normal distribution to the left of $z = 0.854$.
2. Find z to three decimal places if the area to the left of z is
    (a) 0.90.   (b) 0.95.   (c) 0.975.   (d) 0.99.   (e) 0.995.

3. Compare your answers in Exercise 2 to the last row ($r = \infty$) in the Student-t table (Table F).

4. What is the area under a Student-t distribution with 12 degrees of freedom *above* the value $t_{12;0.05}$?

5. Using linear interpolation, if necessary, find:
   (a) $t_{5;0.05}$, $t_{17;0.025}$, $t_{200;0.975}$, $t_{34;0.99}$.
   (b) $\chi^2_{20;0.05}$, $\chi^2_{12;0.99}$, $\chi^2_{8;0.95}$, $\chi^2_{62;0.025}$.
   (c) $F_{4,15;0.95}$, $F_{12,7;0.975}$, $F_{6,12;0.05}$, $F_{18,15;0.99}$.

6. Find p, $t_{12,p}$, and $t_{12,1-p}$ if $P(t_{12,p} < t_{12} < t_{12,1-p})$ equals
   (a) 0.99.   (b) 0.98.   (c) 0.95.   (d) 0.90.

7. Find the mean, variance, and standard deviation for random variable
   (a) t_{10}.   (b) χ^2_8.   (c) $F_{10,6}$.

8. Verify that $\chi^2_{1;0.95} = (z_{0.975})^2$ and that $\chi^2_{1;0.99} = (z_{0.995})^2$. What generalization do these two examples suggest?

9. Verify that $F_{1,10;0.95} = (t_{10;0.975})^2$ and that $F_{1,15;0.99} = (t_{15;0.995})^2$. What generalization do these two examples suggest?

Summary

A *continuous random variable* is a variable (a) whose values are associated with outcomes in a chance experiment, and (b) whose set of possible values (domain) is the set of all real numbers in an interval on the real number line.

A *continuous probability function* for continuous randon variable x, denoted by $f(x)$, is a function on an interval from a to b such that (a) $f(x)$ is always nonnegative, and (b) the total area under the graph of $f(x)$ (i.e., between the graph and the x axis) equals 1.00, and (c) $P(c < x < d)$ equals the area under the graph between $x = c$ and $x = d$ (assuming $c \le d$). Unlike the discrete case, $P(x = c) = 0$ for every individual value c in the interval. The notation "x_p" denotes the value of x below which the area under the graph is p. If $100p$ is a whole number, then we say that x_p is the $(100p)$th *percentile* in $f(x)$. The *mean, variance,* and *standard deviation* for continuous random variables are defined in terms of integral calculus. We assume, therefore, the existence of these parameters but do not define them formally. If random variable x has mean μ_x and standard deviation σ_x, then the transformation $x' = (x - \mu_x)/\sigma_x$ converts the values of x to *standard units*.

The most famous and useful probability function is the *normal probability function*. If x has a normal probability function and $z = (x - \mu_x)/\sigma_x$, then z has a *standard normal probability function* (i.e., a normal probability function with $\mu_z = 0$ and $\sigma_z = 1$). Thus, any normally distributed random variable x can be transformed into the variable z. Areas (probabilities) under the standard normal probability function $n(z)$ are found in Table E in the Appendix. Table E can be used to find probabilities relating to random variables which have normal probability distributions, and it can be used to approximate binomial probability.

Other important probability functions include (a) the Student-t distribution, (b) the chi-square distribution, and (c) the F distribution (see Tables F, G, and H in the Appendix).

part 3

statistical inference

9

Sampling

9.1 Introduction

In Part I we discussed methods of describing data which have been generated from samples selected from populations. We assumed at the time that the samples had already been selected, and we concentrated solely upon the task of defining various sample statistics which can be used to describe characteristics of sample data. In addition, we stated that such descriptions can be used to estimate certain corresponding characteristics of the populations from which the samples were selected. Since any sample provides only partial information concerning the population from which it was selected, it follows that any statement that we might make (based on sample statistics) concerning a characteristic of the population is necessarily subject to error. That is, we cannot predict or infer with 100 percent accuracy the characteristics of a population. Thus any statement that we make about a population *based on sample information* is made with some *uncertainty*. It is for this reason that we dropped the development of descriptive statistics after Chapter 5 and discussed probability in the next three chapters (Part II).

In Part III we now address ourselves to the problems associated with estimating population characteristics and making decisions based on sample information. In doing so we bring together the notions of descriptive statistics and probability to form a theory of *statistical inference*. We shall develop statistical inference procedures from the classical approach (sometimes called the "sampling theory" approach). Generally speaking, in the classical approach we make statistical inferences and decisions *based solely upon sample information*. In recent years an extension of the classical approach has developed which utilizes not only sample information, but other evidence as well. This new approach, which is rapidly gaining new adherents, is based upon Bayes' Theorem (Theorem 6.10), and has therefore become known as the *Bayesian* approach. In this text we shall concentrate on the classical approach exclusively. However, students interested in extending their knowledge to include Bayesian methods are directed to Schmitt (1969) and Hays and Winkler (1970, Chapter 8), which are both excellent introductions to Bayesian inference.

Before we get into statistical inference directly, we need to explore (a) the notion of *random sampling,* (b) some basic *sampling methods,* and (c) the concept of *sampling distribution,* which are fundamental to the basic structure of statistical inference theory.

9.2   Random Sampling

We have established that a sample is simply *any* subset of a population. However, we understand intuitively that not every subset of a population is *representative* of that population with respect to a particular variable. For example, if we were measuring the intelligence of fifth grade students in the U.S., then any fifth grade class in any elementary school could be thought of as a sample of the population of all fifth grade students in the country. However, if we are planning to estimate "average" intelligence or spread of intelligence in the population, then we hope that our sample will be representative enough so that the various sample statistics will approximate closely the corresponding measures in the population. If the sample is very atypical, then any inferences that we draw from that sample will be misleading. For example, if we were to choose fifth graders from a high economic area where most of the parents are professional people with a graduate school education, then the distribution of IQs for this sample would differ significantly from the distribution of IQs in the general fifth grade population. The average IQ would be larger and the spread in IQs would probably be smaller. Such a sample would not be considered "representative" of the population.

9.2.1 Defining a Random Sample

Actually we have no real way of guaranteeing that a sample is representative of a given population. However, we can hope to minimize the probability that a sample will be significantly atypical by selecting the sample in such a way that *every member of the population has an equal chance to be chosen.* Such a sampling procedure is called *random sampling.* And a sample thus obtained is called a *random sample.* Therefore, to get a random sample of fifth graders, we would need to select them in such a way that each fifth grader in the U.S. has an equal chance to be a member of the sample.

We can view the notion of random sampling in a slightly different but equivalent way. If a sample of size n is to be selected from a population of size N, then there are $\binom{N}{n}$ different subsets of n elements that could be chosen. If the selection procedure is *random*, then each of the $\binom{N}{n}$ possible samples must have probability $1/\binom{N}{n}$ of being selected. That is, each possible result must be equally likely.

Definition 9.1: If a sample of size n is selected from a population of size N in such a way that each of the $\binom{N}{n}$ possible subsets of size n has probability $1/\binom{N}{n}$ of being selected, then the sample is called a *random sample.**

Suppose, for example, that we wish to select a sample of size $n = 3$ from a population of size $N = 5$. The elements in the population can be represented by a, b, c, d, and e. Therefore, the number of different subsets of three elements is $\binom{5}{3} = 10$. These 10 possible samples are

1.	$\{a,b,c\}$	2.	$\{a,b,d\}$
3.	$\{a,b,e\}$	4.	$\{a,c,d\}$
5.	$\{a,c,e\}$	6.	$\{a,d,e\}$
7.	$\{b,c,d\}$	8.	$\{b,c,e\}$
9.	$\{b,d,e\}$	10.	$\{c,d,e\}$

*This definition applies strictly to sampling *without replacement from a finite population.* The definition of random sample for sampling from an *infinite* population (or equivalently, from a finite population with replacement) includes the additional assumption of *independence.* (See Remington and Schork (1970, pages 92 and 105), or for a more mathematical treatment, see Freund (1962, pages 176 and 181).)

According to Definition 9.1, each of these possible outcomes must have probability 1/10 in order that the sample thus obtained can be considered a *random* sample. A logical way to proceed would be to write the numbers 1, 2, 3, . . . , 10 on 10 pieces of paper, put them in a hat, shake them up, and then make a "blind-folded" selection. This procedure is intuitively sound, but unfortunately rather inefficient from a practical standpoint. Before we discuss more efficient procedures the reader should note the following: (a) *if n elements are selected from a population in such a way that on each selection the elements existing in the population are equally likely to be selected, then Definition 9.1 will be satisfied,* and (b) *if Definition 9.1 holds, then every element in the population has an equal chance of being selected in the sample.* Let us justify these statements for our example.

First, we can see from the example that if the elements in the population are given an equal opportunity to be chosen on each selection, then the probability that the subset {a,b,c}, for example, is selected equals

$$\left(\frac{1}{5} \cdot \frac{1}{4} \cdot \frac{1}{3}\right)3! = \frac{1}{10}$$

The product $(1/5)(1/4)(1/3)$ is the probability that a, b, and c are selected in a specific order; 3! is the number of orders possible. Thus, each subset (sample) has the probability 1/10 and Definition 9.1 holds. Secondly, we can see from the example that six of the 10 possible outcomes involve an a. Therefore, the probability that a is selected is 6/10. The same is true for each of the elements b, c, d, and e. Thus, each has an equal chance of being included in the sample. Consequently, saying that Definition 9.1 holds implies that the elements in the population are equally likely to be chosen.

It follows from the preceding statements that rather than numbering the possible subsets of size n and selecting one of them, it is legitimate, and certainly more convenient, to number each element in the population and make n selections in succession, making sure that in each successive selection the remaining elements in the population are equally likely to be chosen. This approach is a common random sampling procedure although writing numbers on slips of paper and selecting from a hat is somewhat tedious, especially when the population is large as it usually is (if the population were small, it would not be necessary to sample in the first place). For example, if we wanted to select a sample of four from $N = 50$, then rather than making one selection from 230,300* slips of paper, we could make four successive selections (without replacement) from 50 slips of paper.

---

*$230,300 = \binom{50}{4}$, which is the total number of possible samples of size four.

9.2.2 Random Number Table

An alternative to writing numbers on slips of paper is the use of a *random number table*. Table I, in the Appendix, shows a small part of such a table, taken from RAND (1954). A random number table in its entirety consists of hundreds of pages of digits (0, 1, 2, . . . , 9) which have been randomly recorded, usually by a computer. Each of the two pages displayed in Table I contains 2500 digits, 50 columns by 50 rows. Although the computer generates these digits in a rather sophisticated way, we can imagine that each of the digits from zero through nine was written on a piece of paper and put in a hat. And then each page of the table was produced by taking 2500 "blind-folded" selections *with replacement*, each outcome being recorded in a 50 by 50 matrix. Such a table is extremely useful for the selection of random samples. For example, if we want a random sample of four elements from a population of 50 elements, we begin by assigning the numbers 00, 01, 02, . . . , 49 to the 50 elements in the population. Then we arbitrarily select a page, a row, and a pair of successive columns on which to enter the random number table.* Suppose that we decide to start on the *second* page, row 11, columns 36 and 37 (page 331). The two-digit number at that juncture in the table is 25. Proceeding *down* the table we read off

25, 27, 59, 42, 91, 90, 87, 45

Since we need only four elements in our sample, we need to proceed no further. Ignoring those two-digit numbers which are 50 or greater (since the population elements are numbered 00 to 49), we select the four elements in the population which are labeled 25, 27, 42, and 45.

Suppose instead that we want to select a sample of 10 college freshmen from a total enrollment of 845 freshmen. First, each freshman is assigned a number from 000 to 844. Then we arbitrarily choose a page, a row, and three successive columns for our starting point in the random number table. Suppose that we start this time on the *first* page, row 33, columns 7, 8, and 9 (page 330). Proceeding down the table we find the following three-digit numbers.

644, 955, 540, 611, 860, 323, 245, 201, 217, 401, 915, 033, 651

Since 955, 860, and 915 do not represent freshmen in the population, they are ignored, leaving us with a random sample of 10 three-digit numbers, representing 10 freshmen. Please note that (a) had we exhausted the numbers in columns 7, 8, and 9 before we completed the sample, we would have continued at the top of the page in columns

---

*This "arbitrary" entrance into the table should in theory be made randomly also by some sort of gambling device, especially if several samples are to be selected using the name random number table.

10, 11, and 12, etc., until we had attained the desired sample size, and (b) should the same three-digit number occur a second time, it is ignored.

Exercises

1. How many different samples of size $n = 3$ can be selected from a population of size $N = 10$?

2. If a random sample of four is to be selected from a population of 20, then each possible sample of four has what probability of being selected?

3. Consider a population of size $N = 6$ which contains the elements 1, 2, 3, 4, 5, and 6. We plan to select a random sample of size $n = 4$.
    (a) List the possible samples of size four.
    (b) What is the probability of each possible sample?
    (c) What is the probability that element "5" will be included in the sample chosen?
    (d) Does the probability in (c) apply to each of the other elements in the population?
    (e) What is the probability that elements 4 and 5 will both be included in the sample chosen?
    (f) Does the probability in (e) apply to every pair of elements in the population?

4. If a person selects in his mind a number between 1 and 10 (inclusive), does this sample of size $n = 1$ qualify as a *random* sample? Explain.

5. A population of 4826 airplane pistons are numbered consecutively from 0000 to 4825. Given that we use Table I (beginning in row 6, columns 41–44 of the first page) to select a random sample of eight pistons, list the numbers on the pistons in the sample.

6. In Exercise 5 we have to ignore all four-digit numbers from 4826 to 9999. How could we change the procedure slightly so that only a few numbers are ignored in the selection process?

7. A table of random numbers can be used to *simulate* certain chance experiments for which a particular probability distribution is determined. For example, suppose that we want to flip a balanced coin 10 times, but no coin is available. To simulate this experiment we enter the random number table on a single-digit column and read off 10 digits—regarding even digits (0, 2, 4, 6, 8) as "heads" and odd digits as "tails." List the outcomes of such a simulated experiment, assuming that you enter Table I on the first page, row 27, column 15.

8. Simulate 10 flips of three balanced coins and record the number of heads (x) that occur on each of the 10 flips. Remember that $P(x = 0) = 0.125$, $P(x = 1) = 0.375$, $P(x = 2) = 0.375$, and $P(x = 3) = 0.125$. (Use the second page of Table I and start in row 1, column 1–3.)

9. In a large population of marbles 27 percent are red, 38 percent are blue, and the rest are green. Simulate the selection (without replacement) of 20 marbles, using the second page of Table I, beginning in row 9, columns 11 and 12. List the results.

9.3 Sampling Methods

9.3.1 Simple Random Sampling

We have been describing what statisticians call *simple random sampling*. Earlier we noted that a sample should be representative of the population from which it is selected. If it is not representative, then any descriptive characteristic of the sample would give misleading information regarding the population. Although a random sample does not guarantee representativeness, it does help to avoid the selection of an unusually atypical sample. The method of simple random sampling has two other important advantages.

First, the method eliminates selective bias, either intentional or unintentional, on the part of the investigator. An eager researcher may inadvertently select a sample which would be more likely to support his theories than would a perfectly random sample. Sometimes bias occurs when the sample chooses itself. This happens when, for example, a college history class is used as a sample to investigate variables relating to the entire college student body. In this type of situation the members of the sample have in fact chosen themselves, since each of them chose to take that particular course. It is not unlikely that such a group would be somewhat atypical with respect to the population. The extent of the damage would, of course, depend upon the specific nature of the variable(s) measured and how atypical the sample really was.

Secondly, and most important, since the method guarantees equal probability for each possible sample, it is possible to express our results (i.e., any inferences we might want to make concerning the characteristics of the population) *in terms of probability.* That is, random sampling makes it possible for us to make statements such as "the probability that the population mean has a value between 46.8 and 50.2 is 0.95." Without random sampling we would be limited to statements such as "the sample mean = 28.6," and there would be no way to estimate the probability that the population mean is anywhere near that value. Since we are sampling presumably for the purpose of indirectly discovering certain characteristics of a population, it is absolutely essential that the sampling method we use involve *random selection.*

9.3.2 Stratified Sampling

Despite the intuitive simplicity of the simple random sampling method, there are situations in which certain variations on the method are desirable. For example, in some instances getting a representative sample is especially crucial. These situations are most likely to occur when the variable being measured has considerable dispersion and when the researcher must work with a relatively small sample. In such instances a method of sampling called *stratified sampling* can be more effective.

For example, suppose that an investigator wishes to measure the beliefs held by seniors in a large university on the subject of birth control. Since one's religious conviction is likely to be a significant factor in one's beliefs regarding birth control, it is likely that the researcher would want the various religious groups represented in his sample in the *same proportion* that they occur in the population (the senior class). Suppose that this investigator is restricted to a sample size of 50 and that the senior class breaks down in the category of religious affiliation as follows: Catholics (46 percent), Protestants (24 percent), Jews (10 percent), Other (8 percent), and Uncommitted (12 percent). It is not likely that a random sample of 50 chosen from the senior class would have the same proportional representation for the groups that exist in the population. Therefore, to guarantee the same proportions, we alter the selection method as follows; the sample of 50 is chosen by randomly selecting 23 from the Catholic category, 12 from the Protestants, five from the Jews, four from seniors with other affiliations, and six from the uncommitted group. That is, the population is partitioned into *strata*, and then simple random sampling is used on each stratum. For obvious reasons this stratified sampling method is sometimes called *proportional* sampling. This method produces a random sample according to Definition 9.1.

9.3.3 Cluster Sampling

For the selection of samples over a large geographical area, such as a large city, a state, or an entire country, simple random sampling is virtually impossible since it is often difficult to determine the exact membership of the population. Even if this problem is overcome, the cost and time involved in measuring a sample of elements which are widely scattered over a large geographical area can be prohibitive. To overcome these problems researchers use a method called *cluster sampling* (sometimes called *area sampling*), in which the geographical region is partitioned into smaller regions and a certain number of these subregions are selected randomly. Then each of the selected subregions is again partitioned into still smaller regions and a certain number of these subregions are selected randomly. This procedure can go on until the subregions remaining are small enough to handle efficiently. Then a random sample of elements (people, apple trees, or whatever) is selected from each of those remaining subregions. The collection of these elements constitutes one sample to be used for investigation. The selection of samples for the purpose of predicting voting patterns at the state and national level involves the use of cluster sampling. This method produces a random sample according to Definition 9.1.

9.3.4 Systematic Sampling

A method called *systematic sampling* is another popular sampling technique. This method does not, strictly speaking, produce a random sample according to Definition 9.1. The method, in essence, involves the random selection of one element from the first k elements in an ordered population, followed by the selection of every kth element thereafter, until the desired sample size is obtained. This procedure guarantees equal probability for each individual in the population. However, not all samples of the desired size are equally likely. In fact, many of the samples have zero probability. Why is that so? A slight variation of this method is used by quality control experts who select samples directly off the factory assembly line.

An advantage of this method is its simplicity. However, problems can arise if the population is ordered in such a way that every kth element is different from the rest in some significant way. For example, suppose that a machine has 10 stations on it for putting caps on bottles, but one station has a defect so that every 10th bottle coming off the assembly line is not correctly capped. Thus, if the researcher selects every kth bottle for inspection, where k is any multiple of 10, either the bottles will all be correctly capped or all incorrectly capped! Neither of these cases would accurately reflect the fact that 1/10 of the bottles coming off the line are incorrectly capped. Despite this possible drawback, the systematic sampling method is used with success by experts who know how to avoid its disadvantages.

9.3.5 Multistage Sampling

It is not uncommon for large-scale sampling to involve more than one of the methods mentioned above. Some sampling methods are indeed a complex combination of these methods. Such a combination of methods is called *multistage sampling*. The extensive survey work which precedes every national election, which is designed to poll public opinion and predict voting outcomes, is a perfect example of multistage sampling.

In this relatively short section on sampling methods we have hardly scratched the surface of this extensive subject. Actually each sampling method presents its own special problems with respect to manipulation of the data and interpretation of results. For more information on sampling methodology the reader is directed to Walker and Lev (1969) and Cochran (1963).

For simplicity we shall assume from now on, unless we state otherwise, that any sample under discussion is a sample selected by the simple random sampling method.

Exercises

1. In light of Definition 9.1 determine which of the following examples describes the selection of a *random* sample; and if it does not, explain why.

   (a) The phone company wishes to sample the opinions of its subscribers. So, a number between 01 and 99 is chosen randomly. The number turns out to be 52. Therefore, from an alphabetical listing of the telephone company's subscribers every 52nd name is selected for the sample.

   (b) Suppose that the phone company in (a) instead sent an opinion questionnaire to *all* of its subscribers and got replies from 1382 of them.

   (c) A deck of 52 playing cards is thoroughly shuffled and then 13 cards are dealt to a player.

   (d) Assume the same situation as in (c), but after each card is dealt it is replaced in the deck which is then shuffled thoroughly before the next deal.

   (e) A deck of 52 playing cards is thoroughly shuffled. Then the third card, and every fourth card thereafter are selected to form the sample (i.e., sample contains the 3rd, 7th, 11th, . . . , and 51st cards in the deck).

2. A population is composed of four women with weights of 102, 108, 114, and 126 lb, and two men with weights of 183 and 192 pounds.

   (a) Suppose that a random sample of size three is to be selected from the six, and $\bar{x}$ is to be calculated for the purpose of estimating the mean of the population. What are the values of $\bar{x}$ for each of the possible samples that could be selected?

   (b) Suppose instead that a stratified sample is to be selected (two women and one man). What are the values of $\bar{x}$ for each of the possible samples?

   (c) With reference to your answers to (a) and (b) explain what advantage could be gained here by using stratified sampling rather than simple random sampling.

3. A college professor wants to select a random sample of 50 college sophomores. He wants samples to be as representative of the 1000 sophomores at the college as possible with regard to verbal ability. He notices from the grade records that 160 of the 1000 got As in freshman English, 240 got Bs, 420 got Cs, 140 got Ds, and 40 got Es. He decides to select a stratified sample using the five letter-grade categories as strata. Describe what he must do.

4. A magazine publishes a short questionnaire asking its subscribers their preferences regarding upcoming congressional elections. Assuming that 65 percent of the subscribers answer and return the questionnaire, state briefly any criticism you might have regarding the adequacy of this sampling procedure for determining voter preference for

   (a) all who subscribe to this magazine;

   (b) the general public.

9.4   Parameters of a Finite Population

Let us return again to the basic task in statistical work—that of selecting a sample from a population, describing the sample in various ways, and then using these statistics as a basis for inferences about the population. Our basic model for the distribution of sample values is called a *frequency distribution*, and the basic model for the distribution of population values is called a *probability function*. We talk about a sample of size n, thus implying that the sample is *finite*. Populations, on the other hand, may be thought of as being either *finite* or *infinite*, depending upon the situation. As a matter of fact we cannot tell in discrete cases whether a population is finite or infinite by looking at the probability function. For example, the discrete probability function

x	$f(x)$
1	2/7
2	1/7
3	4/7

gives no clues regarding the *number* of elements in the population. It tells us only the proportional occurrence of each value in the population. The population for which the above probability function is the model could be *infinite* with 2/7 of its values 1s, 1/7 of them 2s, and 4/7 of them 3s, OR, it could be *finite* with the seven values: 1,1,2,3,3,3,3, or the 14 values: 1,1,1,1,2,2,3,3,3,3,3,3,3,3, and so on. If, in fact, a population is finite, we shall use the capital letter N to denote the number of elements in that population.

Suppose that a finite population has the five values: 1,4,7,10, and 13. We can represent this population with a probability function $f(x)$ which assigns equal probability to each value.

x	$f(x)$
1	1/5
4	1/5
7	1/5
10	1/5
13	1/5

Therefore, the parameters μ_x, $\sigma_x{}^2$, and σ_x can be calculated for this population by the usual methods for calculating the mean, variance, and standard deviation of a discrete random variable.

$$\mu_x = \sum_{\text{all } x} x\, f(x) = 7$$

$$\sigma_x{}^2 = \sum_{\text{all } x} x^2 f(x) - (\mu_x)^2 = 67 - 49 = 18$$

$$\sigma_x = \sqrt{18} = 4.24$$

The following theorem suggests an alternative method for obtaining these same results.

Theorem 9.1:   Given a finite population of size N of measurements on variable x, the mean and variance of x are

$$\mu_x = \frac{\sum\limits_{i=1}^{N} x_i}{N} \quad \text{and} \quad \sigma_x^2 = \frac{N \sum\limits_{i=1}^{N} x_i^2 - \left(\sum\limits_{i=1}^{N} x_i\right)^2}{N^2}$$

Applying Theorem 9.1 to the finite population: 1,4,7,10,13, we get

$$\mu_x = \frac{35}{5} = 7 \quad \text{and} \quad \sigma_x^2 = \frac{5(335) - (35)^2}{5} = 18$$

The standard deviation of x is of course $\sqrt{18} = 4.24$. Theorem 9.1, therefore, gives us special formulas for calculating the mean, variance, and standard deviation for a *finite* population. Note the similarity between these formulas and the formulas for $\bar{x}$, s_x^2 and s_x in Chapter 2.

Exercises

1. A population contains the following elements: 0, 1, 2, 3, 4, 5, 6, 7, 8, 9.
    (a)   What is the value of N?
    (b)   Express the population as a probability function.
    (c)   Find the mean, variance, and standard deviation of the population using
        (i)   the methods developed in Chapter 7;
        (ii)   Theorem 9.1.

2. Consider a finite population which contains 50 zeros, 50 "1s," 50 "2s," and so forth, up through 50 "9s," and no other elements.
    (a)   What is the value of N?
    (b)   Express the population as a probability function.
    (c)   Find the mean, variance, and standard deviation of the population using
        (i)   the methods in Chapter 7;
        (ii)   Theorem 9.1.

3. Consider the following set of numbers: 0, 2, 3, 3, 7. Find the variance of this set, given that the set is thought of as a
    (a)   sample;   (b)   population.

4. For the following population of x values: $-1, -3, 2, 0, 1, -1, 6, 0, 1, 3, -2, -1, 3, 0, 2, 1$, find μ_x, σ_x^2, and σ_x.

9.5   The Theoretical Sampling Distribution of $\bar{x}$

In a typical sampling situation a random sample of size n is selected from a population of size N. Suppose that we are interested in estimating the population mean μ_x. To do this we would calculate $\bar{x}$ from the sample. But what can we say about $\bar{x}$ in relation to how *good* an estimate it is likely to be? How much confidence can we have that the difference between μ_x and the $\bar{x}$ we calculate is "tolerably" small? Before we can answer such questions we must know how the statistic $\bar{x}$ behaves. It is clear that the value we get for $\bar{x}$ will depend upon the particular sample we choose. Each possible sample of size n would have its own value of $\bar{x}$. To find out how these various possible values of $\bar{x}$ are distributed *we think of the selection of the random sample as a chance experiment and $\bar{x}$ as a random variable.*

Take, for example, the population we considered earlier: 1, 4, 7, 10, 13. If we plan to choose a sample of size three from this population, then this action is a chance experiment and the sample mean is a random variable defined on the experiment. To determine what values $\bar{x}$ can possibly have, we can list all possible samples of three that could be selected (without replacement) from this population. There are, of course, $\binom{5}{3}$, or 10, such possibilities. Table 9.1 lists these 10 samples, and the value of the sample mean ($\bar{x}$) for each. Since this is a *random* sample, each of these possible samples has equal probability, and therefore each of them has probability 1/10. From Table 9.1 we can construct the *probability function* for $\bar{x}$ (Table 9.2). This probability function is called the *theoretical sampling distribution* for $\bar{x}$. Of course, a theoretical sampling distribution is as abstract as any other probability function. It simply tells us how the probability is distributed for various values of $\bar{x}$, *assuming* certain information about the population. In this demonstration example we are assuming complete information about the population.

Table 9.1. Listing of Possible Samples, Their Means, and Probabilities

Sample	$\bar{x}$	Probability
1,4,7	4	1/10
1,4,10	5	1/10
1,4,13	6	1/10
1,7,10	6	1/10
1,7,13	7	1/10
1,10,13	8	1/10
4,7,10	7	1/10
4,7,13	8	1/10
4,10,13	9	1/10
7,10,13	10	1/10

Table 9.2. The Theoretical Sampling Distribution for $\bar{x}$

$\bar{x}$	$f(\bar{x})$
4	1/10
5	1/10
6	2/10
7	2/10
8	2/10
9	1/10
10	1/10

The most important thing that we learn from studying the theoretical sampling distribution for $\bar{x}$ is the relationship between it and the probability function which represents the population. This relationship can be expressed in part by comparing the means and variances for the two functions.

Theorem 9.2: If x has a finite population of size N with mean $= \mu_x$ and variance $= \sigma_x^2$, then for the theoretical sampling distribution of $\bar{x}$,

$$\mu_{\bar{x}} = \mu_x \text{ and } \sigma_{\bar{x}}^2 = \frac{\sigma_x^2}{n}\left(\frac{N-n}{N-1}\right)$$

In other words, the distribution of x in the population and the theoretical sampling distribution of $\bar{x}$ have the same mean and their variances differ by a factor of

$$\frac{1}{n}\left(\frac{N-n}{N-1}\right)$$

In our example we have already established that $\mu_x = 7$ and $\sigma_x^2 = 18$ (page 218). We can calculate $\mu_{\bar{x}}$ and $\sigma_{\bar{x}}^2$ directly from the information in Table 9.3 as follows:

$$\mu_{\bar{x}} = \frac{70}{10} = 7 \quad \text{and} \quad \sigma_{\bar{x}}^2 = \frac{520}{10} - (7)^2 = 3$$

Table 9.3. Sums Needed for the Calculation of $\mu_{\bar{x}}^2$ and $\sigma_{\bar{x}}^2$

$\bar{x}$	$f(\bar{x})$	$\bar{x}f(\bar{x})$	$\bar{x}^2 f(\bar{x})$
4	1/10	4/10	16/10
5	1/10	5/10	25/10
6	2/10	12/10	72/10
7	2/10	14/10	98/10
8	2/10	16/10	128/10
9	1/10	9/10	81/10
10	1/10	10/10	100/10
		70/10	520/10

Therefore, for our example $\mu_x = \mu_{\bar{x}} = 7$, which is consistent with Theorem 9.2. In addition, $\sigma_x^2 = 18$ and $\sigma_{\bar{x}}^2 = 3$, which is also consistent with Theorem 9.2 because

$$\sigma_{\bar{x}}^2 = \frac{\sigma_x^2}{n}\left(\frac{N-n}{N-1}\right) = \frac{18}{3}\left(\frac{5-3}{5-1}\right) = 3$$

Note that in our example we consider the case in which a sample is drawn *without replacement* from a *finite* population. This situation is the typical one in statistical sampling. However, in most practical sampling situations, although the population is usually finite it is typically quite large (if it were not, there would be no need to sample). For convenience, statisticians typically think of large populations as infinite populations. And it can be proved that if the population is infinite in size, $\sigma_{\bar{x}}^2$ becomes simply σ_x^2/n.

> **Theorem 9.3*:** If x has an *infinite* population with mean $= \mu_x$ and variance $= \sigma_x^2$, then for the theoretical sampling distribution of $\bar{x}$,
>
> $$\mu_{\bar{x}} = \mu_x \text{ and } \sigma_{\bar{x}}^2 = \sigma_x^2/n$$

Compare Theorem 9.3 to Theorem 9.2 and note that the only differences are the replacement of "finite" by "infinite" and the elimination of the factor $(N - n)/(N - 1)$ from the variance of $\bar{x}$. The latter difference is consistent with our intuitive reasoning, for as N approaches infinite size, the factor $(N - n)/(N - 1)$ must approach 1! In practice, populations are rarely infinite, per se; however, as a rule of thumb, if the sample size is less than 5 percent of the population size, then we shall drop the $(N-n)/(N - 1)$ factor (i.e., use Theorem 9.3). The standard deviation of $\bar{x}$, denoted by $\sigma_{\bar{x}}$, is, of course, equal to $\sqrt{\sigma_{\bar{x}}^2}$ in both the finite and infinite cases. This parameter is sometimes called the *standard error of the mean*.

> **Problem:** Given that x has an infinite population with mean equal to 50 and a variance of 36, find the mean, variance, and standard deviation of $\bar{x}$ for a sample of size 25.
>
> **Solution:** According to Theorem 9.3,
>
> $$\mu_{\bar{x}} = \mu_x = 50$$
>
> $$\sigma_{\bar{x}}^2 = \frac{\sigma_x^2}{n} = \frac{36}{25} = 1.44$$
>
> and therefore,
>
> $$\sigma_{\bar{x}} = \sqrt{1.44} = 1.2$$

For obvious reasons it is impossible to demonstrate by example the construction of a theoretical sampling distribution of $\bar{x}$ in the case where the population is infinite (or even large). There is enough work involved constructing such a distribution for the finite case: $N = 5$ and $n = 3$. We can, however, *simulate* the construction of a theoretical sampling distribution for the infinite case by using the same finite example, but assuming that the sample will be selected *with replacement*. Such an exercise is suggested in Exercise 2, page 224.

---

* The proof of this theorem (and the remaining theorems in this chapter) carries the assumption that the sample whose mean is $\bar{x}$ is a random sample defined for *infinite* population sampling (see footnote on page 209).

Theorem 9.3 is in fact very consistent with common sense. The reader should note that if n is large $\sigma_{\bar{x}}$ is necessarily small! Since $\sigma_{\bar{x}}$ measures spread in the theoretical sampling distribution, it follows that a large sample produces a very "narrow" theoretical sampling distribution. And since the mean of the theoretical sampling distribution $(\mu_{\bar{x}})$ is equal to the mean of the population (μ_x), it is clear that a large sample is more likely to have a sample mean which is close to μ_x than a small sample. This fact makes sense, since a larger sample gives us more information about the population than a small sample, and therefore is more likely to have a sample mean $(\bar{x})$ close in value to the population mean (μ_x). We shall capitalize on this fact when we use $\bar{x}$ to estimate μ_x (Chapter 10).

Up to this point we have found that the theoretical sampling distribution of $\bar{x}$ has a mean and variance which can be expressed in terms of the population mean and variance. The next two theorems tell us something about the general shape of this theoretical sampling distribution.

Theorem 9.4: If the population is normally distributed, then the theoretical sampling distribution of $\bar{x}$ is also normally distributed.

Theorem 9.4 says that the theoretical sampling distribution of $\bar{x}$ has a normal distribution provided that the population from which the sample is drawn is itself normally distributed. Certainly this theorem makes intuitive sense. However, since we rarely know the shape of the population, Theorem 9.4 is inadequate for most practical purposes. Theorem 9.5, *The Central Limit Theorem*, may be used when we do not know the shape of the population distribution.

Theorem 9.5: The Central Limit Theorem   Regardless of the shape of the population distribution, the theoretical sampling distribution of $\bar{x}$ can be *approximated* by a normal probability function. The larger the sample size (n), the closer the approximation.

The Central Limit Theorem is indeed a powerful statement, and probably the most important single theorem in statistical theory. Even though the shape of the population distribution may be unknown, this theorem allows us to assume that $\bar{x}$ has an approximately normal distribution, as long as n is large enough. There is no hard and fast rule regarding how large n must be to guarantee that a normal probability function closely fits the theoretical sampling distribution. In most practical cases a sample size of 20 is large enough for a good approximation, although smaller samples are sufficient when the population is nearly normal. Clearly, the closeness of the approximation depends not only on the size of n, but also on how close the population itself comes to being normally distributed.

In summary, we have established that the random variable $\bar{x}$, the mean of a random sample of size n selected from an infinite (or very large) population with mean $= \mu_x$ and variance $= \sigma_x^2$, has an approximately normal distribution with mean $= \mu_x$ and variance $= \sigma_x^2/n$. Thus, the standard deviation of $\bar{x}$ is $\sigma_x/\sqrt{n}$. It follows from these statements that the random variable

$$\frac{\bar{x} - \mu_x}{\sigma_x/\sqrt{n}}$$

has a *standard normal distribution* (by Theorem 8.1). We shall find many applications for this bit of statistical theory in later chapters. For example, suppose that a large population has a mean of 50 and variance equal to 36, and we plan to select a random sample of size 25 and calculate the sample mean. What is the probability that the sample mean will have a value greater than 51.8? Translating this problem into notation, we know that $\mu_x = 50$, $\sigma_x^2 = 36$, and $n = 25$, and we want to calculate $P(\bar{x} > 51.8)$. From the Central Limit Theorem we know that $\bar{x}$ has (approximately) a normal distribution with $\mu_{\bar{x}} = 50$ and $\sigma_{\bar{x}} = 6/\sqrt{25}$. Therefore,

$$P(\bar{x} > 51.8) = P\left(z > \frac{51.8 - 50}{6/\sqrt{25}}\right)$$

$$= P(z > 1.5)$$

$$= 1 - 0.9332$$

$$= 0.0668$$

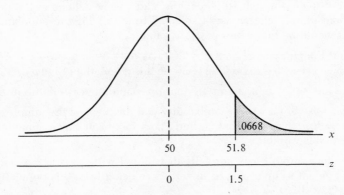

Figure 9.1. Application of the central limit theorem to find $P(\bar{x} > 51.8)$.

Exercises

1. Keeping the same finite population used in the text: 1, 4, 7, 10, 13, let $\bar{x}$ be defined as the mean of a random sample of size $n = 2$ selected from this population.
    (a) How many different samples of size two are possible?
    (b) Construct the theoretical sampling distribution of $\bar{x}$.
    (c) Calculate $\mu_{\bar{x}}$ and $\sigma_{\bar{x}}^2$ from the sampling distribution in (b).
    (d) Calculate $\mu_{\bar{x}}$ and $\sigma_{\bar{x}}^2$, using Theorem 9.2.
    (e) Find the standard error of $\bar{x}$.

2. Using the same finite population: 1, 4, 7, 10, 13, assume that $\bar{x}$ is the mean of a sample of size $n = 2$ selected randomly from this population *with replacement*.
    (a) How many different samples of size two are possible?
    (b) Construct the theoretical sampling distribution of $\bar{x}$.
    (c) Calculate $\mu_{\bar{x}}$ and $\sigma_{\bar{x}}^2$ from the sampling distribution in (b).
    (d) Calculate $\mu_{\bar{x}}$ and $\sigma_{\bar{x}}^2$, using Theorem 9.3.
    (e) Why is Theorem 9.3 applicable in (d), rather than Theorem 9.2?

3. For the population of size $N = 10$ in Exercise 1, page 218, assume that a random sample of size $n = 4$ is to be selected from the population. Find the mean, variance, and standard deviation of the theoretical sampling distribution of $\bar{x}$.

4. For the population of size $N = 500$ in Exercise 2, page 218, assume that a random sample of size $n = 4$ is to be selected from the population. Find the mean, variance, and standard deviation of the theoretical sampling distribution for $\bar{x}$, and compare your answers to those obtained in Exercise 3 above.

5. If we had assumed in Exercises 3 and 4 that the selections would be made with replacement, what would $\mu_{\bar{x}}$, $\sigma_{\bar{x}}^2$, and $\sigma_{\bar{x}}$ be in each case?

6. If an infinite population contained 10 percent 0s, 10 percent 1s, 10 percent 2s, and so on, through 10 percent 9s, and $\bar{x}$ were defined as the mean of a random sample of size $n = 4$ selected from this infinite population, then what would be the values of $\mu_{\bar{x}}$, $\sigma_{\bar{x}}^2$ and $\sigma_{\bar{x}}$?

7. A finite population of size $N = 200$ has $\mu_x = 50$ and $\sigma_x^2 = 64$. A random sample of size 25 is to be selected from this population. Find $\mu_{\bar{x}}$, $\sigma_{\bar{x}}^2$, and $\sigma_{\bar{x}}$.

8. Do Exercise 7 again, assuming that the population is infinite in size.

9. In Exercise 8 what is the probability that the mean of the sample selected will be less than 52? That is, find $P(\bar{x} < 52)$. (You may assume that x is a continuous variable.)

10. A random variable x is normally distributed with $\mu_x = 12.6$ and $\sigma_x = 4.8$. If a random sample of size $n = 16$ is selected from this population, find
    (a) $P(\bar{x} = 13.8)$
    (b) $P(\bar{x} > 14.0)$
    (c) $P(11.4 < \bar{x} < 13.2)$

11. When the sample size is changed from 240 to 15, what effect does this have on the size of $\sigma_{\bar{x}}$ (the standard error of the mean)?

12. If a population is normally distributed, the population mean and median are equal. The theoretical sampling distribution for the *sample median* has mean equal to population mean and

$$\text{variance} = \left(\frac{\pi}{2}\right)\frac{\sigma_x^2}{n}$$

In other words, assuming a normal population, *the sample median is more variable from sample to sample than the sample mean.*
(a) Show that the sample median is more variable from sample to sample than $\bar{x}$ in the text example (page 219), by constructing the theoretical sampling distribution for the sample median, calculating the variance of that distribution and comparing the result to $\sigma_{\bar{x}}^2$.
(b) Why is the answer in (a) not exactly equal to
$\left(\frac{\pi}{2}\right)\frac{\sigma_x^2}{n}$? ($\pi/2$ = approximately 1.57)

13. Assume that in the general population IQ is normally distributed with mean 100 and standard deviation 16.
(a) If we were to randomly select one person from this population, what is the probability that he would have an IQ greater than 102?
(b) If we were to select a random sample of 25 people from the general population, what is the probability that the mean IQ in the sample is greater than 102?
(c) Do (b) again, assuming $n = 100$.

14. A soft drink company produces a cola in 12-ounce bottles. The number of ounces per bottle actually varies slightly from bottle to bottle. However, when the bottling machines are working correctly, the number of fluid ounces per bottle is normally distributed with $\mu_x = 12$ ounces and $\sigma_x = 0.2$ ounces. Every so often a sample of 16 bottles is randomly selected, and the mean volume ($\bar{x}$) of the contents is measured. If a sample mean differs from 12 ounces by more than 1.5 $\sigma_{\bar{x}}$ ounces, then production stops and the machines are repaired; otherwise, production continues as usual.
(a) Find $\mu_{\bar{x}}$ and $\sigma_{\bar{x}}$.
(b) What action takes place if $\bar{x} = 11.86$ ounces?
(c) If the distribution of the population is in fact normal with $\mu_x = 12$ and $\sigma_x = 0.2$, what is the probability that a given sample check will stop production?

15. In Exercise 2, page 216, which of the two sampling methods is more likely to produce the smaller $\sigma_{\bar{x}}$?

9.6 Sampling Distributions for Some Other Statistics

Every sample statistic has a theoretical sampling distribution. These sampling distributions are crucial to the study of statistical inference. Every sample statistic in some way estimates a population parameter. Unless we know how a sample statistic is distributed, we can make no meaningful probability statements with respect to how good an estimate that statistic is. In Chapter 10 we shall discuss in detail the kinds of statements that we can make concerning the estimation of population parameters.

However, before we proceed to Chapter 10 we shall consider briefly some of the more useful sample statistics and their theoretical sampling distributions. The sample statistic $\bar{x}$ is probably the most "celebrated" of all statistics, and therefore, we have used it as the prototype for discussion. We know that $\bar{x}$ has an approximately normal distribution with mean $= \mu_x$ and variance σ_x^2/n (assuming N large and n moderately large). Therefore, the transformed statistic

$$\frac{\bar{x} - \mu_x}{\dfrac{\sigma_x}{\sqrt{n}}}$$

has a standard normal distribution. This statement can be shortened in the following way.

$$\frac{\bar{x} - \mu_x}{\dfrac{\sigma_x}{\sqrt{n}}} \longrightarrow z \qquad\qquad \text{SD1}$$

In a practical problem the statistician typically does not know the standard deviation (σ_x) of the population. In this event he calculates s_x from the sample and substitutes it for σ_x, whenever σ_x is needed. Statement SD1 above is useful only if σ_x is known. If σ_x is unknown and s_x is substituted for it, then statement SD1 becomes

$$\frac{\bar{x} - \mu_x}{\dfrac{s_x}{\sqrt{n}}} \longrightarrow t_{n-1} \qquad\qquad \text{SD2}$$

That is, when we substitute s_x for σ_x, the transformed statistic no longer has a standard normal distribution, but instead it has a *Student-t distribution with n − 1 degrees of freedom.*

The theoretical sampling distribution for s_x^2 (the variance of a random sample of size n) is not particularly useful to us. However, if we assume that the random sample is selected from a *normally distributed* population, then

$$\frac{(n-1)s_x^2}{\sigma_x^2} \longrightarrow (\chi_{n-1}^2) \qquad\qquad \text{SD3}$$

The above statement will be useful to us for making inferences about σ_x^2 and σ_x.

In a binomial experiment where x is the number of successes in n independent trials and p is the probability of success on each trial, the sample statistic x/n, denoted by $\hat{p}$, the *proportion of successes in the n trials*, can be used to estimate p. Since x is approximately normally distributed with $\mu_x = np$ and $\sigma_x{}^2 = np(1-p)$, it follows that $\tilde{p}$ is approximately normally distributed with

$$\mu_p = \mu_{x/n} = \frac{1}{n}\,\mu_x = \frac{1}{n}\,(np) = p$$

and

$$\sigma_p{}^2 = (\sigma_{x/n})^2 = \left(\frac{1}{n}\right)^2 np(1-p) = \frac{p(1-p)}{n}$$

Therefore,

$$\frac{\tilde{p} - p}{\sqrt{\dfrac{p(1-p)}{n}}} \longrightarrow z \qquad\qquad\qquad \text{SD4}$$

When the true correlation between two random variables $(\rho)^*$ is zero, then the theoretical sampling distribution for the sample correlation $(r)^*$ is symmetric and it can be shown that

$$\frac{r\sqrt{n-2}}{\sqrt{1-r^2}} \longrightarrow t_{n-2} \qquad\qquad\qquad \text{SD5}$$

When the true correlation is not zero, the theoretical sampling distribution of r is skewed to the left if ρ is positive and to the right if ρ is negative. The skewness becomes acute when ρ is near -1 or $+1$. A transformation introduced by R. A. Fisher, namely,

$$w_r = \frac{1}{2}\log_e \frac{1+r}{1-r}$$

converts r into the statistic w_r which is normally distributed with mean $= w_\rho$ and variance $= 1/(n-3)$. Therefore, the statistic

$$\frac{w_r - w_\rho}{\sqrt{\dfrac{1}{n-3}}}$$

has a standard normal probability distribution, or equivalently,

$$(w_r - w_\rho)\sqrt{n-3} \longrightarrow z \qquad\qquad\qquad \text{SD6}$$

In this section we have considered only a few of the more important sample statistics and their corresponding sampling distributions. In Chapters 10 and 11 we shall use this theory to estimate certain population parameters and to test hypotheses regarding these parameters. Other useful sampling distributions will be introduced when needed.

---

*The subscript "xy" has been omitted from ρ_{xy} and r_{xy} for convenience.

9.7 Degrees of Freedom

The expression "degrees of freedom" was first introduced in Chapter 8 and then again in the previous section. Whenever we use the Student-t, chi-square, or F distributions, the number of degrees of freedom determines the exact probability function under consideration. The term actually refers to a characteristic of a statistic.

> **Definition 9.2:**   The *number of degrees of freedom* of a statistic is the number of observations used in its calculation minus the number of restrictions* imposed upon them.

The reader is not expected to fully appreciate this definition. A thorough understanding of this concept is dependent upon a more advanced study of theoretical statistics. However, a couple of examples may help to encourage intuitive understanding. For instance, the sample variance (s_x^2) is defined as

$$s_x^2 = \frac{\sum_{i=1}^{n} (x_i - \bar{x})^2}{n-1}$$

The number of observations is n and the statistic $\bar{x}$ is a *linear restriction* on s_x^2. That is, the value of s_x^2 always depends upon the value of $\bar{x}$. Since we have n observations and one restriction, the statistic s_x^2 is said to have $n-1$ degrees of freedom. The sample standard deviation (s_x) also has $n-1$ degrees of freedom. Thus, the statistics

$$\frac{\bar{x} - \mu_x}{\frac{s_x}{\sqrt{n}}} \quad \text{and} \quad \frac{(n-1)s_x^2}{\sigma_x^2}$$

both have a distribution with $n-1$ degrees of freedom.

Another approach to an intuitive understanding of degrees of freedom is to consider the number of sample measurements which are free to vary. For example, let us suppose that the sample mean of five measurements is 8.0. This means that $\Sigma x = 40$. There are many sets of five measurements which could add up to 40. However, if any four of the measurements are determined, the fifth is not free to vary, because the sum must be 40. In other words, if the mean of a sample of size n is specified, the number of degrees of freedom is $n-1$, because only $n-1$ of the measurements are free to vary.

*Actually the number of "algebraically independent linear restrictions" to quote Roscoe (1969, page 161).

Summary

The method of *simple random sampling* involves the selection of n objects from a population of objects such that every possible sample of size n has an equal probability of selection. A *table of random numbers* is used to facilitate the job of selecting random samples. Other sampling methods such as *stratified sampling, cluster sampling,* and *systematic sampling* are variations on the basic method—applicable in special situations.

Parameters μ_x, σ_x^2, and σ_x for a finite population of size N can be calculated from the usual formulas for the mean, variance, and standard deviation of a discrete random variable, *or* from special formulas involving N, which are similar to the formulas for $\bar{x}$, s_x^2, and s_x.

The selection of a sample from a population may be thought of as a chance experiment, and a sample statistic, such as $\bar{x}$, may be thought of as a random variable. The *theoretical sampling distribution* for a sample statistic is the probability function for that statistic. If a population has mean μ_x and variance σ_x^2, then the theoretical sampling distribution for $\bar{x}$ (the mean of a random sample of size n selected from that population) is approximately normal with mean μ_x and variance σ_x^2/n. Every sample statistic has a theoretical sampling distribution, including s_x^2, $\tilde{p}$, and r_{xy}.

The *number of degrees of freedom* of a statistic is the number of measurements used in its calculation minus the number of restrictions imposed upon them.

10

Estimation and Hypothesis Testing

10.1 Point Estimation

When a sample statistic is used to estimate a population parameter, we call that statistic a *point estimate*. For example, we call $\bar{x}$ (the sample mean) a point estimate of μ_x (the population mean). The sample variance s_x^2 is a point estimate of σ_x^2, s_x is a point estimate of σ_x, and $\tilde{p}$ is a point estimate of p. One important characteristic of a statistic is that it be an *unbiased estimator* of the parameter it estimates.

> **Definition 10.1:** The sample statistic q is called an *unbiased estimator* of population parameter θ if the mean of the theoretical sampling distribution of q is equal to θ.

Since the mean of the theoretical sampling distribution of $\bar{x}$ is μ_x, it follows, according to Definition 10.1, that $\bar{x}$ is an unbiased estimator of μ_x.

It can be proved that the mean of the theoretical

sampling distribution of s_x^2 is σ_x^2, and therefore s_x^2 is an unbiased estimator of σ_x^2. The reader probably recalls from Chapter 2 that s_x^2 is thought of as the "average squared deviation from $\bar{x}$." And yet, rather than divide $\Sigma\,(x_i - \bar{x})^2$ by n to get the mean of the squared deviations, we choose to divide by $n - 1$ instead. The reason for dividing by $n - 1$ is that we want s_x^2 to be an unbiased estimator of σ_x^2. If s_x^2 were defined as

$$\frac{\sum\limits_{i=1}^{n} (x_i - \bar{x})^2}{n}$$

then it would *not* be an unbiased estimator of σ_x^2. Each value of s_x^2 would be slightly smaller, and therefore the mean of the sampling distribution would slightly underestimate σ_x^2.

Oddly enough s_x is not an unbiased estimator of σ_x. The reason for this fact is mathematical; it can be explained intuitively by noting that in general *the mean of a set of numbers is not equal to the positive square root of the mean of the squares of that set of numbers.* For example, if we used the set of numbers: 1, 2, 3, 4, it is easy to show that

$$\frac{1 + 2 + 3 + 4}{4} \neq \sqrt{\frac{1^2 + 2^2 + 3^2 + 4^2}{4}}$$

10.2 Interval Estimation

The estimation of a population parameter by a point estimate is a useful technique, especially when the sample size is large, since we know that under this condition the estimate is likely to be close in value to the parameter. For example, we know that the variance of $\bar{x}$ is σ_x^2/n, which means that when n is large, the theoretical sampling distribution for $\bar{x}$ is very narrow, compared to the spread of the population distribution. It follows, therefore, that when n is large, any value of $\bar{x}$ is likely to be close to $\mu_{\bar{x}}$, which we know is equal to μ_x. Therefore, when n is large, we can feel confident that the mean of the sample is a close approximation of the population mean.

Since a point estimate is almost never exactly equal to the parameter value, however, we must always assume that some error exists when we use a point estimate. A disadvantage in point estimation is the fact that the estimate itself does not indicate how much error is likely to exist or how confident we can be that the estimate is close in value to the parameter we are estimating. To avoid this disadvantage we can use another method of estimating parameters called *interval estimation.* In this method we calculate a point estimate as before, and then we use the estimate to construct an interval within which we can be confident (to a specified degree) that the true parameter value lies. The upper and lower boundaries of such an interval are called *confidence limits*. The interval itself is called a *confidence interval*, and the degree of confidence is expressed in terms of probability.

10.2.1 Confidence Interval for μ_x

As an illustration let us derive what is called a *95 percent confidence interval for* μ_x. We know that for large n the transformed statistic

$$\frac{\bar{x} - \mu_x}{\frac{\sigma_x}{\sqrt{n}}}$$

has a standard normal distribution. We also know that in a standard normal distribution,

$$P(-1.96 < z < 1.96) = 0.95$$

which says that the probability is 0.95 that a random variable which has a standard normal distribution has a value between -1.96 and 1.96. (The reader should verify this statement by referring to Table E in the Appendix.) Since

$$\frac{\bar{x} - \mu_x}{\frac{\sigma_x}{\sqrt{n}}}$$

has a standard normal distribution, it follows that

$$P\left(-1.96 < \frac{\bar{x} - \mu_x}{\frac{\sigma_x}{\sqrt{n}}} < 1.96\right) = 0.95$$

Using the appropriate algebraic manipulations, we can convert the above expression to

$$P\left(\bar{x} - 1.96 \frac{\sigma_x}{\sqrt{n}} < \mu_x < \bar{x} + 1.96 \frac{\sigma_x}{\sqrt{n}}\right) = 0.95$$

This expression says the probability is 0.95 that μ_x has a value between

$$\bar{x} - 1.96 \frac{\sigma_x}{\sqrt{n}} \quad \text{and} \quad \bar{x} + 1.96 \frac{\sigma_x}{\sqrt{n}}$$

Therefore, we say that

$$\bar{x} - 1.96 \frac{\sigma_x}{\sqrt{n}} < \mu_x < \bar{x} + 1.96 \frac{\sigma_x}{\sqrt{n}}$$

is a 95 percent confidence interval for the parameter μ_x. For example, suppose that a random sample of size 100 is selected from a population which has a variance of 25. The sample mean is calculated and found to be 35.6. Thus, $n = 100$, $\sigma_x = 5$, and $\bar{x} = 35.6$, so the 95 percent confidence interval for μ_x is

$$35.6 - 1.96\left(\frac{5}{\sqrt{100}}\right) < \mu_x < 35.6 + 1.96\left(\frac{5}{\sqrt{100}}\right)$$

which reduces to

$$34.62 < \mu_x < 36.58$$

Thus, we are 95 percent confident that the true value of μ_x is somewhere between 34.62 and 36.58.

We are not limited to 95 percent confidence intervals. The following theorem generalizes the method of interval estimation for μ_x.

Theorem 10.1: If a large random sample of size n is selected from a population with unknown mean μ_x and known standard deviation σ_x, and the mean of the sample is $\bar{x}$, then

$$\bar{x} - z_{(1+c)/2}\,\frac{\sigma_x}{\sqrt{n}} < \mu_x < \bar{x} + z_{(1+c)/2}\,\frac{\sigma_x}{\sqrt{n}}$$

is called a *100c percent confidence interval for* μ_x.

Suppose, for example, that we want to find a 99 percent confidence interval for μ_x in the same example. The only thing that changes is the value of z used in the calculation. For 99 percent confidence, $c = 0.99$ and therefore $(1 + c)/2 = 0.995$. Thus, we want $z_{0.995}$, rather than $z_{0.975}$, which means that we use 2.58 rather than 1.96. Therefore, a 99 percent confidence interval for μ_x is

$$35.6 - 2.58\left(\frac{5}{\sqrt{100}}\right) < \mu_x < 35.6 + 2.58\left(\frac{5}{\sqrt{100}}\right)$$

which reduces to

$$34.31 < \mu_x < 36.89$$

Note that by increasing the level of confidence we have increased the length of the interval. This makes logical sense since a large interval is more likely to include μ_x than a small interval.

Obviously the most desirable situation is one in which the level of confidence is high and the interval is as short as possible. There are three variables involved in the calculation of the confidence interval for μ_x, which directly affect the length of the interval: $z_{(1+c)/2}$, σ_x, and n. Note that the interval is formed by starting with a point estimate $\bar{x}$ and then adding and subtracting the quantity

$$z_{(1+c)/2}\,\frac{\sigma_x}{\sqrt{n}}$$

If this quantity is large, the interval is large, and vice versa. Therefore, it is easy to see that a given interval can be shortened by (a) *decreasing* the value of $z_{(1+c)/2}$ (which would unfortunately decrease our confidence in the result), or (b) *decreasing* the size of σ_x (which cannot be done in practice), or (c) *increasing* the size of the sample. Therefore, if we want to decrease the length of the confidence interval, but keep the confidence level the same, the *only* alternative is to select a larger sample. Again we see that a larger sample size leads to more desirable results — a fact which by now should be no surprise to the reader.

We noted in Chapter 9 that in practice the value of the population standard deviation (σ_x) is not usually known. In the event that σ_x is unknown, the sample standard deviation s_x is calculated and substituted for σ_x in the confidence interval calculation. As we noted earlier, the statistic

$$\frac{\bar{x} - \mu_x}{\dfrac{s_x}{\sqrt{n}}}$$

has a Student-*t* distribution with $n - 1$ degrees of freedom.

Theorem 10.2: If a large random sample of size n is selected from a population with unknown mean μ_x and *unknown* standard deviation, and the mean and standard deviation of the sample are $\bar{x}$ and s_x, respectively, then

$$\bar{x} - t_{n-1;(1+c)/2}\,\frac{s_x}{\sqrt{n}} < \mu_x < \bar{x} + t_{n-1;(1+c)/2}\,\frac{s_x}{\sqrt{n}}$$

is called a *100c percent confidence interval for* μ_x.

Suppose, for example, that a random sample of size 16 is selected from a large population with unknown mean and variance. The sample mean turns out to be 65.80 and the sample variance is 72.25. And suppose that we wish to find a 90 percent confidence interval for μ_x. We know that $\bar{x} = 65.80$, $n = 16$, and $s_x = \sqrt{72.25} = 8.5$. Since $c = 0.90$, $(1+c)/2 = 0.95$, and therefore $t_{n-1;(1+c)/2} = t_{15;0.95} = 1.75$. Thus, according to Theorem 10.2,

$$65.80 - 1.75\left(\frac{8.5}{\sqrt{16}}\right) < \mu_x < 65.80 + 1.75\left(\frac{8.5}{\sqrt{16}}\right)$$

and therefore, this 90 percent confidence interval for μ_x is

$$62.08 < \mu_x < 69.52$$

We shall now consider the notion of *100c percent confidence* from an intuitive standpoint. Just what does it mean to say, for example, that we have 95 percent confidence that the true mean has a value between a and b? The interval from a to b is determined by measurements in a random sample. If we were to select two different samples of size n, the two 95 percent confidence intervals determined by them would no doubt be different. In fact, if we think for a moment about all possible samples of size n that could be selected from the population, we can imagine a large collection of 95 percent confidence intervals, each determined by a different sample. By "95 percent confidence" we mean that 95 percent of all of those possible intervals do in fact "bracket" or "capture" the true mean within their limits. Since 95 percent of them do bracket μ_x, we can feel 95 percent confident that any one 95 percent confidence interval brackets μ_x.

10.2.2 Determination of n for a Given Error

It is clear that a point estimate is almost always in error by (hopefully) a small amount. And it is also clear that the size of the sample (n) is very important with regard to the precision with which we can expect to estimate μ_x with a single value $(\bar{x})$. We note from Theorem 10.1 that the limits of the confidence interval for μ_x are found by adding the quantity

$$z_{(1+c)/2} \frac{\sigma_x}{\sqrt{n}}$$

to $\bar{x}$ (to get the upper limit) and then subtracting the same quantity from $\bar{x}$ (to get the lower limit). In other words, the point estimate $\bar{x}$ is at the very center of the confidence interval for μ_x. Thus, *if μ_x is in fact in the* confidence interval, then the difference between $\bar{x}$ and μ_x is *at most E*, where E equals the quantity mentioned above. It follows, therefore, that

$$E = \frac{z_{(1+c)/2}\,\sigma_x}{\sqrt{n}}$$

By multiplying both sides of this equation by $\sqrt{n}/E$ we get

$$\sqrt{n} = \frac{z_{(1+c)/2}\,\sigma_x}{E}$$

And finally, by squaring both sides of the equation, we get

$$n = \left[\frac{z_{(1+c)/2}\,\sigma_x}{E}\right]^2$$

This equation permits us to determine how large a sample to select to guarantee (within the appropriate degree of confidence) that the point estimate $\bar{x}$ will be in error by no more than the quantity E.

Theorem 10.3:   A population has unknown mean μ_x and known standard deviation σ_x. If we want to have $100c$ percent confidence that the mean $\bar{x}$ of a large random sample of size n will differ from μ_x by at most the quantity E, then n must be at least

$$\left[\frac{z_{(1+c)/2}\,\sigma_x}{E}\right]^2$$

Note that we cannot apply Theorem 10.3 unless we know or can approximate σ_x. Suppose, for example, that we wish to estimate the mean IQ for a large population of high school seniors—the scores determined by the Stanford-Binet Intelligence Test. Let us assume that the standard deviation of the population of IQ scores is 16 (based on earlier experience with such scores). How large a sample is needed if we want to be 95 percent confident that the sample mean will not differ from the population mean by more than 2.5? Since $c = 0.95$, $z_{(1+c)/2} = z_{0.975} = 1.96$. Therefore, n must be at least

$$\left[\frac{(1.96)(16)}{2.5}\right]^2 = (12.544)^2 = 157.351936$$

Thus, with a sample size of at least 158 we can be 95 percent confident that the sample mean is within 2.5 IQ points of the exact mean IQ of the population.

10.2.3 Confidence Intervals for Other Parameters

We have discussed two methods for finding a $100c$ percent confidence interval for μ_x, which depend upon whether or not σ_x is known.

$$\bar{x} - z_{(1+c)/2}\,\frac{\sigma_x}{\sqrt{n}} < \mu_x < \bar{x} + z_{(1+c)/2}\,\frac{\sigma_x}{\sqrt{n}} \qquad \text{CI1}$$

and

$$\bar{x} - t_{n-1;(1+c)/2}\,\frac{s_x}{\sqrt{n}} < \mu_x < \bar{x} + t_{n-1;(1+c)/2}\,\frac{s_x}{\sqrt{n}} \qquad \text{CI2}$$

We can, of course, calculate confidence intervals for many other population parameters. With reference to Section 9.6 in the previous chapter, it can be shown that a $100c$ percent *confidence interval for the population variance* σ_x^2 is given by

$$\frac{(n-1)s_x^2}{\chi_{n-1;(1+c)/2}^2} < \sigma_x^2 < \frac{(n-1)s_x^2}{\chi_{n-1;(1-c)/2}^2} \qquad \text{CI3}$$

where s_x^2 is the variance of a random sample selected from a normally distributed population. A *confidence interval for* σ_x (the population standard deviation) is found by taking the positive square root of each term in CI3. Suppose, for example, that a sample of 30 measurements has a variance of 36.0, and we want a 90 percent confidence interval

$(c = 0.90)$ for σ_x^2. It follows that $\chi_{n-1;(1+c)/2}^2 = \chi_{29;0.95}^2 = 42.56$ and $\chi_{n-1;(1-c)/2}^2$
$= \chi_{29;0.05}^2 = 17.71$. Therefore, a 90 percent confidence interval for σ_x^2 is

$$\frac{(29)(36)}{42.56} < \sigma_x^2 < \frac{(29)(36)}{17.71}$$

which reduces to

$$24.53 < \sigma_x^2 < 58.95$$

And by taking the positive square root of each term, we find that the corresponding 90 percent confidence interval for σ_x is

$$4.95 < \sigma_x < 7.68$$

Earlier we noted that the statistic $\tilde{p}$, which equals x/n (where x is the number of successes in a random sample of size n) has an approximately normal distribution (assuming that n is large) with a mean of p and variance $p(1-p)/n$, where p is the true proportion of successes in the population. Therefore, a $100c$ percent confidence interval for p, assuming a large n, is given by

$$\tilde{p} - z_{(1+c)/2}\sqrt{\frac{\tilde{p}(1-\tilde{p})}{n}} < p < \tilde{p} + z_{(1+c)/2}\sqrt{\frac{\tilde{p}(1-\tilde{p})}{n}} \qquad \text{CI4}$$

The reader may question the fact that $\tilde{p}$ has replaced p in the expression for the variance. This is done by necessity; since the variance is expressed in terms of the unknown parameter p, we use $\tilde{p}$ to estimate it. It is therefore very important that the sample size be large, especially if it is suspected that p is far from 0.50.

For example, a sample of 400 car registrations are randomly selected from the state motor vehicle records and the cars are inspected. It is found that 168 of the cars cannot pass the inspection. Thus, $\tilde{p} = x/n = 0.42$. Therefore, the 95 percent confidence interval for the true proportion (p) of registered cars in the state which cannot pass inspection is

$$0.42 - 1.96\sqrt{\frac{(0.42)(0.58)}{400}} < p < 0.42 + 1.96\sqrt{\frac{(0.42)(0.58)}{400}}$$

which reduces to

$$0.372 < p < 0.468$$

Constructing a confidence interval for the population correlation ρ, using r as the point estimate, is a somewhat more complicated procedure because we must use Fisher's transformation,

$$w_r = \frac{1}{2}\log_e\frac{1+r}{1-r}$$

to convert the sample correlation r to the statistic w_r, which is normally distributed with mean w_ρ and variance $1/(n-3)$. Therefore, we first construct a $100c$ percent confidence interval for w_ρ, and then convert the limits back in terms of r to give us a 100 percent confidence interval for ρ.

For example, suppose that a random sample of 124 people have their weights and heights measured and the sample correlation between the two variables turns out to be 0.74. Let us find the 95 percent confidence interval for ρ, the true correlation between height and weight in the population from which the random sample was selected. First, we convert 0.74 to $w_{0.74}$, using Table J in the Appendix. The value of $w_{0.74}$ is 0.950. Therefore, the confidence interval for w_ρ is

$$0.950 - \frac{1.96}{\sqrt{124-3}} < w_\rho < 0.950 + \frac{1.96}{\sqrt{124-3}}$$

which simplifies to

$$0.772 < w_\rho < 1.128$$

We then convert 0.772 in 1.128 back in terms of r, using the same table, but in reverse,* to get the 90 percent confidence interval for ρ;

$$0.65 < \rho < 0.81$$

Exercises

1. A population has variance equal to 36.0. A random sample of size 225 is selected from this population. The sample mean is calculated and found to be 46.82. Find a 95 percent confidence interval for μ_x.

2. The following set of numbers represents a random sample of 16 measurements selected from a population which has $\sigma_x{}^2 = 6.25$. Find:
    (a) a point estimate for μ_x
    (b) a 99 percent confidence interval for μ_x

23.6	27.8	31.1	24.4
28.1	25.1	30.0	25.0
27.2	22.5	26.3	19.6
21.0	18.4	20.6	22.2

3. A random sample of 100 college freshmen from a large university have a mean IQ score of 124.6. The standard deviation for these IQ scores is 8.2.
    (a) Find a 90 percent confidence interval for the mean IQ score of the entire freshman class (the population).
    (b) Find a 90 percent confidence interval for the variance of the population.

4. A sample of size 144 is selected from a large population of tropical fish of a

---

*The value of w_r in the table which is closest to 0.772 is 0.775, which corresponds to $r = 0.65$. We do not need to interpolate unless we want the answer correct to three or more decimal places.

certain species. The mean length of the fish in the sample is 5.26 cm. Find a 98 percent confidence interval for μ_x,

(a) given that the population variance is 1.44 cm;

(b) given that the population variance is unknown, but the variance for the sample is 1.69.

5. For a random sample of 36 measurements on variable x,

$$\sum_{i=1}^{36} x_i = 120.0 \quad \text{and} \quad \sum_{i=1}^{36} x_i^2 = 500.0$$

Find a 95 percent confidence interval for μ_x, σ_x^2, and σ_x.

6. Find 99 percent confidence intervals for μ_x, σ_x^2, and σ_x, given the sample data in Exercise 4, page 32.

7. In Exercise 2 what can we say with 96 percent confidence about the possible size of the error involved in estimating μ_x by $\bar{x}$?

8. A factory manager wants to estimate the average time (in minutes) that it takes to assemble a certain product. He plans to select a random sample of n workers, measure the time (x) each takes to assemble the product, and then calculate $\bar{x}$. How large must n be in order that the manager can be 95 percent confident that $\bar{x}$ differs from the true mean time (μ_x) by no more than half a minute? (You may assume that $\sigma_x = 2.2$ minutes.)

9. You want to estimate the mean number of ounces (μ_x) of coffee dispensed from a certain coffee machine. You know from past experience that the amount of coffee dispensed varies slightly from cup to cup. In fact, $\sigma_x = 0.2$ ounces. How large a sample is needed so that you can be 99 percent confident that the mean of the sample differs from μ_x by no more than 0.1 ounces?

10. A random sample of 20 cigarettes of a particular brand has a mean nicotine content of 15.8 milligrams with a variance of 4.50 milligrams. Find a 98 percent confidence interval for

(a) the true mean nicotine content (for that brand of cigarette);

(b) the true variance in nicotine content;

(c) the true standard deviation in nicotine content.

11. A random sample of 100 measurements on variable x is selected from a population of such measurements whose variance is 9.61. The mean of the sample is 25.40. With what degree of confidence can we assert that the mean of the population is a value between 24.84 and 25.96?

12. The standard deviation of the lifetimes of Brand X television tubes is assumed to be 50 hours. If you use the mean lifetime of n television tubes (selected randomly from a large population of tubes) to estimate the true mean lifetime for Brand X, how large must n be in order that you can be 90 percent confident that the error in estimation is no more than 10 hours?

13. Given a confidence interval for μ_x, what effect would each of the following changes make on the length of the interval?

(a) Increase the level of confidence.

(b) Increase the standard deviation.

(c) Increase the size of the sample.

14. In a sample poll of 256 voters chosen at random from all registered voters in a given county, 56 percent are in favor of candidate A. Find a 95 percent confidence interval for p, the proportion of registered voters in the county who are in favor of candidate A.

15. An urn contains 100,000 marbles—some red and some black. Given that we randomly select 144 of these marbles and find that 48 of them are red, find a 97 percent confidence interval for the true proportion of red marbles in the urn.

16. A sociologist studying ethnic problems selects a random sample of 225 names from the registered voter rolls in a large city and finds that 45 are Puerto Rican.
    (a) Find a point estimate for the true proportion p of registered Puerto Rican voters.
    (b) Find a 90 percent confidence interval for p.

17. In a systematic sample of 625 parts coming off the assembly line, a quality control expert discovers that 10 percent are defective. Find a 95 percent confidence interval for the true proportion of defective parts being produced.

18. For a large sample size the statistic s_x has a theoretical sampling distribution which is approximately normal with mean σ_x and standard deviation $\sigma_x / \sqrt{2n}$. Thus, it can be shown that a *large sample 100c percent confidence interval for σ_x is*

$$\frac{s_x}{1 + \dfrac{z_{(1+c)/2}}{\sqrt{2n}}} < \sigma_x < \frac{s_x}{1 - \dfrac{z_{(1+c)/2}}{\sqrt{2n}}}$$

Use the interval above to again find a 95 percent confidence interval for σ_x in Exercise 5.

19. Find a 90 percent confidence interval for ρ_{xy}, which represents the true correlation between verbal aptitude (x) and mathematics achievement (y), given the sample data on page 71 (i.e., given that $n = 15$ and $r_{xy} = 0.84$).

20. Find a 95 percent confidence interval for ρ_{xy}, given that $n = 124$ and $r_{xy} = 0.67$.

10.3 Testing a Statistical Hypothesis—An Example

10.3.1 Introduction

So far, the extent of our activity in the area of statistical inference has been the calculation of point and interval estimates. That is, we have been content to estimate a population parameter by calculating a particular sample statistic and either (a) reporting the value of that statistic as a *point estimate*, or (b) using the point estimate to generate an interval within which the parameter in question is likely to lie—such an interval being called a *confidence interval*. We turn now to the task of making *decisions* based upon sample information. This process, called *hypothesis testing*, is a logical extension of estimation. The sample data provides an estimate of a particular population parameter. The estimate is used as the basis for making a decision. A *statistical hypothesis* is usually a statement about one or more population parameters, such as μ_x, σ_x^2, or ρ_{xy}, which is either accepted or rejected on the basis of information provided by the sample data. Such an acceptance or rejection constitutes a *statistical decision.*

Suppose, for example, that we find an unusual looking coin, and we wonder whether or not it is balanced (i.e., heads and tails equally likely). To help us decide, we plan to gather evidence by flipping the coin 400 times and counting the number (x) of times heads occur. If the coin is balanced, then the average number of heads occurring in 400 flips should be 200, since $\mu_x = np^* = 400(1/2) = 200$. If the coin is not balanced, then $\mu_x \neq 200$.

10.3.2 The Null and Alternate Hypotheses

Since a coin is either balanced or not balanced, the decision is a simple choice between the hypothesis $\mu_x = 200$ and hypothesis $\mu_x \neq 200$. We shall call the former hypothesis a *null hypothesis* and the latter an *alternate hypothesis*. Thus, we set up the hypothesis test by stating

> Null Hypothesis: $\mu_x = 200$
> Alternate Hypothesis: $\mu_x \neq 200$

Note that the alternate hypothesis includes all possible values of μ_x which are not included in the null hypothesis. By formulating the alternate hypothesis in this way, we have indicated that we wish to perform a *two-sided test*. That is, if the coin is not balanced, then it could be unbalanced in favor of heads OR in favor of tails. Thus, we allow both possibilities in the formulation of the alternate hypothesis. On the other hand, if, for some reason, we either strongly suspect that the coin is unbalanced in one particular direction, or we are not concerned about the possibility that the imbalance favors one particular side, we can state the alternate hypothesis either as $\mu_x > 200$ or $\mu_x < 200$, whichever one fits the situation. Such an alternate hypothesis indicates a *one-sided test*. For example, suppose that we have been involved in a gambling game in which the coin in question has been flipped many times to determine the outcomes. And suppose that we have noticed that the coin comes up heads many more times than tails. If we have been losing in this game, we may suspect that the coin is unbalanced

---

*The number of heads (x) occurring in n flips of the coin is a binomially distributed random variable with $\mu_x = np$ and $\sigma_x^2 = np(1 - p)$.

in favor of heads. And, therefore, we might be motivated to test the null hypothesis that $\mu_x = 200$ (balanced) against the alternate hypothesis that $\mu_x > 200$ (unbalanced in favor of heads). Note that we keep the null hypothesis *simple*. That is, the null hypothesis states that the parameter in question has one specific value. The reason for this practice should become clear later in this chapter.

As we follow through this coin example, we shall assume that the test of the null hypothesis, $\mu_x = 200$, is two-sided (i.e., the alternate hypothesis is $\mu_x \neq 200$). To test this null hypothesis we must collect sample data (by flipping the coin 400 times and recording the number of heads that occur), and, on the basis of the results, either accept the null hypothesis or reject it in favor of the alternate hypothesis. But before we do this, we must determine how many heads can occur in 400 flips of the coin and what decision we will make for each possible result.

10.3.3 The Test Criterion

Obviously, the number of heads (x) can vary anywhere from 0 to 400, although if the null hypothesis is true, then x will probably be in the neighborhood of 200. The big question is, "What set of values of x would be suitable as a *criterion* for accepting the null hypothesis?" Obviously, even if the coin is perfectly balanced it is not likely that we will get exactly 200 heads in 400 flips. If we were to make 400 flips of a balanced coin on several different occasions, the number of heads occurring on each occasion would fluctuate around 200 due to the element of chance. Thus, the question is, "How far from 200 must x be before we are likely to say that we do not believe the null hypothesis?" The acceptance/rejection criterion is an arbitrary choice on the part of the investigator. Suppose that we decide on the following test criterion.

Accept the NH if $185 \leq x \leq 215$
Reject the NH if $x < 185$ or $x > 215$

10.3.4 Type I Error

In hypothesis testing we make a decision to either accept or reject a null hypothesis. Since the decision is based upon information supplied by a sample, we realize that whatever decision we make, there is a chance that we have committed an error. Since we have two choices—to *accept* or *reject* the null hypothesis—there are two types of errors possible. The first type can occur only when the null hypothesis is rejected.

Definition 10.2: When a *true* null hypothesis is *rejected*, we say that a Type I error has been committed.

In our example, a Type I error is committed if we decide to reject the null hypothesis when in fact the coin is balanced. That is, if in the 400 flips the number of heads occurring is either less than 185 or more than 215, then the null hypothesis must be rejected. And if the coin is in fact balanced, such a rejection of the null hypothesis is clearly an error— namely, a Type I error. We hope that the probability of the occurrence of a Type I error is not high. We can, in fact, calculate this probability; all we need to find is the probability that the number of heads obtained in the 400 flips is not in the interval from 185 to 215, *given that the null hypothesis is true.* Under the null hypothesis, x is binomially distributed with $\mu_x = 400(1/2) = 200$ and $\sigma_x = \sqrt{(400)(1/2)(1/2)} = 10$. Therefore, we want to find $1 - P(185 \leq x \leq 215 | \mu_x = 200)$, and we can approximate this binomial probability by finding the area to the left of 184.5 plus the area to the right of 215.5 under the graph of a normal distribution with mean equal to 200 and standard deviation 10. Note in Figure 10.1 that the vertical lines through 184.5 and 215.5 cut the area under the graph into three regions: an acceptance region and two rejection regions. The sum of the areas in the rejection regions equals the *probability of a Type I error.* Therefore,

$$P(x < 184.5) = P\left(z < \frac{184.5 - 200}{10}\right) = P(z < -1.55) = 0.0606$$

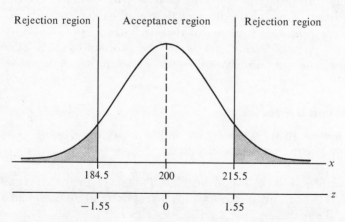

Figure 10.1. *P*(Type I error): represented by the shaded area.

Since the area to the right of 215.5 is also 0.0606, P(Type I error) = 0.1212. In other words, as the test criterion is now set up, if the coin is, in fact, perfectly balanced, the probability that this statistical test will lead us to the conclusion that the coin is *not* perfectly balanced is

0.1212. Perhaps this probability is too high to suit us. Perhaps we do not want there to be that much of a chance that a Type I error will occur. The size of this probability is determined by the test criterion which we specified arbitrarily. If we feel that 0.1212 is too high, we might, for example, decide to change the test criterion to

Accept the NH if $180 \le x \le 220$
Reject the NH if $x < 180$ or $x > 220$

With this new test criterion the probability of a Type I error should be smaller, because the new criterion decreases the size of the rejection regions. The probability is in fact only 0.0404, which the reader should verify.

Definition 10.3:   The *probability of a Type I error* is called the *level of significance*. If $P(\text{Type I error}) = \alpha$, then we say that the null hypothesis is being tested at the α level of significance.

In statistical research the level of significance is usually set in the vicinity of 0.05 or less. The reason for this custom is not entirely clear, except that it appears to be human nature to take a chance on being right when the probability of being wrong is less than 0.05. Of course, an individual researcher must decide what level is appropriate for his particular purpose. His decision is based upon many considerations, not the least of which is the fact that as the level of significance decreases, the probability increases that he will accept the null hypothesis when in fact it is false. That is, as the probability of a Type I error decreases, the probability increases that another type of error will occur, known as a Type II error.

10.3.5   Type II Error

Definition 10.4:   When a *false* null hypothesis is *accepted*, we say that a *Type II error* has been committed.

If the criterion is changed so that $P(\text{Type I error})$ decreases, then $P(\text{Type II error})$ must increase! This statement makes logical sense since a decrease in the significance level increases the acceptance region. And since a Type II error can occur only when a null hypothesis is accepted, it follows that the probability of a Type II error must increase when the probability of a Type I error decreases. Thus, a researcher cannot afford to make the level of significance so small that the probability of a Type II error becomes too great.

Calculating the probability of a Type II error is somewhat more difficult than calculating the probability of a Type I error, due to the fact that a Type II error occurs only when the null hypothesis is *false*.

In our coin example, a Type II error would occur if we accepted the null hypothesis ($\mu_x = 200$) when in fact $\mu_x \neq 200$. But when $\mu_x \neq 200$, the probability distribution for x is *not specifically determined*. Thus, we cannot calculate P(Type II error) *per se!* But we can calculate P(Type II error), *given* that μ_x is a specific value other than 200. For example, suppose that $\mu_x = 208$, which is equivalent to saying that the probability of getting a head is 0.52. When $p = 0.52$, $\sigma_x = \sqrt{(400)(0.52)(0.48)} = 9.99$. Thus, x is binomially distributed and can be approximated by a normal distribution with $\mu_x = 208$ and $\sigma_x = 9.99$. To aid our calculations, we picture a graph which has shifted to the right of the graph pictured in Figure 10.1. The rejection and acceptance regions are still determined by the vertical lines through 184.5 and 215.5 (assuming the original test criterion). Thus, P(Type II error$|\mu_x = 208$) is represented by the area in the acceptance region under the shifted graph (see Figure 10.2).

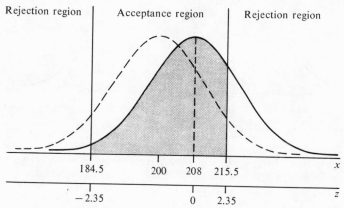

Figure 10.2. P(Type II error$|\mu_x = 208$): represented by the shaded area.

Therefore,

$$P(\text{Type II error}|\mu_x = 208) = P(184.5 < x < 215.5)$$

$$= P(x < 215.5) - P(x < 184.5)$$

$$= P\left(z < \frac{215.5 - 208}{9.99}\right) - P\left(z < \frac{184.5 - 208}{9.99}\right)$$

$$= P(z < 0.75) - P(z < -2.35)$$

$$= 0.7734 - 0.0094$$

$$= 0.7640$$

This calculation tells us that if the true mean is 208 (i.e., if $p = 0.52$) then the probability that we will accept the null hypothesis is 0.7640. In other words, if the true mean is close to the hypothesized mean (208 compared to 200) there is a strong chance (probability = 0.7640) that we may erroneously conclude that the hypothesized mean is the true one. We would hope that if the true mean were not quite so close in value to the hypothesized mean of 200, that the probability of a Type II error would be much smaller. For example, let us calculate $P(\text{Type II error}|\mu_x = 220)$. If $\mu_x = 220$, this means that $p = 0.55$ and therefore $\sigma_x = \sqrt{(400)(0.55)(0.45)} = 9.96$. Thus, $P(\text{Type II error}|\mu_x = 220)$ is represented by the area between 184.5 and 215.5 under the graph of a normal distribution which has a mean of 220 and $\sigma_x = 9.96$ (see Figure 10.3).

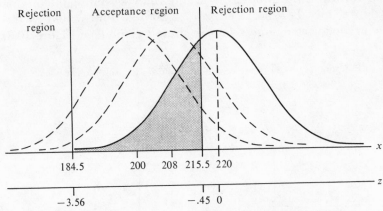

Figure 10.3. $P(\text{Type II error}|\mu_x = 220)$: represented by the shaded area.

Therefore,

$$P(\text{Type II error}|\mu_x = 220) = P(184.5 < x < 215.5)$$
$$= P(x < 215.5) - P(x < 184.5)$$
$$= P\left(z < \frac{215.5 - 220}{9.96}\right) - P\left(z < \frac{184.5 - 220}{9.96}\right)$$
$$= P(z < 0.45) - P(z < -3.56)$$
$$= 0.3264 - 0.0000^{\circ}$$
$$= 0.3264$$

As expected, this result (0.3264) is lower than the previous probability (0.7640). When the null hypothesis is not true, μ_x can have any value

---

°We know from Table E that the area to the left of $z = -3.56$ is less than 0.0002, but how much less we do not know, because the table goes out only as far as $z = -3.49$. Therefore, we say that the area = 0.0000.

except 200. Each one of these values has associated with it a value: P(Type II error$|\mu_x$). Thus, for a given hypothesis test, P(Type I error) $= \alpha$, a *constant* value, but P(Type II error$|\mu_x$) is a *variable* which is a function of μ_x. If we plot each value of μ_x against its corresponding Type II error probability, we generate a graph called the *operating characteristic curve* (OCC). Figure 10.4 shows the OCC for our example. The probability values we calculated are labelled on the vertical axis, and they correspond to 208 and 220 on the horizontal axis. Note that the probabilities for $\mu_x = 208$ and $\mu_x = 192$ are equal. This is also true for $\mu_x = 220$ and $\mu_x = 180$. Why? Note also that there is a break in the OCC at $\mu_x = 200$. The reason for this break is that each height represents a probability of "error" when accepting the null hypothesis. There is no error committed if $\mu_x = 200$ and we accept the null hypothesis. The value on the vertical axis which corresponds to $\mu_x = 200$ is 0.8788 which equals $1 - P$(Type I error). Why?

Since a null hypothesis is either true or false, and in either case we may accept or reject it, there are actually four distinct situations involved (see Table 10.1). We have just examined two of these situations —the two which constitute erroneous decisions. The other two situations, namely, (a) *accepting* a *true* null hypothesis, and (b) *rejecting* a *false* null hypothesis, involve no error. They are correct decisions.

Table 10.1. Summary of the Types of Error Possible When a Null Hypothesis (NH) Is Accepted or Rejected

	Accept NH	Reject NH
NH is true	No error	Type I error
NH is false	Type II error	No error

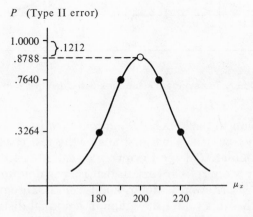

P (Type II error)

Figure 10.4. Operating characteristic curve.

10.3.6 *Protecting Against Error*

In the actual test situation it is virtually impossible to protect adequately against *both* types of errors, since a decrease in the probability that one type will occur is accompanied by an increase in the probability that the other type will occur. To meet this problem, the researcher tries to state his null and alternate hypotheses in such a way that a *rejection of the null hypothesis is what he hopes to conclude.* Since he is likely, he feels, to *reject* the null hypothesis, he can sufficiently protect himself against the Type I error and not worry about the Type II error, since a Type II error can occur only when the null hypothesis is *accepted.*

Let us assume in our coin example that, in fact, we do hope to find support for the theory that the coin is not balanced. That is, we hope to reject the null hypothesis that $\mu_x = 200$. To help protect ourselves against a Type I error, we arbitrarily decide what value of α we are willing to accept as a reasonable probability for a Type I error. As we mentioned earlier, many factors enter into this decision. In a business situation, for example, wrong decisions usually cost money. Therefore, the size of α should reflect how much money may be lost should the decision to reject the null hypothesis be in error. Suppose that we choose $\alpha = 0.05$. When $\alpha = 0.05$, we say that we are testing at the 0.05 *level of significance.*

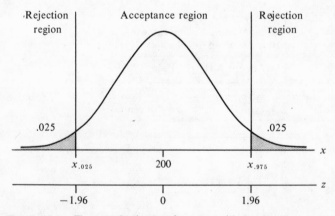

Figure 10.5. The standard setup for a two-sided test with $\alpha = .05$.

The decision to make $\alpha = 0.05$ *automatically determines the test criterion.* In Section 10.3.4 we first stated the test criterion which we then used to calculate P(Type I error), now called α. In actual practice, however, this procedure is reversed; that is, α is determined arbitrarily and then the test criterion is determined from α. In our coin example, we know that the area under the graph of a normal distribution (which approximates the binomial) must be divided into an acceptance region and two rejection regions—one on each end (since the test is two-sided). We choose to make the two rejection regions equal in area. Therefore, since $\alpha = 0.05$, the area of each rejection region must be 0.025 (half of 0.05). Thus the two values of x which determine the three regions—and therefore determine the test criterion—are $x_{0.025}$ and $x_{0.975}$. In terms of these values the test criterion reads as follows:

Accept NH if $x_{0.025} \leq x \leq x_{0.975}$
Reject NH if $x < x_{0.025}$ or $x > x_{0.975}$

In order to calculate $x_{0.025}$ and $x_{0.975}$, we note from Figure 10.5 that these values correspond to the standard normal values $z_{0.025}$ and $z_{0.975}$ in the transformation $(x - \mu_x)/\sigma_x$, where $\mu_x = 200$ and $\sigma_x = 10$. Since $z_{0.025} = -1.96$ and $z_{0.975} = 1.96$, it follows that

$$-1.96 = \frac{x_{0.025} - 200}{10} \quad \text{and} \quad 1.96 = \frac{x_{0.975} - 200}{10}$$

Consequently, $x_{0.025} = 180.4$ and $x_{0.975} = 219.6$, and therefore the test criterion may be expressed as shown below.

Accept NH if $180.4 \leq x \leq 219.6$
Reject NH if $x < 180.4$ or $x > 219.6$

Then, when we collect the sample data (flip the coin 400 times), we count the number of heads obtained and make the decision to accept or reject the null hypothesis based on this test criterion. For example, if we get 228 heads in the 400 flips, we would *reject* the null hypothesis.

In actual practice, however, the test criterion is not expressed in terms of the statistic (x), but instead in terms of the transformed statistic $(z$, in this case). Thus, in our coin example, the test criterion would be expressed as

Accept NH if $-1.96 \leq z \leq 1.96$
Reject NH if $z < -1.96$ or $z > 1.96$

Then, when the sample is taken and number of heads turns out to be 228, we convert 228 to standard units (z),

$$z = \frac{228 - 200}{10} = 2.80$$

and then apply the test criterion. Since 2.80 is greater than 1.96, the decision is to *reject* the null hypothesis.

10.3.7 Reserving Judgment

But what happens when the sample data is such that the null hypothesis is accepted? How do we protect against the Type II error? Since P(Type

II error) is an entire set of values (represented by the OCC), and is therefore somewhat difficult to interpret, some statisticians prefer to *reserve judgment* when the test fails to reject the null hypothesis. In this way the researcher can avoid the decision to *accept* the null hypothesis and therefore avoid making a Type II error. This strategy may seem to the reader to be somewhat dishonest. However, from a practical standpoint reserving judgment is a realistic alternative.

First, in so many statistical tests the researcher is hoping to reject the null hypothesis, which quite often is a statement which in the past has been generally accepted as true. The researcher hopes that his new theory is a preferable alternative. Therefore, he sets up the test so that rejection of the null hypothesis is equivalent to rejection of the old in favor of the new. If the sample data is such that rejection of the null hypothesis is not supportable, it can be argued that the old theory will continue to prevail without the researcher coming out in favor of it by stating, "I accept the null hypothesis." Moreover, the researcher is not motivated to publish the fact that he failed to reject the null hypothesis, since his real interest lies in the belief that the null hypothesis is false. Nor will a publisher be likely to publish such a result, since it would essentially reaffirm an already established fact.

For example, suppose that a professor of mathematics has developed a revolutionary method of teaching mathematics, and he wishes to test his new method against the old "traditional" method. So he gets two samples of students, teaches one group in a traditional fashion and the other with his new method (both groups getting the same material), and then examines them for a comparison of their comprehension. His null hypothesis would be stated as, "Both methods are of equal effectiveness;" and his (one-sided) alternative hypothesis would be, "The new method is superior to the traditional method." If he finds from the results of the examination that those receiving the new method did not do significantly better on the exam than those in the traditional group, then he is likely to reserve judgment on his new method rather than say that the old method is just as good.

Secondly, and much more important, a researcher may wish to avoid accepting a null hypothesis because it is a fact that if the measuring instrument is not reliable, then rejection of a null hypothesis is almost impossible to attain, even though the null hypothesis is, in fact, false. For instance, when the difference in effectiveness of two teaching methods is measured by an examination, as in the previous example, there is a real possibility that the examination is not reliable enough (i.e., does not measure with sufficient precision) to detect real differences in comprehension. If, for example, the examination were too easy, the group with the higher level of comprehension might not have sufficient opportunity to display that superiority. Thus, even though the methods might produce significantly different results, if the differences are not reliably measured, the null hypothesis will not be rejected on the basis of the sample data. Therefore, the act of reserving judgment is neither unreasonable nor dishonest.

Generally speaking then, the decision to accept a null hypothesis should be avoided if possible. We should always try to state null and alternate hypotheses in such a way that we hope to reject the null hypothesis in favor of the alternate. Then, if it turns out that, on the basis of the sample data, we cannot reject the null hypothesis, we should consider reserving judgment, *unless* we can show evidence that (a) the OCC is relatively steep (thus indicating relatively small probability of a Type II error), (b) the reliability of the measuring instrument is high, and (c) the same results hold up in repeated experiments (this condition would in part reflect high measurement reliability).

10.4   Outline of the Basic Procedure

The testing procedure described in the previous section is commonly called a *significance test*. There exist many different significance tests, many more in fact than we shall discuss in this text. In Chapter 11 we cover some of the more common tests, all of which involve the basic concepts presented in Section 10.3. Although each test is different in many ways, each involves basically the same procedure. This procedure is outlined as follows:

(a) *Formulate the hypothesis* for which you hope to find supporting evidence.

(b) *Determine the size of the sample* you wish to select and what *sample statistic* you will use as the basic evidence.

(c) *Formulate a null hypothesis* in terms of a population parameter and in such a way that the probability distribution of the sample statistic is known, or can be closely approximated, if the null hypothesis is true.

(d) *Formulate an alternate hypothesis* — either one-sided or two-sided, depending upon the situation. Rejection of the null hypothesis implies that the alternate hypothesis is accepted. (Hopefully a rejection of the null hypothesis is the desired result; that is, rejection of the null hypothesis will provide support for the hypothesis formulated in (a).)

(e) *Specify α, the probability of a Type I error* (also called the *level of significance*).

(f) *Construct the test criterion.*

(g) *Collect sample data and calculate the appropriate statistic(s).*

(h) *Make the decision to reject or accept the null hypothesis*, (or possibly to *reserve judgment*).

Figure 10.6 shows a diagram of the decision-making alternatives and consequences.

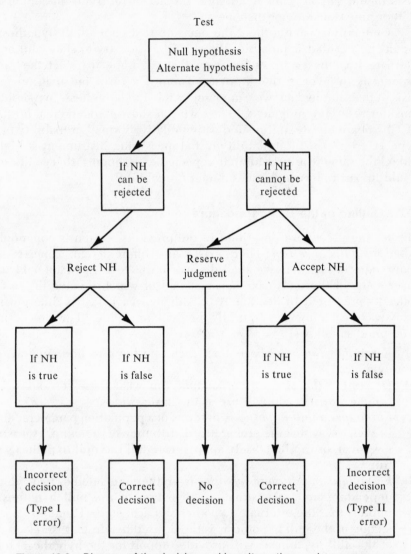

Figure 10.6. Diagram of the decision-making alternatives and consequences.

Let us apply this procedure to another example. Suppose that candidates Smith and Jones are running for mayor of a large city, and Smith's campaign manager wants to find evidence to support the hypothesis that the majority of the voting public will vote for Smith at the polls. He plans to select a random sample from the voting public and ask each member of the sample whom he plans to vote for. He proceeds as follows:

(a) He hopes to support the hypothesis that Smith will win the election.
(b) He decides to select randomly 100 registered voters from the general public and count the number of voters in the sample who say that they plan to vote for Smith.
(c) The number in the sample who favor Smith can be thought of as a binomially distributed random variable x, where p is the proportion of voters in the population who favor Smith and $n = 100$. If $p = 0.5$, then Smith and Jones have an equal chance to win and $\mu_x = np = 100(0.5) = 50$. Thus, the null hypothesis is: $\mu_x = 50$.*
(d) The alternate hypothesis must be one-sided, since he wants to test the hypothesis that Smith will win the election (refer back to (a)). Thus, the alternate hypothesis is: $\mu_x > 50$.
(e) Let $\alpha = 0.01$.
(f) Under the null hypothesis the probability distribution of x can be closely approximated by a normal distribution with $\mu_x = 50$ and $\sigma_x = 5$. Thus z, where $z = (x - 50)/5$, has a standard normal distribution. And since $z_{0.99} = 2.33$, the test criterion can be expressed as

Accept NH (or reserve judgment) if $z \leq 2.33$
Reject NH if $z > 2.33$

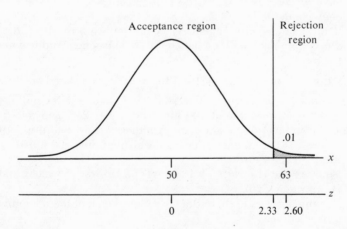

Figure 10.7. Graphical representation of the test criterion (one-sided test), showing the sample value in the rejection region.

*The reader may wonder why the null hypothesis is not $\mu_x \leq 50$ or $\mu_x < 50$. The reason why we use $\mu_x = 50$ is that we need a *simple* null hypothesis (refer back to Section 10.3.2), one which specifies the exact distribution of x.

(g) He selects the sample of 100 and finds that 63 people favor Smith. Therefore, the z value which corresponds to $x = 63$ is

$$z = \frac{63 - 50}{5} = \frac{13}{5} = 2.60$$

(h) Since $z = 2.60$ is larger than $z = 2.33$, his decision is to *reject* the null hypothesis in favor of the alternate hypothesis at the 0.01 level of significance. That is, he can state (with probability 0.01 that he is wrong) that Smith will win the election over Jones.

In Chapter 11 we shall use this same general procedure for each of several different types of hypothesis tests involving many different population parameters, sample statistics, and probability functions.

Exercises

1. A manufacturer wants to test the null hypothesis that 95 percent of the equipment turned out is free from defect, against the alternate hypothesis that less than 95 percent is defect-free.
    (a) If, in fact, only 92 percent of the equipment is defect-free, and the manufacturer decides, on the basis of some sample information, to reject the null hypothesis, what type of error has he committed?
    (b) If, in fact, 95 percent of the equipment is defect-free, and the manufacturer decides to reject the null hypothesis, what type of error has been committed?
    (c) If the manufacturer cannot reject the null hypothesis on the basis of the sample data, under what condition(s) can he avoid making a Type II error?

2. Verify that $P(\text{Type I error}) = 0.0404$ for the criterion stated on page 244.

3. In the coin example, suppose that you decide to accept the null hypothesis ($\mu_x = 200$) if in a sample of 400 flips $190 \leq x \leq 210$, and reject the null hypothesis in favor of the alternate hypothesis ($\mu_x \neq 200$) if $x < 190$ or $x > 210$ (where x is the number of heads occurring in the 400 flips).
    (a) Find $P(\text{Type I error})$.
    (b) Find $P(\text{Type II error}|\mu_x = 188)$, $P(\text{Type II error}|\mu_x = 208)$, and $P(\text{Type II error}|\mu_x = 220)$.
    (c) Use the answers to (b) to draw a rough sketch of the operating characteristic curve.
    (d) Why are the Type II error probabilities for $\mu_x = 208$ and $\mu_x = 220$ smaller here than they are in Figure 10.4?

4. Suppose that you wish to test the null hypothesis that a particular coin is balanced, against the alternate hypothesis that the coin is not balanced, at the five percent level of significance. You decide to select a sample of size 100 and count the number of heads (x) which occur.
    (a) State the null and alternate hypotheses (in terms of μ_x).
    (b) State the test criterion in terms of z.

(c) State the test criterion in terms of x.

(d) What is P(Type I error)?

(e) Find $P(\text{Type II error}|\mu_x = 60)$.

(f) If you get 59 heads in the 100 tosses, what decision do you make with regard to the null hypothesis?

5. Repeat Exercise 4 assuming the one-sided alternate hypothesis that the coin is unbalanced in favor of heads (i.e., the probability of getting a head is greater than 0.50).

6. You wish to test the null hypothesis that a pair of dice are balanced, against the alternate hypothesis that they are not, at the 0.02 level of significance. You base your decision on the number (y) of times a total of 7 occurs in 120 tosses of the pair of dice.

(a) State the null and alternate hypotheses.

(b) State the test criterion in terms of z.

(c) State the test criterion in terms of y.

(d) What is P(Type I error)?

(e) Find $P(\text{Type II error}|\mu_y = 15)$.

(f) What decision would be made if the 100 tosses of the pair of dice yields sixteen "7s"?

7. A state department of motor vehicles claims that 60 percent of the cars on the road are in violation of the state inspection laws. Test this (null) hypothesis at the 0.05 level of significance, against the alternate hypothesis that less than 60 percent violate inspection laws, given that in a random sample of 900 car inspections 500 fail to pass.

8. In the type of hypothesis test presented in this section, one involving a binomially distributed random variable x, we have been expressing the hypothesis in terms of the parameter μ_x, the mean *number* of successes in n trials, rather than the parameter p, the mean *proportion* of successes in n trials. For instance, in the text example, the hypothesis could have been expressed as

$$\text{Null Hypothesis:}\quad p = \frac{1}{2}$$

$$\text{Alternate Hypothesis:}\quad p \neq \frac{1}{2}$$

Express the null and alternate hypotheses in Exercises 5, 6, and 7 in terms of p.

Summary

We use sample statistics to aid in the estimation of population parameters. One type of estimate is called a *point estimate*, where a single value of a statistic is used to estimate a parameter. A sample statistic q is considered an *unbiased estimator* of a parameter θ if the mean of the sampling distribution of q is θ. Another type of estimate is called an *interval estimate*, where a statistic is used to generate a *confidence interval* within which a given parameter is likely to lie.

A process called *hypothesis testing* is a logical extension of estimation which requires that a *decision* be made between two hypotheses called the *null* and *alternate* hypotheses. The decision to accept or reject a statistical hypothesis is based upon statistical information supplied by a sample randomly selected from the population. If the decision is to *reject* a true *null* hypothesis in favor of the alternate hypothesis, then we say that a *Type I error* has been committed. The probability of a Type I error, denoted by α, is called the *level of significance*. If a *false* null hypothesis is *accepted*, then we say that a *Type II error* has been committed. The probability of a Type II error depends upon the true value of the parameter in question. The *operating characteristic curve* is the graph of possible parameter values vs. Type II error probabilities. The *test criterion* states what decision will be made relative to the possible results obtained from the sample. It is often preferable to *reserve judgment* rather than accept a null hypothesis.

11

Classical Significance Tests

11.1 Introduction

Now that we understand the fundamental concepts underlying hypothesis testing, the remainder of this text should present no new conceptual difficulties. In the latter part of Chapter 10 we covered the basic notions involved in hypothesis testing, without attempting to survey the various types of hypothesis tests available to us. In this chapter we shall discuss several types of hypothesis tests, which we shall call *tests of significance*, or simply *significance tests*. In each test we determine whether or not the observed data is *significantly* different from data we might expect were the null hypothesis true. If we decide that the difference is *significant* (too great), then we reject the null hypothesis. We shall concentrate on those significance tests most frequently used in applied areas.

The tests included in this chapter, which we shall call *classical significance tests*, involve assumptions regarding the shape of the population distribution, the values of certain population parameters, and the manner in which the sample data is collected. Tests which, for the most part, do not involve such assumptions we shall call *distribution-free* (or *nonparametric*) tests. We deal with distribution-free tests exclusively in Chapter 12.

The key in hypothesis testing is *knowing the distribution of the appropriate sample statistic* (or transformation thereof) *under the assumption that the null hypothesis is true.* Therefore, knowing the theoretical sampling distribution of various statistics is absolutely essential. In Chapter 9 we described theoretical sampling distributions for several statistics (SD1–SD6, pages 226–227). We saw in Chapter 10 how this information can be used to develop confidence intervals. The same information, and more like it, comprises the basic mathematical structure for the development of significance tests. We shall define some new statistics, most of them variations on $\bar{x}$, s_x^2, r_{xy}, and others, each of which has a known theoretical sampling distribution (given certain assumptions).

11.2   Tests Concerning Proportions

11.2.1   One Proportion

Many practical problems involve questions regarding populations which are conceptually partitioned into two subsets determined by a dichotomous trait. For example, newborn babies are either boys or girls, machine parts are either defective or nondefective, a student either passes or fails an examination, a person does or does not have type AB blood, or a heart attack victim lives or dies. The basic mathematical theory underlying these situations is the binomial probability function. In fact, our demonstration example in Chapter 10 depends upon the fact that x, the *number* of heads occurring in n flips of the coin—where the probability of getting a head on any given flip is p—is binomially distributed with mean equal to np and variance $np(1-p)$. We can think of the population as a virtually infinite collection of heads and tails, in which the *proportion* of heads is p. We select a sample from this population by flipping the coin n times and recording the results. We can estimate the population parameter p by dividing the number of heads that occur (x) by the size of the sample (n). This estimate (x/n) is denoted by $\tilde{p}$.

In short, the occurrence of heads in the sample can be expressed either in terms of the *number* of heads (x), or the *proportion* of heads $(x/n = \tilde{p})$. The statistic x has an approximately normal distribution with mean np and variance $np(1-p)$, and the statistic $\tilde{p}$ is also approximately normally distributed, but with mean p and variance $p(1-p)/n$. Therefore, when we are testing a hypothesis having to do with the value of a population proportion (p), we have a choice in how we express the null and alternate hypotheses. For instance, in the coin example we hypothesize that the coin is balanced, against the alternate hypothesis that it is not balanced. Suppose that we choose to test at the 0.05 level of significance and that a sample of 400 flips results in 228 heads. The two approaches may be compared as follows:

"Number" Approach	"Proportion" Approach
NH: $\mu_x = 200$	NH: $p = 0.5$
Alt. H: $\mu_x \neq 200$	Alt. H: $p \neq 0.5$

$$\frac{x - np}{\sqrt{np(1 - p)}} \longrightarrow z \qquad\qquad \frac{\tilde{p} - p}{\sqrt{\dfrac{p(1 - p)}{n}}} \longrightarrow z$$

Accept NH if $-1.96 \leq z \leq 1.96$
Reject NH if $z < -1.96$ or $z > 1.96$ (same)

$$x = 228 \qquad\qquad\qquad p = \frac{228}{400} = 0.57$$

$$z = \frac{228 - (400)(0.5)}{\sqrt{400(0.5)(0.5)}} = 2.80 \qquad\qquad z = \frac{0.57 - 0.50}{\sqrt{\dfrac{(0.5)(0.5)}{400}}} = 2.80$$

Reject NH, since $2.80 > 1.96$ (same)

From now on we shall use the proportion approach exclusively. Let us consider a different example. A congressman claims that 90 percent of the adult male population in his state are employed. Test this claim against the alternate hypothesis that the percentage is lower than 90 (using $\alpha = 0.01$), given that, in a random sample of 1600 adult males, only 1400 are employed. The hypothesis should read

Null Hypothesis: $p = 0.90$
Alternate Hypothesis: $p < 0.90$

And we must reject the null hypothesis if $z < -2.33$. If we cannot reject, then we either accept or reserve judgement. Since the number of employed in the sample of 1600 is 1400, it follows that $\tilde{p} = 0.875$. Therefore,

$$z = \frac{0.875 - 0.90}{\sqrt{\dfrac{(0.90)(0.10)}{1600}}} = -3.33$$

Since -3.33 is less than -2.33, we conclude that the congressman's claim is false (see Figure 11.1). That is, we conclude that less than 90 percent of the state's adult male population is employed. We recognize, of course, that the probability is 0.01 that we are making an incorrect decision.

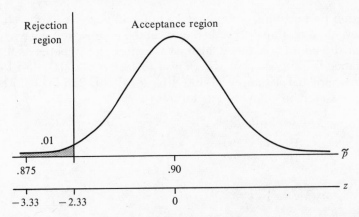

Figure 11.1. Rejection of the null hypothesis ($p = .90$) in the employment example.

11.2.2 *Two Proportions*

The researcher often finds himself trying to decide whether an observed difference between two sample proportions (or percentages) is simply due to chance or due to the fact that the two corresponding population proportions are in fact different. That is, he wants to determine if an observed difference is *significant* (due to real differences in the population proportions) or *not significant* (due only to chance). For example, suppose that we suspect that candidate A, who is running for public office, will get more support at the polls from women than men. And suppose that we interview 200 female voters and 300 male voters and find that 64 percent of the female voters and 56 percent of the male voters favor candidate A. The question is, "In the general voting public, is candidate A favored more by women than men, or is he equally favored by both sexes?" In other words, is the observed difference between proportions 0.64 and 0.56 due simply to chance fluctuations from one sample to another, or does it reflect a real difference between the proportion (p_1) of women in the population who support candidate A and the proportion (p_2) of men in the population who support candidate A?

The key to the answer is the theoretical sampling distribution of the statistic $\tilde{p}_1 - \tilde{p}_2$, the difference between two sample proportions from *independent* samples; $\tilde{p}_1 = x_1/n_1$ and $\tilde{p}_2 = x_2/n_2$, where x_1 and x_2 are the number of voters in each group who favor candidate A, and n_1 and n_2 are the two sample sizes. It can be shown that for large samples the statistic $\tilde{p}_1 - \tilde{p}_2$ is approximately normally distributed with

$$\text{mean} = p_1 - p_2 \quad \text{and} \quad \text{variance} = \frac{p_1(1 - p_1)}{n_1} + \frac{p_2(1 - p_2)}{n_2}$$

Thus, we may conclude that

$$\frac{(\tilde{p}_1 - \tilde{p}_2) - (p_1 - p_2)}{\sqrt{\dfrac{p_1(1 - p_1)}{n_1} + \dfrac{p_2(1 - p_2)}{n_2}}} \longrightarrow z \qquad \text{SD7}$$

In our example, the hypotheses are

Null Hypothesis: $p_1 = p_2$
Alternate Hypothesis: $p_1 > p_2$

Since under the null hypothesis the two population proportions are equal, let us assume that they are equal to p. Therefore, *under the assumption that the null hypothesis is true*, we can rewrite SD7 in the following way.

$$\frac{\tilde{p}_1 - \tilde{p}_2}{\sqrt{p(1-p)\left(\frac{1}{n_1} + \frac{1}{n_2}\right)}} \longrightarrow z \qquad\qquad \text{SD7}'$$

Since p is unknown, we estimate it from the sample data by the formula

$$p(\text{est}) = \frac{x_1 + x_2}{n_1 + n_2} \qquad \begin{array}{r}\text{(the overall proportion}\\ \text{in the two samples who}\\ \text{favor candidate }A)\end{array}$$

In our example, since $\tilde{p}_1 = 0.64$, $\tilde{p}_2 = 0.56$, $n_1 = 200$, and $n_2 = 300$, it follows that $x_1 = 128$ and $x_2 = 168$. Therefore, our estimate of p is

$$p(\text{est}) = \frac{128 + 168}{200 + 300} = 0.59$$

Thus,

$$z = \frac{0.64 - 0.56}{\sqrt{(0.59)(0.41)\left(\frac{1}{200} + \frac{1}{300}\right)}} = 1.25$$

Assuming that we are testing at the 0.05 level of significance, we cannot reject the null hypothesis. Why? Therefore, we conclude that the observed difference between sample proportions (0.08) is apparently due to chance alone (i.e., the difference is not "significant"). Thus, the data does not support the notion that candidate A has greater support from women voters.

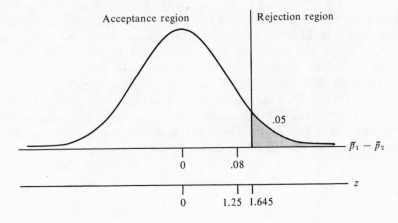

Figure 11.2.   Failure to reject the null hypothesis ($p_1 = p_2$) in the voter preference vs. sex example.

This type of significance test, as well as the type involving only one proportion, can be either one-sided or two-sided. The reader is reminded that the decision to have a one-sided or two-sided alternate hypothesis is properly done *before* we look at the sample data.

This significance test for two proportions is based upon the assumption that the two samples are independently selected. We must be careful not to apply this test when the proportions are *correlated*. Correlated proportions occur whenever the same sample (or part of it) is used to generate both sample proportions. For example, if a sample of fifth grade students is given two different types of geometry exams, then the proportion $(\tilde{p}_1)$ of those who pass one exam and the proportion $(\tilde{p}_2)$ of those who pass the other exam are considered correlated. That is, the responses cannot be assumed to be independent of one another. The sample proportions are more likely to be close in value than a comparable pair of proportions generated from two independent samples. When the assumption of independence is violated, a different test is used (see McNemar, 1962, page 227).

11.2.3   k Proportions

In some situations we must decide whether the observed differences among more than two sample proportions reflect true differences in the population proportions or only chance fluctuation due to sampling. To do this we set up the following hypotheses.

Null Hypothesis: $p_1 = p_2 = \ldots = p_k$
Alternate Hypothesis: Not all p_is are equal

For example, suppose that candidate A's campaign manager wants to know whether or not people with different religious preferences differ in their support of candidate A. Let p_1 represent the true proportion of Protestants who favor candidate A, p_2 the true proportion of Catholics who favor candidate A, and p_3 the true proportion of Jews who favor candidate A. The manager decides to test at the 0.05 level the null hypothesis that p_1, p_2, and p_3 are equal, against the alternate hypothesis that they are not all equal. He interviews a total of 400 voters. The results appear in Table 11.1. From the table we can calculate the sample proportions, which are $\tilde{p}_1 = 90/170 = 0.53$, $\tilde{p}_2 = 110/165 = 0.67$, and $\tilde{p}_3 = 40/65 = 0.62$. The question is, "Do the observed differences in $\tilde{p}_1$, $\tilde{p}_2$, and $\tilde{p}_3$ reflect differences in p_1, p_2, and p_3, or are they simply due to chance fluctuation?" If the null hypothesis is true, it is implied that the true proportion for each religious group is equal to the overall proportion for the combined group in the population. An estimate of that population proportion can be calculated by dividing the total number in the sample who favor candidate A by 400; that is, $p(\text{est.}) = 240/400 = 0.60$. If (a) 0.60 is the overall proportion in the population who favor candidate A, and (b) the null hypothesis is true, then we can *expect* 60 percent of the Protestants in the sample to favor candidate A. That is, we can expect $(0.60)(170) = 102$ Protestants in the sample to favor candidate A. And, therefore, we can *expect* $170 - 102 = 68$ Protestants not to favor candidate A. Similarly, we can *expect* $(0.60)(165) = 99$ Catholics to favor candidate A, and so on. These *expected frequencies* are shown in Table 11.2. If we let the six *observed* frequencies in Table 11.1 be labelled $o_1, o_2, \ldots, o_6$, and the corresponding *expected* frequencies in Table 11.2 be labelled $e_1, e_2, \ldots, e_6$, then the statistic

$$\sum_{i=1}^{6} \frac{(o_i - e_i)^2}{e_i}$$

has a chi-square distribution with $k - 1$ degrees of freedom. Since this is a new statistic, we shall state its sampling distribution formally below.

$$\sum_{i=1}^{2k} \frac{(o_i - e_i)^2}{e_i} \longrightarrow \chi_{k-1}^2 \qquad \text{SD8}$$

Table 11.1. Survey Results—Voter Preference for Candidate A, by Religion

	Protestant	Catholic	Jew	Total
For Candidate A:	90	110	40	240
Not for Candidate A:	80	55	25	160
Total:	170	165	65	400

Table 11.2. "Expected" Frequencies, Assuming That the Null Hypothesis Is True

	Protestant	Catholic	Jew	Total
For Candidate A:	(102)	(99)	(39)	240
Not for Candidate A:	(68)	(66)	(26)	160
Total:	170	165	65	400

Intuitively it is clear that if the observed and expected frequencies for each of the six cells in the table differ very little, then we have evidence to support the null hypothesis. On the other hand, if the observed and expected frequencies differ greatly, it is likely that the null hypothesis is false. A close look at the statistic

$$\sum_{i=1}^{2k} \frac{(o_i - e_i)^2}{e_i}$$

reveals that its value is small when the differences between o_i and e_i are small, and large when the differences are large. Therefore, if the value of this statistic is large enough, we can reject the null hypothesis. If the level of significance is α, then we accept the null hypothesis if the value of the statistic is less than or equal to $\chi^2_{k-1,1-\alpha}$ and reject the null hypothesis for any other value. This type of test (for k proportions) is *always one-sided*.

Getting back to our example, the computed χ^2 value is

$$\chi^2 = \frac{(90 - 102)^2}{102} + \frac{(110 - 99)^2}{99} + \frac{(40 - 39)^2}{39}$$

$$+ \frac{(80 - 68)^2}{68} + \frac{(55 - 66)^2}{66} + \frac{(25 - 26)^2}{26}$$

$$= 6.77$$

Since $\chi^2_{2,0.95} = 5.99$ (from Table G in the Appendix), we can reject the null hypothesis. That is, we conclude that not all of the population proportions are equal, which means that *at least two* differ. Note that we are *not* saying that the population proportions are all different!

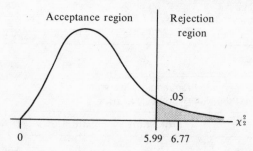

Figure 11.3. Rejection of the null hypothesis ($p_1 = p_2 = p_3$) in the voter preference vs. religion example.

For small proportions and small sample sizes this particular test is not very accurate. Therefore, it is best not to use it if any of the expected frequencies is less than five. Remember also that this chi-square test is based upon the assumption that the k proportions are independent. A significance test for k correlated proportions can be found in McNemar (1962, page 227). Finally, it is interesting to note that if we apply this test for k proportions *when* $k = 2$, the computed χ^2 value is exactly the *square* of the computed z value we would get if we used the test for two proportions discussed in Section 11.2.2 (see Exercise 13, page 266).

Exercises

1. Test the null hypothesis in Exercise 7, page 255, again, but this time in terms of the population proportion p.

2. An eminent authority claims that in two out of every three families the husband has primary responsibility for the handling of financial matters. Test this claim at the 0.05 level of significance (using a two-sided test), given that in 360 of 500 randomly selected families, the husband has primary responsibility for financial matters.

3. A company produces and sells an ointment which it claims removes warts in 90 percent of the cases. An independent firm wishes to test this claim against the counter-claim that the product is overrated. In a random sample of 64 people who have used the product, 48 report that the ointment successfully removed the warts. What should be the firm's decision with regard to the company's claim? (Test at the 0.01 level of significance.)

4. Experience at State College has shown that typically 30 percent of the freshman class fail to pass the college's entrance examination in mathematics. However, this year's freshman class of 900 had only 24 percent fail the examination. Perform a significance test to determine whether or not this year's freshmen are better prepared than usual in mathematics. (Use $\alpha = 0.01$.)

5. To test the effectiveness of a drug designed to cure a disease, a researcher injects the drug into 50 mice who have the disease. As a control, he has a second sample of diseased mice who do not receive the drug. What can he conclude about the effectiveness of the drug if, after one week, 32 mice in the treated sample and 24 mice in the control sample have fully recovered from the disease? (Use a one-sided test at $\alpha = 0.05$.)

6. In the assembly department of a toy factory the foreman wants to find out whether or not there is a difference between the quality of work done during the day shift as opposed to work done at night. He randomly samples 400 toys from each shift, checking each toy for defective assembly. Given that 56 of the day-shift toys and 44 of the night-shift toys are found to have assembly defects, test (at $\alpha = 0.01$) the null hypothesis that there is no difference in the quality of work done in the two shifts.

7. In a random sample of 100 sorority girls, 39 say they are in favor of open housing. In a random sample of 200 nonsorority girls, 60 are in favor of open housing. Use a two-sided test at the 0.05 level of significance to determine if there is a real difference of opinion between sorority and nonsorority girls on the issue of open housing.

8. A safety expert believes that women are more accident-prone than men. He administers a test which measures "degree of accident-proneness" to

random samples of 100 men and 150 women. Sixteen men and 30 women were declared accident-prone by the test results. Test the safety expert's belief at the 0.03 level of significance.

9. In a random sample of 250 psychotic patients who receive psychotherapy, 200 show improvement. In a random sample of 180 psychotic patients who receive shock therapy, 150 show improvement. Test at $\alpha = 0.05$ the hypothesis that the two types of therapy are not equally effective for the treatment of psychotic patients.

10. Random samples of 100 medical doctors, 150 college professors, and 200 trial lawyers take a psychological examination which measures aggressiveness. Results show that 80 of the doctors, 100 of the professors, and 180 of the lawyers are classified as "aggressive." Test at $\alpha = 0.01$ the hypothesis that the members of the three professions do not significantly differ in aggressiveness.

11. A random sample of 50 students is chosen from each of the four classes in a large high school. The number of students in each class who smoke cigarettes regularly is summarized in the following table.

	Freshman	Sophomore	Junior	Senior
Smoker:	18	24	22	16
Nonsmoker:	32	26	28	34

From this evidence, what can be concluded regarding the true proportions of smokers in the four classes? (Let $\alpha = 0.05$.)

12. A farmer has three types of fertilizer. He wants to determine if the fertilizers differ at all in their effectiveness in promoting germination of corn seed. Therefore, he selects three samples of 100 corn seeds each, plants them, and applies a different fertilizer to each sample. After an appropriate time interval, he counts the number of seeds which have germinated in each sample.

	Fertilizer		
	A	B	C
Germinated:	62	72	76
Not germinated:	38	28	24

Perform the appropriate significance test to determine if any real differences exist among the three types of fertilizers with respect to their effect on germination.

13. Rework Exercise 7 using the χ^2 statistic, and show that the obtained χ^2 value equals the square of the z value previously obtained. (This relationship between χ^2 and z was first mentioned in Exercise 8, page 202.)

14. An equivalent form of the χ^2 statistic in SD8 is

$$\sum_{i=1}^{2k} \left(\frac{o_i^2}{e_i}\right) - n$$

where n is the total number of observations. Use this alternate form to recalculate the χ^2 value in Exercise 10.

11.3   Tests Concerning Means

In this section we shall consider significance tests which apply to hypotheses concerning a single population mean, the difference between two population means, and, finally, differences among k population means.

11.3.1   One Mean

A bulb manufacturer claims that his 100-watt bulbs have a mean life of 150 hours. However, a random sample of 100 bulbs is tested, and it is discovered that the mean life for the sample is 146.2 hours. Is the manufacturer's claim in error or is the observed difference between 150 and 146.2 due merely to chance?

To answer questions of this type we need to know the theoretical sampling distribution for the sample mean $(\bar{x})$. We know from Chapter 9 that this sampling distribution is normally distributed (if the population is normal) or approximately normally distributed (if the shape of the population is unknown, but N and n are sufficiently large). Therefore, we know that in most cases $\bar{x}$ is at least approximately normally distributed with mean equal to μ_x (the population mean) and variance equal to σ_x^2/n (where σ_x^2 is the population variance). Therefore, according to SD1 (page 226), the transformed statistic

$$\frac{\bar{x} - \mu_x}{\dfrac{\sigma_x}{\sqrt{n}}}$$

has a standard normal distribution. In order to use this fact we must know the value of σ_x. In the bulb example, let us assume that $\sigma_x = 12.6$ hours. To determine whether or not the observed difference between the claimed μ_x and the observed $\bar{x}$ is likely to be due to chance, we set up a significance test with the following hypotheses.

Null Hypothesis: $\mu_x = 150$
Alternate Hypothesis: $\mu_x < 150$

If we test at the 0.05 level of significance, then we must reject the claim if $z < -1.645$. If $z \geq -1.645$, then we either accept the claim or reserve judgement. Next we convert the observed mean to standard units and compare the computed z value with -1.645. Since $n = 100$, $\bar{x} = 146.2$, and $\sigma_x = 12.6$, therefore

$$z = \frac{146.2 - 150}{\dfrac{12.6}{\sqrt{100}}} = \frac{-3.8}{1.26} = -3.02$$

Consequently, we conclude that the true mean lifetime of the population of 100-watt bulbs is less than 150 hours. Intuitively, we are saying that it would be too unlikely that we would get $\bar{x} = 146.2$, *given that* $\mu_x = 150$;

therefore, we conclude that μ_x must be less than 150, *since we did in fact get $\bar{x} = 146.2$.*

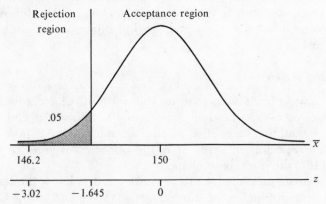

Figure 11.4. Rejection of the null hypothesis ($\mu_x = 150$) in the bulb example ($\sigma_x = 12.6$).

Too often laymen look at a difference, such as the difference between 150 and 146.2, and conclude that it is either insignificantly small or significantly large, based on some sort of informal intuition. However, the decision to accept or reject the null hypothesis ($\mu_x = 150$) depends upon more than just the observed difference between 150 and 146.2. There are, in fact, two other very important determining factors: (i) the size of the sample (n), and (ii) the size of the population standard deviation (σ_x). If σ_x had been much larger, or if n had been much smaller (or a combination of the two), the resulting decision could have been quite different. For example, if σ_x had been 28.2 (instead of 12.6), then the calculated z value would have been

$$z = \frac{146.2 - 150}{\dfrac{28.2}{\sqrt{100}}} = \frac{-3.8}{2.82} = -1.35$$

Or, if n had been 25 (instead of 100), then the calculated z value would have been

$$z = \frac{146.2 - 150}{\dfrac{12.6}{\sqrt{25}}} = \frac{-3.8}{2.52} = -1.51$$

In both cases, even though the observed difference is still -3.8, we would not have rejected the null hypothesis.

More often than not the value of σ_x is unknown, in which case we calculate s_x from the sample data and substitute it for σ_x. The only effect this substitution has is a slight change in the shape of the theoretical sampling distribution for the transformed statistic. We noted this effect in Chapter 9 when we stated in SD2 (page 226) that the statistic

$$\frac{\bar{x} - \mu_x}{\dfrac{s_x}{\sqrt{n}}}$$

as a Student-t distribution with $n-1$ degrees of freedom. The effect of this change is to make the test slightly more conservative (i.e., makes rejection more difficult), because we are estimating σ_x with s_x. Suppose that σ_x had been unknown in the bulb example. Therefore, with $n = 100$ and $\alpha = 0.05$, the criterion would have been to reject the null hypothesis if $t < -1.66$ (see Table F in the Appendix), rather than $z < -1.645$. Note that the change in the criterion is very slight for this example because n is large. Suppose that s_x (calculated from the sample data) is 13.2. Therefore,

$$t = \frac{146.2 - 150}{\dfrac{13.2}{\sqrt{100}}} = \frac{-3.8}{1.32} = -2.88$$

Since -2.88 is less than -1.66, we would reject the null hypothesis.

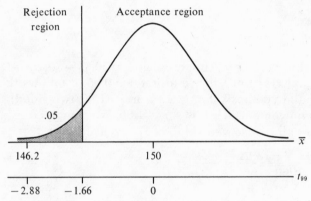

Figure 11.5.   Rejection of the null hypothesis ($\mu_x = 150$) in the bulb example (σ_x unknown).

11.3.2 Two Means

Suppose that two different manufacturers produce 100-watt bulbs. An independent firm wishes to determine whether or not there is a real difference between the true mean life of one brand as opposed to the other. That is, if μ_1 is the mean life of brand 1, and μ_2 is the mean life of brand 2, then the independent firm wishes to test the following hypothesis.

> Null Hypothesis: $\mu_1 = \mu_2$
> Alternate Hypothesis: $\mu_1 \neq \mu_2$

The firm plans to select a sample of each brand (of sizes n_1 and n_2) and calculate the sample mean lives $\bar{x}_1$ and $\bar{x}_2$. The difference between the two sample means, $\bar{x}_1$ and $\bar{x}_2$, is the statistic of primary interest. After $\bar{x}_1 - \bar{x}_2$ has been calculated, the question will be, "Is this observed difference large enough so that we can conclude that μ_1 and μ_2 are not equal, or is $\bar{x}_1 - \bar{x}_2$ small enough so that we can conclude that $\mu_1 = \mu_2$ and that the observed difference is due only to chance?" To perform this type of test we must know the theoretical sampling distribution of the statistic $\bar{x}_1 - \bar{x}_2$. If the two samples are independently selected, and if the two populations can be assumed to be normally distributed, then we can assume that $\bar{x}_1 - \bar{x}_2$ has a normal distribution with

$$\text{mean} = \mu_1 - \mu_2 \quad \text{and} \quad \text{variance} = \frac{\sigma_1^2}{n_1} + \frac{\sigma_2^2}{n_2}$$

Therefore, if we transform this normally distributed statistic into standard normal form, we can state the following relationship.

$$\frac{(\bar{x}_1 - \bar{x}_2) - (\mu_1 - \mu_2)}{\sqrt{\frac{\sigma_1^2}{n_1} + \frac{\sigma_2^2}{n_2}}} \longrightarrow z \qquad\qquad \text{SD9}$$

If the null hypothesis $\mu_1 = \mu_2$ is true, then $\mu_1 - \mu_2 = 0$. Therefore, we find the z value which corresponds to $\bar{x}_1 - \bar{x}_2$ by computing

$$\frac{\bar{x}_1 - \bar{x}_2}{\sqrt{\frac{\sigma_1^2}{n_1} + \frac{\sigma_2^2}{n_2}}}$$

In the bulb example, suppose that $n_1 = 50$, $n_2 = 30$, $\sigma_1^2 = 60$, $\sigma_2^2 = 66$, $\bar{x}_1 = 142.5$ and $\bar{x}_2 = 139.1$. If we test at $\alpha = 0.05$, it follows that we shall reject the null hypothesis if $z < -1.96$ or $z > 1.96$ (two-sided), and accept (or reserve judgement) if $-1.96 \leq z \leq 1.96$. We compute the z value corresponding to the observed value of $\bar{x}_1 - \bar{x}_2$ as follows:

$$z = \frac{142.5 - 139.1}{\sqrt{\frac{60}{50} + \frac{66}{30}}} = \frac{3.4}{\sqrt{3.4}} = 1.85$$

Since 1.85 falls between -1.96 and 1.96, we cannot reject the null hypothesis.

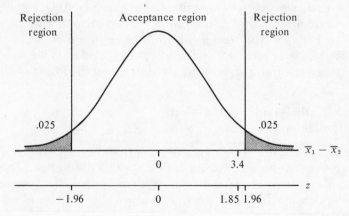

Figure 11.6. Failure to reject the null hypothesis ($\mu_1 = \mu_2$) in the bulb example (σ_1^2 and σ_2^2 known).

In the previous example we assumed that σ_1^2 and σ_2^2 are known. This is not usually the case in practice. Usually the only information we have relating to σ_1^2 and σ_2^2 are the statistics s_1^2 and s_2^2 obtained from the sample data. When the population variances are unknown, the question arises, "Are they equal or unequal?" If σ_1^2 and σ_2^2 can be assumed to be equal, then we can estimate each of them with the so-called *pooled estimate*, denoted by s^2, which is defined as

$$s^2 = \frac{(n_1 - 1)s_1^2 + (n_2 - 1)s_2^2}{n_1 + n_2 - 2}$$

Then s^2 can be substituted into SD9 for σ_1^2 *and* σ_2^2. This substitution changes the theoretical sampling distribution from standard normal to Student-t with $n_1 + n_2 - 2$ degrees of freedom. That is,

$$\frac{(\bar{x}_1 - \bar{x}_2) - (\mu_1 - \mu_2)}{\sqrt{\dfrac{s^2}{n_1} + \dfrac{s^2}{n_2}}} \longrightarrow t_{n_1 + n_2 - 2} \qquad \text{SD10}$$

Let us consider an example in which SD10 can be applied. Suppose that a new hormone is developed which is supposed to promote weight-increase in hogs. To test the effectiveness of this discovery, two samples of 15 hogs each are randomly selected. One sample gets regular injections of the hormone. The other sample gets no hormone injections. After three weeks the sample getting the hormone shows a mean increase in weight ($\bar{x}_1$) of 15.6 pounds with a variance (s_1^2) of 6.5. In the other sample $\bar{x}_2 = 12.8$ and $s_2^2 = 6.3$. Thus, we want to test

Null Hypothesis: $\mu_1 = \mu_2$
Alternate Hypothesis: $\mu_1 > \mu_2$

Let us test at the 0.01 level of significance. We assume that the population variances are equal, and then we proceed to calculate the pooled variance, as follows:

$$s^2 = \frac{14(6.5) + 14(6.3)}{15 + 15 - 2} = 6.4^*$$

Therefore, according to SD10

$$t = \frac{(15.6 - 12.8) - (0)}{\sqrt{\dfrac{6.4}{15} + \dfrac{6.4}{15}}} = \frac{2.80}{0.92} = 3.04$$

Since $\alpha = 0.01$, the value of t which separates the rejection and accep-

---

*Note that when $n_1 = n_2$, s^2 is simply the mean of s_1^2 and s_2^2; and in general s^2 is the *weighted* mean of s_1^2 and s_2^2, using $(n_1 - 1)$ and $(n_2 - 1)$ as weights.

tance regions is $t_{28;0.99}$, which is 2.47. Therefore, we reject the null hypothesis.

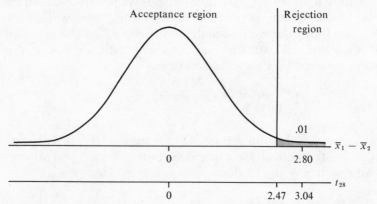

Figure 11.7. Rejection of the null hypothesis ($\mu_1 = \mu_2$) in the hormone example (σ_1^2 and σ_2^2 unknown).

In this example we have *assumed* that $\sigma_1^2 = \sigma_2^2$. This assumption need not be made on pure guesswork. It is possible to formally test the hypothesis that $\sigma_1^2 = \sigma_2^2$ (see Section 11.4.2). If conditions are such that we cannot assume that $\sigma_1^2 = \sigma_2^2$, then neither of the two procedures described in this section can be used legitimately. There are other alternatives depending upon the nature of the data, sample size, and so forth. (See Exercise 14, page 276.)

11.3.3   k Means (Analysis of Variance)

We turn now to a generalization of the significance test just discussed. That is, we shall describe a procedure which tests the null hypothesis $\mu_1 = \mu_2 = \ldots = \mu_k$, against the alternate hypothesis that not all of the μ_is are equal. We shall assume that each population is normally distributed and that the k variances are equal to the same value, which we shall call σ^2. To perform this test we estimate σ^2 from the sample data in two different ways and then compare the two estimates. For simplicity we shall assume that the k samples are of equal size (n).

First, we estimate σ^2 by calculating the mean of the sample variances. We call this estimate the *variation within samples*, and we denote it by V_w. This estimate is comparable to the "pooled" estimate described in the previous section. That is,

$$V_w = \frac{\sum\limits_{i=1}^{k} s_i^2}{k}$$

Secondly, we estimate σ^2 by calculating the variance of the sample means $(s_{\bar{x}}^2)$ and multiplying the result by n. We call this estimate the *variation among samples*, and we denote it by V_a. V_a makes a sensible estimate of σ^2 assuming that $\mu_1 = \mu_2 = \cdots = \mu_k$, because under that assumption (plus the assumptions of equal variances and normality) we can think of our k samples as independent samples from the same normally distributed population with variance equal to σ^2. Thus, we know from Theorem 9.3 that $\sigma^2 = n\sigma_{\bar{x}}^2$. Since $V_a = ns_{\bar{x}}^2$, it is an appropriate estimate of σ^2—*assuming that the null hypothesis is true*. Therefore,

$$V_a = n \left[\frac{\sum_{i=1}^{k} (\bar{x}_i - \bar{\bar{x}})^2}{k - 1} \right]$$

($\bar{\bar{x}}$ = the mean of the means)

If the null hypothesis is true, then the two estimates V_a and V_w should be close in value, since they are both good estimates of σ^2. However, if the null hypothesis is not true, V_a will overestimate σ^2, since V_a is very sensitive to variation among the sample means. Therefore, assuming that the null hypothesis is true, the quotient V_a/V_w fluctuates around a mean value of 1.00 (approximately), but if the null hypothesis is not true, V_a/V_w has a mean value greater than 1.00. According to statistical theory, when $\mu_1 = \mu_2 = \cdots = \mu_k$, the statistic V_a/V_w has an F probability distribution with $k - 1$ and $k(n - 1)$ degrees of freedom. That is,

$$\frac{V_a}{V_w} \longrightarrow F_{k-1, k(n-1)}$$

SD11

Therefore, to test at the α level of significance the hypothesis that k population means are equal (against the alternate hypothesis that they are not all equal), we calculate V_a/V_w and see if the value obtained is near 1.00 or is much larger. If it is so large that it is greater than $F_{k-1,k(n-1);1-\alpha}$, then we reject the null hypothesis. Otherwise, we accept the null hypothesis (or reserve judgement).

For example, suppose that an ecologist wishes to investigate the relative levels of mercury pollution in three major lakes. He catches five lake trout from each lake and measures the concentration of mercury present in each fish. The measurements for each sample, in parts per million (ppm) are

Sample 1: 2.2, 3.4, 3.0, 2.6, 3.8
Sample 2: 4.0, 3.2, 3.7, 3.0, 3.6
Sample 3: 2.6, 2.2, 3.0, 2.4, 1.8

Thus, the means and variances for the amount of mercury present in the samples are

$\bar{x}_1 = 2.8$ $\bar{x}_2 = 3.5$ $\bar{x}_3 = 2.4$
$s_1^2 = 0.20$ $s_2^2 = 0.16$ $s_3^2 = 0.20$

We must answer the question, "Do the observed differences among the means reflect actual differences among μ_1, μ_2, and μ_3, *or* is it true that $\mu_1 = \mu_2 = \mu_3$ and that the observed differences are due simply to chance?" Our formal hypotheses are

Null Hypothesis: $\mu_1 = \mu_2 = \mu_3$
Alternate Hypothesis: Not all of the μ_is are equal

First we estimate σ^2 by calculating V_w.

$$V_w = \frac{s_1^2 + s_2^2 + s_3^2}{k} = \frac{0.20 + 0.16 + 0.20}{3} = 0.19$$

Then we estimate σ^2 again by calculating V_a.

$$V_a = n\, s_{\bar{x}}^2 = 5 \left[\frac{\sum\limits_{i=1}^{3} (\bar{x}_i - 2.9)^2}{3 - 1} \right] = 5(0.31) = 1.55$$

Therefore, the calculated F value is

$$F = \frac{V_a}{V_w} = \frac{1.55}{0.19} = 8.16$$

If we are testing at $\alpha = 0.01$, then we must reject the null hypothesis if $V_a/V_w > F_{2,12;0.99}$. According to Table H in the Appendix, $F_{2,12,0.99} = 6.93$. Thus, our conclusion is to reject the null hypothesis. That is, we conclude that the observed means differ so much that we cannot believe that the true population means are all equal. Therefore, we have some good evidence that the three lakes are not equally polluted with mercury. This test, incidentally, by its nature is always one-sided.

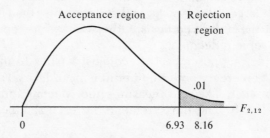

Figure 11.8. Rejection of the null hypothesis ($\mu_1 = \mu_2 = \mu_3$) in the mercury pollution example.

This F-test is designed as a first step in determining whether or not any of k population means differ from each other. A rejected null hypothesis simply indicates that not all of the k population means are equal; it does *not* tell us which ones are different from others. Therefore, if the null hypothesis is rejected, the next step is to find out exactly which population means differ. Special methods devised for this purpose have been developed by Duncan, Scheffé, and others. Winer (1962) summarizes and compares these methods. If, on the other hand, the F-test is not rejected, then we conclude that none of the population means differ, and no further testing is necessary.

In this section we have only scratched the surface of a widely-used testing procedure called *analysis of variance*. We have discussed only the very simplest case—a "one-way" analysis designed to test k means from *independent* samples of *equal* size. The sample sizes in the mercury pollution example (and in the exercises to come) are somewhat unrealistic. Samples this small do not usually yield reliable information about a population. However, analysis of variance procedures require considerable computational effort, and, for the sake of learning basic concepts, small samples serve our purpose well. One last note—when *two* sample means from samples of equal size are tested with this procedure, the calculated F value turns out to be the *square* of the t value you would get if you applied the t-test in Section 11.3.2 (see Exercise 19, page 277).

$$\therefore\ t = \sqrt{F}$$

Exercises

1. A large population has variance equal to 36 and an unknown mean μ_x. Test the hypothesis that $\mu_x = 30$, against the alternate hypothesis that $\mu_x \neq 30$, given that a random sample of size 225 has a sample mean of 29.0. Let $\alpha = 0.01$.

2. In Exercise 1 find $P(\text{Type II error} \mid \mu_x = 32)$.

3. Given the data in Exercise 3, page 238, test the hypothesis that the freshman class has a mean IQ of 120, against the alternate hypothesis that the mean IQ is greater than 120. Let $P(\text{Type I error}) = 0.01$.

4. A sample of 25 tropical fish of a certain species is selected from a large population. The investigator wants to find evidence to support his theory that this particular species has mean length greater than 5.00 centimeters. He plans to test at the 0.05 level of significance. If the sample has mean length of 5.25 centimeters, what is the investigator's decision regarding the mean length for the species if
    (a) the population standard deviation is 0.75 centimeters?
    (b) the population standard deviation is unknown and the sample standard deviation is 0.75 centimeters?

5. Given the data in Exercise 10, page 239, test the hypothesis that the true mean nicotine content for that brand of cigarette is 14.5 milligrams. Use a two-sided test with $\alpha = 0.02$.

6. According to the norms published for a certain intelligence test, college freshmen are expected to average 65 points with a standard deviation of 10 points. A sample of 80 freshmen at State U. average 68.2 points on the test. Test at $\alpha = 0.01$ the hypothesis that State U. freshmen are more intelligent than average college freshmen.

7. The same aptitude test is given to two different groups of people with the following results:

$$\bar{x}_1 = 82.6 \qquad s_1 = 5.6 \qquad n_1 = 50$$
$$\bar{x}_2 = 80.2 \qquad s_2 = 5.2 \qquad n_2 = 65$$

Test at $\alpha = 0.05$ the hypothesis that the means are not significantly different.

8. The makers of Brand A toothpaste claim that children who use Brand A have fewer cavities than children who use any other brand. Test this claim at the $\alpha = 0.01$ level of significance, given that 10 children use Brand A for six months and 10 more use Brand B for the same period, with the following results (in terms of number of cavities).

$$\text{Brand } A: \quad \bar{x}_1 = 2.32 \qquad s_1^2 = 0.25$$
$$\text{Brand } B: \quad \bar{x}_2 = 2.86 \qquad s_2^2 = 0.36$$

9. In the previous exercise what would be the conclusion if the same data had been collected from samples twice the size?

10. One typist makes an average of 2.8 errors per page on a 20-page assignment. Another typist makes an average of 2.5 errors per page on a 12-page assignment. Test to see whether or not there is a significant difference between the two performances. Use $\alpha = 0.05$ and assume that $\sigma_1^2 = \sigma_2^2 = 0.50$.

11. A sample of 20 subjects is asked to estimate the height of a tree with one eye closed. The mean estimate is 15.6 feet with a standard deviation of 4.2 feet. A second sample of 25 subjects is asked to perform the same task, but with both eyes open. Their mean estimate is 12.2 feet with a standard deviation of 3.8 feet. Test at $\alpha = 0.01$ the hypothesis that the true mean of estimated heights is the same whether people have one or both eyes open.

12. A random sample of 28 fifth graders take a standardized achievement test immediately after an hour of recess. A second random sample of 35 fifth graders take the same test after an hour of rest period. The recess group scored, $\bar{x}_1 = 56.5$, $s_1 = 6.0$; and the rest group scored, $\bar{x}_2 = 62.2$, $s_2 = 12.6$. Test at $\alpha = 0.01$ the hypothesis that the rest group did significantly better than the recess group.

13. What major assumption must be made regarding σ_1^2 and σ_2^2 in Exercises 11 and 12?

14. In Section 11.3.2 we considered a test for the case when σ_1^2 and σ_2^2 are known and a test for the case when σ_1^2 and σ_2^2 are unknown but equal. A third case occurs when σ_1^2 and σ_2^2 are unknown and we are not willing to assume that they are equal. In this case, if the samples are large enough (each at least 25), then

$$\frac{(\bar{x}_1 - \bar{x}_2) - (\mu_1 - \mu_2)}{\sqrt{\dfrac{s_1^2}{n_1} + \dfrac{s_2^2}{n_2}}} \longrightarrow z$$

Do Exercise 12 again, assuming that σ_1^2 and σ_2^2 are not equal.

15. Another assumption underlying the test of significance for two means is that the samples involved are *independent*. Therefore, if a sample of subjects were examined *before* and *after* receiving instruction, we cannot use the conventional t-test to compare the means for the two examinations. However, to make the comparison we can simply find the difference (d)

between the two examination scores for each subject and then test the null hypothesis that the mean difference is 0 ($\mu_d = 0$), against the alternate hypothesis that $\mu_d > 0$, using the t-test described in Section 11.3.1. Use this theory to test (at $\alpha = 0.05$) the null hypothesis that instruction did not significantly increase the scores on the retest, given the following data.

Subject:	1	2	3	4	5	6	7	8
Score Before Instr.:	50	60	55	72	85	78	65	90
Score After Instr.:	56	70	60	70	82	84	68	88

16. A sample of 10 non-German-speaking subjects are asked to copy a passage written in German. The number of errors made is recorded. The group is given a six-week crash course in German. Finally, the 10 are asked to copy over the same German passage, and the number of errors made is recorded. Given the following data, test the hypothesis that the mean number of errors made before and after the six-week training session are not significantly different. (Use $\alpha = 0.01$.)

Subject:	1	2	3	4	5	6	7	8	9	10
# Errors Before:	10	6	8	7	7	12	4	0	7	6
# Errors After:	6	4	5	3	5	8	0	2	5	4

17. Three brands of gasoline are tested to see if they differ at all with respect to gas mileage. In four test runs on each brand, the number of miles driven on one gallon of gas is recorded.

  Brand 1: 15, 20, 21, 16
  Brand 2: 22, 18, 19, 21
  Brand 3: 18, 24, 22, 20

Test the hypothesis that the three brands have equal effect on gas mileage. (Let $\alpha = 0.05$.)

18. Four different methods of instruction are used to cover a unit in high school biology — each method used on a different random sample of five students. All 20 students take a common examination with the following results (scores).

  Method 1: 50, 65, 40, 50, 55
  Method 2: 60, 60, 50, 45, 65
  Method 3: 75, 85, 70, 80, 70
  Method 4: 65, 80, 60, 65, 60

Test at $\alpha = 0.01$ the hypothesis that no method shows any superiority over the others.

19. A random sample of five football players and five soccer players are randomly selected. All 10 run 100 yards and are timed (in seconds). Results are

  Football: 11.5, 11.1, 10.8, 12.4, 12.2
  Soccer: 11.6, 10.5, 10.3, 11.2, 11.4

Test (at $\alpha = 0.05$) the hypothesis that football and soccer players run equally fast, using
(a) the methods of Section 11.3.2.
(b) the methods of Section 11.3.3. (Note that the calculated F value equals the square of the calculated t value in (a).)

11.4   Tests Concerning Variances

In this section we shall consider significance tests which apply to hypotheses concerning a single population variance, the difference between two population variances, and finally, differences among k population variances. These tests are used whenever we wish to check on the consistency or uniformity of a process or product. In addition, the tests for $k = 2$ and $k > 2$ are used in conjunction with the t-test described in Section 11.3.2 and the analysis of variance test described in Section 11.3.3 for the purpose of justifying the equal-variance assumption.

11.4.1   One Variance

A manufacturer of flashlight batteries claims that the variance of battery lives is 16 hours. ("Battery life" means the number of hours of use before the battery dies.) Let us assume that we want to test this claim against the alternate hypothesis that $\sigma_x^2 \neq 16$. We select a random sample of 15 batteries and find that the sample variance of the battery lives is 18.5. We must answer the question, "Is $s_x^2 = 18.5$ *likely* to occur in a sample of size 15 when in fact $\sigma_x^2 = 16$, *or*, is this result very *unlikely* to occur under the given conditions?" If this result is an unlikely one, assuming $\sigma_x^2 = 16$, then we must conclude that $\sigma_x^2 \neq 16$. To answer this question we need to know something about how s_x^2 varies from sample to sample — in other words, we need to know the theoretical sampling distribution of s_x^2, or a transformation of s_x^2. Earlier in Chapter 9, page 226, we stated in SD3 that the quantity

$$\frac{(n-1)s_x^2}{\sigma_x^2}$$

has a chi-square distribution with $n - 1$ degrees of freedom (assuming that the population is normally distributed). This quantity obviously varies from sample to sample depending upon the value of s_x^2, since n and σ_x^2 are constants. Therefore, this quantity gives us an excellent means for comparing the observed s_x^2 to the hypothesized σ_x^2. If there is a great difference, then the computed χ^2 value will be either so large or so small that it will fall into one of the rejection regions. If they are close in value, we cannot reject the null hypothesis. In our example, the hypotheses are

Null Hypothesis: $\sigma_x^2 = 16$
Alternate Hypothesis: $\sigma_x^2 \neq 16$

(It should be noted that at the same time we are testing the null hypothesis that $\sigma_x = 4$, against the alternate hypothesis that $\sigma_x \neq 4$.) If we decide to test at $\alpha = 0.05$, then the values of χ^2 which separate the acceptance region from the two rejection regions are $\chi^2_{14;0.025}$ and $\chi^2_{14;0.975}$, which are

5.63 and 26.12, respectively. Therefore, the computed χ^2 value in our example is

$$\chi^2 = \frac{(16-1)(18.5)}{16} = 17.34$$

Since 17.34 falls in the acceptance region (see Figure 11.9), we cannot reject the claim that the true variance is 16. One-sided tests are also possible, in which case there is only one rejection region with area equal to α. Remember, however, that the decision to have a one-sided or two-sided alternative hypothesis must be made *before* the sample data is collected.

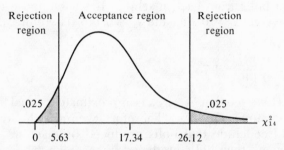

Figure 11.9.   Failure to reject the null hypothesis ($\sigma_x^2 = 16$) in the battery example.

11.4.2 Two Variances

Two machines produce bolts which have a mean length of five inches. However, it is suspected that the lengths of bolts produced by machine #1 vary more than the lengths of bolts produced by machine #2. That is, given that σ_1^2 and σ_2^2 are the actual variances for the two populations of bolt lengths, it is suspected that $\sigma_1^2 > \sigma_2^2$. To test this suspicion we set up the following hypotheses.[*]

Null Hypothesis: $\sigma_1^2 = \sigma_2^2$
Alternate Hypothesis: $\sigma_1^2 > \sigma_2^2$

Intuitively, it is reasonable that we should base our eventual decision on the relative sizes of sample variances calculated from a sample of bolts produced by machine #1 and a second sample produced by machine #2. Thus, we need to develop some additional theory regarding the theoretical sampling distribution involving two sample variances. Let s_1^2 and s_2^2 represent variances calculated from two independent samples. In order to compare the two variances we could consider the difference $s_1^2 - s_2^2$, as we do in comparing sample means. An alternate procedure would be to consider the ratio s_1^2/s_2^2. If s_1^2 and s_2^2 are equal, then the ratio equals one. If they are not equal, the ratio is either less than one or greater than one. If the ratio is nowhere near one, it is likely

[*]Equivalent to NH: $\sigma_1 = \sigma_2$, against Alt. H: $\sigma_1 > \sigma_2$.

that the population variances are not equal. However, to use this ratio as our test statistic, we need to know its theoretical sampling distribution *under the assumption that* $\sigma_1^2 = \sigma_2^2$. It can be proved that if two *independent* random samples of size n_1 and n_2 are selected from two *normally distributed* populations which have *equal variances*, then the statistic s_1^2/s_2^2 has an F probability distribution with $n_1 - 1$ and $n_2 - 1$ degrees of freedom. In other words,

$$\frac{s_1^2}{s_2^2} \longrightarrow F_{n_1 - 1, n_2 - 1} \qquad\qquad \text{SD12}$$

In our example, therefore, let us set α at 0.05 and then select a sample of 60 bolts from each machine. Let us suppose that from the sample data we get $s_1^2 = 0.252$, and $s_2^2 = 0.146$. Therefore, the computed F value is

$$F = \frac{s_1^2}{s_2^2} = \frac{0.252}{0.146} = 1.73$$

From Table H we see that $F_{59, 59; 0.95}$ is approximately equal to 1.53 (which is the value given for $F_{60, 60; 0.95}$). Therefore, we reject the hypothesis that $\sigma_1^2 = \sigma_2^2$ and conclude that bolts produced by machine #1 are more variable in length than bolts produced by machine #2.

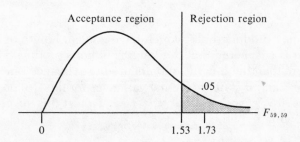

Figure 11.10. Rejection of the null hypothesis $(\sigma_1^2 = \sigma_2^2)$ in the bolt example.

This F-test need not always be one-sided; this decision is arbitrary. One particular application of this test, however, *always* has a two-sided alternative—namely, when we are testing two sample variances for the purpose of justifying the equal-variances assumption in the t-test described in Section 11.3.2. For example, in that section, we tested to see if the difference between $\bar{x}_1 = 15.6$ and $\bar{x}_2 = 12.8$ is significant (hormone example, page 271). To do this we assumed that $\sigma_1^2 = \sigma_2^2$. Is that assumption legitimate in this case? The sample variances are $s_1^2 = 6.5$ and $s_2^2 = 6.3$, for samples of size 15 each. To test the null hypothesis that $\sigma_1^2 = \sigma_2^2$, against $\sigma_1^2 \neq \sigma_2^2$, at the 0.05 level of significance, we calculate

$$F = \frac{s_1^2}{s_1^2} = \frac{6.5}{6.3} = 1.03$$

which we must compare to $F_{14,14;0.025}$ and $F_{14,14;0.975}$; these values are 0.34 and 2.98, respectively. Since the calculated F value lies between 0.34 and 2.98, it falls in the acceptance region. Therefore, we accept the null hypothesis: $\sigma_1^2 = \sigma_2^2$. This is one of the few occasions when we hope to accept a null hypothesis. To help protect against a Type II error, therefore, it is wise to set $\alpha = 0.05$ *or larger*.

11.4.3 k Variances

The significance test for k variances is a generalization of the test for two variances described in the previous section. It is especially useful as a preliminary test to the analysis of variance procedure. This test, called *Bartlett's test for homogeneity of variance* involves considerable calculation, and so shall be omitted from this text. Examples of this procedure can be found in Cornell (1956, page 292) and Edwards (1960, page 125).

Exercises

1. A random sample of 25 measurements selected from a normal population has a sample variance equal to 12.85. Test at $\alpha = 0.05$ the hypothesis that the population variance is 8.00, against the alternate hypothesis that it is greater than 8.00, and answer the following questions:
   (a) What value of χ^2 determines the boundary between the acceptance and rejection regions?
   (b) What is the calculated χ^2 value?
   (c) What is your decision?

2. Test the hypothesis that $\sigma_x = 5.6$, given that $s_x = 6.2$ for a sample of size 100. Use a two-sided test at the 0.05 level of significance.

3. John and Bill bowl 30 games apiece, both averaging 150 pins per game. However, John's scores have a standard deviation of 14.3 pins, while Bill's score standard deviation is 8.7 pins. Test at $\alpha = 0.01$ the hypothesis that Bill is a more consistent bowler than John.

4. A factory makes a certain computer part for NASA which, according to specifications, must have a mean length of 1.5 centimeters with a variance no greater than 0.0050 square centimeters. In a random sample of 25 parts it is discovered that $\bar{x} = 1.472$ centimeters and $s_x^2 = 0.0081$ square centimeters.
   (a) Test the hypothesis that $\mu_x = 1.5$ ($\alpha = 0.05$, two-sided).
   (b) Test the hypothesis that $\sigma_x^2 = 0.0050$ ($\alpha = 0.05$, two-sided).
   (c) What improvement would you make in this quality control procedure?

5. An investigator wishes to test the null hypothesis that $\mu_1 = \mu_2$, against the alternate hypothesis that $\mu_1 \neq \mu_2$. He collects a random sample of size 25 from each population and finds that $\bar{x}_1 = 50.6$, $\bar{x}_2 = 54.2$, $s_1^2 = 42.72$, and $s_2^2 = 20.24$.
   (a) Test to see if it is reasonable to assume that $\sigma_1^2 = \sigma_2^2$, at $\alpha = 0.05$.
   (b) Depending upon the results in (a), use the appropriate statistic to test the significance between the means, at $\alpha = 0.05$.

6. In Exercises 11 and 12, page 276, test the assumption that $\sigma_1^2 = \sigma_2^2$. Let $\alpha = 0.05$.

11.5 Tests Concerning Correlations

In Chapter 5 we defined r_{xy}, the sample linear correlation coefficient between variables x and y. This statistic measures the extent to which x and y are related in a linear fashion. However, even though $|r_{xy}|$ may be greater than zero, it may be that the true linear correlation between x and y, denoted by ρ_{xy}, is actually zero! It is quite likely that two variables which, in fact, are not correlated at all (i.e., $\rho_{xy} = 0$) may have a sample correlation coefficient of, say, $r_{xy} = 0.18$ for a particular set of sample data—especially if the sample is small. The larger the sample, the more confidence we can have that the value we get for r_{xy} is close to the true value of ρ_{xy}.

11.5.1 One Correlation

We correlate two variables to determine, first of all, if there is any linear relationship between them. A logical test of significance can be set up with the following hypotheses (dropping the subscripts for convenience).

Null Hypothesis: $\rho = 0$
Alternate Hypothesis: $\rho \neq 0$

We learned in Chapter 9 (SD5, page 227) that when $\rho = 0$ the theoretical sampling distribution of the statistic

$$\frac{r\sqrt{n-2}}{\sqrt{1-r^2}}$$

is a Student-t probability function with $n-2$ degrees of freedom. Therefore, suppose that in a sample of 18 third graders we correlate their weights against their IQ scores and get $r = 0.30$. To test the null hypothesis that $\rho = 0$ at the 0.05 level of significance, we compute

$$t = \frac{0.30\sqrt{18-2}}{\sqrt{1-(0.30)^2}} = 1.26$$

With a two-sided alternate hypothesis, the values of t_{16} which separate the acceptance and rejection regions are $t_{16;0.025} = -2.12$ and $t_{16;0.975} = 2.12$ (see Figure 11.11). Therefore, we cannot reject the null hypothesis in this instance. In other words, we do not have sufficient evidence to support the hypothesis that the two variables are linearly correlated. It is interesting to note that if we replace $n = 18$ by $n = 50$ in this example, we would reject the null hypothesis. Why? This fact again points up the importance of large samples with respect to statistical reliability.

The theoretical sampling distribution of r is skewed when $\rho \neq 0$. Therefore, if we wish to test, for example, the null hypothesis that $\rho = 0.75$, we must use Fisher's transformation (first introduced on page 227), which converts r to w_r. The variable w_r is normally distributed

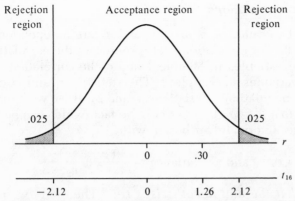

Figure 11.11.   Failure to reject the null hypothesis ($\rho = 0$) in the weight vs. IQ example.

with a mean of w_ρ and variance $1/(n-3)$. Therefore, according to SD6 (page 227),

$$(w_r - w_\rho)\sqrt{n-3} \longrightarrow z$$

Suppose that in a sample of 67 athletes the correlation between height and weight is $r = 0.84$, and suppose we want to test the following hypotheses.

Null Hypothesis: $\rho = 0.75$
Alternate Hypothesis: $\rho > 0.75$

First, we convert r to w_r and ρ to w_ρ, using Table J in the Appendix. For $r = 0.84$, we get $w_r = 1.221$, and for $\rho = 0.75$, we get $w_\rho = 0.973$. Therefore, we calculate

$$z = (1.221 - 0.973)\sqrt{67 - 3} = 1.98$$

With $\alpha = 0.05$, we reject the null hypothesis, since 1.98 is larger than 1.645 (see Figure 11.12), thus concluding that the true correlation between height and weight in the population is larger than 0.75.

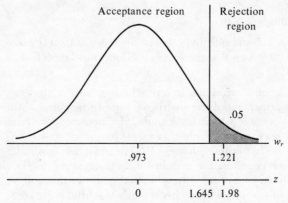

Figure 11.12.   Rejection of the null hypothesis ($\rho = .75$) in the height vs. weight example.

11.5.2 Two Correlations

Given that two samples, of sizes n_1 and n_2, are independently selected from two different populations, let r_1 represent the correlation between two specific variables in Sample 1 and r_2 the correlation between the same two variables in Sample 2. The statistics r_1 and r_2 estimate corresponding population correlations ρ_1 and ρ_2. If we wish to test the null hypothesis that $\rho_1 = \rho_2$, then we use the fact (which can be proved) that $w_{r_1} - w_{r_2}$ has a normal distribution with

$$\text{mean} = 0 \quad \text{and} \quad \text{variance} = \frac{1}{n_1 - 3} + \frac{1}{n_2 - 3}$$

assuming that the null hypothesis is true. Therefore, we can state formally that if $\rho_1 = \rho_2$, then

$$\frac{w_{r_1} - w_{r_2}}{\sqrt{\dfrac{1}{n_1 - 3} + \dfrac{1}{n_2 - 3}}} \longrightarrow z \qquad\qquad \text{SD13}$$

Suppose, for example, that we want to show that the correlation between aggressiveness and anxiety is greater among men than among women. In a sample of $n_1 = 30$ men $r_1 = 0.54$, and in a sample of $n_2 = 40$ women $r_2 = 0.20$. From Table J, $w_{r_1} = 0.604$ and $w_{r_2} = 0.203$. Therefore,

$$z = \frac{0.604 - 0.203}{\sqrt{\dfrac{1}{27} + \dfrac{1}{37}}} = \frac{0.401}{0.253} = 1.58$$

If we test at $\alpha = 0.05$, then we cannot reject the null hypothesis, since 1.58 is less than $z_{0.95} = 1.645$.

Exercises

1. Test at $\alpha = 0.05$ the null hypothesis: $\rho = 0$, against the alternate hypothesis: $\rho \neq 0$, if
   (a) $n = 30$, $r = 0.25$
   (b) $n = 10$, $r = -0.70$
   (c) $n = 200$, $r = 0.15$

2. Test at $\alpha = 0.05$ the null hypothesis: $\rho = 0.60$, against the alternate hypothesis: $\rho \neq 0.60$, given that the corresponding correlation from a sample of size 63 is $r = 0.48$.

3. In a sample of 85 students in a psychology course, students who, after the course was over, rated the instructor high in teaching ability tended to be those who received high final grades. In fact, the correlation between the two variables for the sample was $r = 0.65$.
   (a) Test the hypothesis that $\rho = 0.50$ (against $\rho > 0.50$) at $\alpha = 0.05$.
   (b) Can we consider the conclusion in (a) as fairly good evidence (within the limits of α) that getting a good grade in this course influences the student's opinion of his instructor's teaching ability?

4. A sociologist claims that success of students in college and the annual income of their parents are uncorrelated. Test this claim at $\alpha = 0.01$, given that in a random sample of 100 college students, grade index and parent's income correlate at $r = 0.38$.

5. In Exercise 4 test the null hypothesis that $\rho = 0.20$ against the alternate hypothesis that $\rho > 0.20$ at the same level of significance.

6. Suppose that it is claimed that the correlation between the number of hours spent studying for a certain multiple-choice examination and the number of questions incorrectly answered on the exam is -0.60. Test this claim (at $\alpha = 0.05$) against the hypothesis that the relationship between the two variables is higher than claimed, given that for a sample of 48 examinees $r = -0.74$. (When converting r to w_r enter the Table J with the corresponding positive correlation.)

7. A sample of 50 neurotics and 50 "normal" individuals are examined on several psychological variables. For the neurotic sample the correlation between feelings of inadequacy and number of errors made in attempting to solve a puzzle is $r_1 = 0.72$, and the corresponding correlation for the normal group is $r_2 = 0.34$. Do these results indicate a true difference between ρ_1 and ρ_2? (Test at $\alpha = 0.01$, two-sided.)

11.6 Tests Concerning "Independence" and "Goodness of Fit"

In this section we discuss two commonly used tests of significance involving hypotheses which, unlike those in the previous sections, are not expressed in terms of specific population parameters. Both of these tests involve essentially the same chi-square statistic introduced in Section 11.2.3, with differences only in the manner in which we compute the expected frequencies and determine the number of degrees of freedom.

11.6.1 Independence

Suppose that a college professor distributes to his students in a large lecture course (200 students) a questionnaire which solicits student opinion regarding the effectiveness of his teaching. One particular question reads:

How do you rate your teacher in his ability to explain difficult concepts?

_____   Above Average

_____   Average

_____   Below Average

The results on this question are summarized in Table 11.3 by college class (no freshmen are taking the course).

Table 11.3. Summary of the Responses (Observed Frequencies) to the Question on Teacher-Ability

		Class			
		Sophomore	Junior	Senior	Total
Rating	Above Average:	34	36	30	100
	Average:	12	24	24	60
	Below Average:	4	20	16	40
		50	80	70	200

Two-way tables, such as Table 11.3, are called *contingency tables*. They are usually constructed for the purpose of determining whether or not two variables of classification have a relationship or "dependence" upon each other. For example, the professor may wonder whether or not the ratings he gets on this question depend at all upon the student's class affiliation. Do seniors tend to rate him differently from sophomores, for example? Or is the *probability* that a student rates him in a particular category the same regardless of that student's class affiliation? If the latter case is true, then we say that the variables "rating" and "class" are *independent* in the probability sense. If, on the other hand, the probability that he is given a particular rating depends upon the rater's class affiliation, then we say that the two variables are *dependent*. To test this question of independence vs. dependence, we set up the following hypotheses.

Null Hypothesis: The two variables are *independent*
Alternate Hypothesis: The two variables are *dependent*

The next step involves the calculation of expected frequencies (e_i) based on the assumption that the variables are independent (i.e., the null hypothesis is true). To find a particular expected frequency, we apply a basic rule of probability (Theorem 6.7, page 125) which applies only to *independent events* E and F; namely,

$$P(E \cap F) = P(E) \cdot P(F)$$

For example, if the variables are independent, then the probability that a student is both a sophomore *and* he rates the professor "above average," equals the probability that the student is a sophomore multiplied by the probability that he rates the professor "above average." Using the total of the first column (50) and the sample size (200), we can estimate the probability that a student in this course is a sophomore (50/200). Similarly, using the total of the first row (100), we can estimate the probability that a student in this course rates the professor "above average" (100/200). Therefore, our estimate of the probability that both events occur is $(50/200)(100/200)$. Thus, in a sample of 200 students, we would *expect* to have

$$(200)\left(\frac{50}{200}\right)\left(\frac{100}{200}\right) = 25$$

sophomores who rate the professor above average. The other eight expected frequencies can be calculated in the same way.

This procedure may seem somewhat tedious at first glance; however, there are two significant short cuts. First, note in the above calculation that we can immediately cancel a factor of 200 in both the numerator and denominator—leaving $(50)(100) \div (200)$, which is the *product of the column and row totals for the cell in question divided by the total sample size*. This formula works for each cell in the table. Secondly, we need to apply the formula only four times in this three-by-three contingency table. The reason for this is that the expected frequencies must add up to the same row and column totals as the observed frequencies. Therefore, the last expected frequency in each row (or column) can be found by subtracting the other expected frequencies in that row (or column) from the row (or column) total. In general, in an r-by-k contingency table we need to apply the formula only $(r-1)(k-1)$ times. In our example, therefore, we need to apply it only four times.

After we have determined the expected frequencies (see Table 11.4), we proceed as we did in Section 11.2.3 to calculate

$$\sum_{i=1}^{9} \frac{(o_i - e_i)^2}{e_i}$$

Table 11.4.   Expected Frequencies (in Parentheses) Corresponding to the Observed Frequencies From Table 11.3

		Class			
		Sophomore	Junior	Senior	Total
Rating	Above Average:	34 (25)	36 (40)	30 (35)	100
	Average:	12 (15)	24 (24)	24 (21)	60
	Below Average:	4 (10)	20 (16)	16 (14)	40
	Total:	50	80	70	200

If the two variables are in fact independent, this statistic has a chi-square theoretical sampling distribution with four degrees of freedom. In general, for an r-by-k contingency table the number of degrees of freedom is $(r-1)(k-1)$, which makes sense with reference to the discussion in the previous paragraph and the comments regarding degrees of freedom in Section 9.7. Formally, therefore, for an r-by-k contingency table, with expected frequencies calculated as described above, *if the two variables are in fact independent*, then

$$\sum_{i=1}^{rk} \frac{(o_i - e_i)^2}{e_i} \longrightarrow \chi^2_{(r-1)(k-1)} \qquad \text{SD14}$$

For our example, the calculated χ^2 value is

$$\chi^2 = \frac{(34-25)^2}{25} + \frac{(36-40)^2}{40} + \ldots + \frac{(16-14)^2}{14} = 10.27$$

If we test at $\alpha = 0.05$, then $\chi^2_{4;0.95}$ separates the acceptance and rejection regions (this test is always one-sided). Since $\chi^2_{4;0.95} = 9.49$, we must reject the null hypothesis — see Figure 11.13. That is, we conclude that there exists a dependence between the two variables.

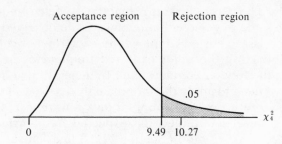

Figure 11.13. Rejection of the null hypothesis (independence) in the questionnaire example.

11.6.2 Goodness of Fit

Another type of significance test, which uses essentially the same chi-square statistic, is used to determine whether or not a random variable has a particular probability distribution, based on an observed frequency distribution. For example, suppose that we wish to test a particular die to determine whether or not it is balanced. If the die is balanced, then when the die is tossed, each of the faces, labelled 1, 2, 3, 4, 5, and 6, has equal probability of occurrence. That is, if random variable x is the number showing when a *balanced* die is tossed, then the theoretical probability distribution for x is

x:	1	2	3	4	5	6
$f(x)$:	1/6	1/6	1/6	1/6	1/6	1/6

Therefore, if the die were balanced and we were to toss it 300 times, ideally we would *expect* a frequency of 50 for each of the values 1 through 6. Suppose that we actually toss the die 300 times and get the following *observed* frequencies:

x_i:	1	2	3	4	5	6
f_i:	45	42	50	64	56	43

We note that there are several departures from the "expected" frequencies of 50 for each x value, and we do expect some chance fluctuations even if the die is balanced. The question is, "Are these observed fluctuations due only to chance (i.e., die is balanced), or are they possibly due *also* to the fact that the die is not balanced?" We can test the null hypothesis that the die is balanced against the alternate hypothesis that it is unbalanced by using the chi-square statistic to compare the observed frequencies to the frequencies expected under the assumption that the die is balanced. In this type of test, if the null hypothesis is true, the chi-square statistic has $k - r$ degrees of freedom, where k is the number of observed frequencies and r the number of quantities (restrictions) obtained from the observed data used in calculating the expected frequencies. Formally, therefore, we say that

$$\sum_{i=1}^{k} \frac{(o_i - e_i)^2}{e_i} \longrightarrow \chi^2_{k-r} \qquad\qquad \text{SD15}$$

In the die example, we use the fact that the total frequency is 300 to calculate the expected frequencies. Therefore, in our example we have five degrees of freedom ($k = 6$ and $r = 1$). If we test at the 0.05 level of significance, the value of χ^2 which separates the acceptance and rejection regions is $\chi^2_{5;0.95} = 11.07$. Since our computed χ^2 value is only 7.40 (see Table 11.5), we cannot reject the null hypothesis. That is, we either accept the hypothesis that the die is balanced, or we reserve judgment.

Table 11.5. Calculation of the χ^2 Value in the Die Example

x_i	o_i	e_i	$(o_i - e_i)^2$	$(o_i - e_i)^2/e_i$
1	45	50	25	.50
2	42	50	64	1.28
3	50	50	0	.00
4	64	50	196	3.92
5	56	50	36	.72
6	43	50	49	.98
	300	300		$7.40 = \chi^2$

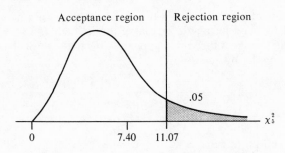

Figure 11.14. Failure to reject the null hypothesis (good fit) in the die example.

This type of chi-square significance test, called a *goodness-of-fit* test, has wide application in a variety of fields. It is especially useful for testing the hypothesis that a population is normally distributed, which is an assumption that underlies most of the classical significance tests. Unfortunately, testing the normality hypothesis with the goodness-of-fit procedure is somewhat laborious, and is therefore omitted here. Examples of such a test can be found in Remington and Schore (1970, page 234) and McNemar (1962, page 232).

The reader should keep in mind that a major assumption which underlies both the "independence" and "goodness-of-fit" chi-square tests is that all of the observations are independently collected. In other words, data which has been generated by measuring the same individuals more than once, such as we find in "before and after" studies, cannot be analyzed with these tests. A second assumption underlying both tests is that the distribution of the difference $o_i - e_i$ is normal. Violations of this assumption can occur when the expected frequencies are very small. Therefore, as a general rule it is wise to avoid using these tests whenever an expected frequency is less than five.

Exercises

1. In the text example (page 285) another question reads, "How do you rate your instructor in his command of the subject? _____ Above average, _____ Average, or _____ Below average." The following contingency table summarizes the joint responses to the two questions.

| | | Rating #2 (Command of Subject) | | |
		Above Average	Average	Below Average
Rating #1 (Explain Concepts)	Above Average	48	30	22
	Average	26	20	14
	Below Average	16	10	14

Test the hypothesis ($\alpha = 0.05$) that the two variables are independent.

2. The members of a community vote on a bond issue for a new high school. Each voter is asked to decide on *one* of the following: Site #1, Site #2, or No Site. The results of the voting appear below in the form of a three-by-four contingency table, by age group.

| | | Age Group | | | |
		18–25	26–35	36–45	Over 45
Decision	Site #1:	120	170	160	50
	Site #2:	60	80	105	55
	No Site:	20	50	85	45

(a)   How many degrees of freedom are there in this contingency table?
(b)   If we test at $\alpha = 0.01$ the hypothesis that voter age and voter decision are not independent variables, what value of χ^2 separates the acceptance and rejection regions?
(c)   What is the calculated χ^2 value?
(d)   What is your decision?

3.   What is the difference between the kind of relationship tested in the "independence" chi-square test and the significance test for the hypothesis $\rho = 0$, against the hypothesis $\rho \neq 0$?

4.   A special case of the chi-square test for an r-by-k contingency table is the test for a two-by-k table—which in fact is the chi-square test described in Section 11.2.3 for k proportions. Verify the above statement by applying the contingency table test to the data in Exercise 11, page 266.

5.   For a two-by-two contingency table it is considered good practice to "correct" the chi-square statistic with a slight adjustment which is comparable to the correction for continuity used in approximating a binomial distribution with a normal distribution. The corrected χ^2 is defined as follows:

$$\chi^2(\text{corrected}) = \sum_{i=1}^{4} \frac{(|o_i - e_i| - 0.5)^2}{e_i}$$

Apply this adjustment, called *Yates's correction*, to the calculation of χ^2 for the data in Exercise 7, page 265, and compare the result of the χ^2 computed in Exercise 13, page 266.

6.   A coin is flipped four times and the total number of heads (x) is recorded. This procedure is repeated 63 more times, with the following results;

x_i:	0	1	2	3	4
f_i:	6	13	23	21	2

Does this data show any evidence that the coin is unbalanced? (Test at $\alpha = 0.05$.)

7.   Select 500 digits out of Table I in the Appendix and test (at $\alpha = 0.01$) to see if the numbers in the table are in fact random (i.e., compare observed frequencies for 0, 1, 2, . . . , 9 to the frequencies you would expect under the assumption of perfect randomization).

8.   Test at $\alpha = 0.05$ the null hypothesis that discrete random variable x has the following probability distribution:

x:	1	2	3	4	5	6
$f(x)$:	0.05	0.26	0.32	0.18	0.12	0.07

A sample of 200 values of x yields the following frequency distribution:

x_i:	1	2	3	4	5	6
f_i:	4	44	58	42	34	18

9.   According to genetic theory, when *tall, yellow* pea plants are crossed with *short, green* pea plants, they produce plants of four basic types: (i) tall and green, (ii) tall and yellow, (iii) short and green, and (iv) short and yellow, in

a 9:3:3:1 proportion. If in a sample of 320 such crosses the observed frequencies for each of these types are

Tall, Green	Tall, Yellow	Short, Green	Short, Yellow
172	55	72	16

then is there sufficient reason to doubt the theory? (Test at $\alpha = 0.05$.)

Summary

Classical significance tests involve assumptions regarding (a) the shape of the population (e.g., the population is normally distributed), (b) certain population parameters (e.g., the population variances are equal), or (c) the manner in which the data is collected (e.g., all observations are independent). The more popular of these tests of significance test hypotheses concerning one or more population parameters including *proportions, means, variances,* and *correlations.* In addition, we have procedures designed to test *independence* and *goodness of fit.*

12

Distribution-Free Significance Tests

12.1 Introduction

The terms "distribution-free" and "nonparametric" (mentioned earlier, page 258) are not really synonymous. However, they are used interchangeably to describe a particular group of significance tests which, under certain conditions, are used in lieu of classical tests. The major advantages of distribution-free tests (as we shall call them) over classical tests are (a) they involve less stringent assumptions, (b) they are usually easier to understand, and (c) the computations are often less laborious. However, distribution-free tests suffer from some disadvantages, including (a) they waste information, and (b) they tend to be too conservative; that is, they tend to lead to acceptance of the null hypothesis more often than they should. Thus, classical tests are preferable *when the underlying assumptions can be met*. But if the situation is such that it is not reasonable to assume, for example, that the population is normally distributed, or that the population variances are equal, then possibly an appropriate distribution-free test can be used as a substitute. Distribution-free tests are especially useful when samples are small, since many of the classical tests assume that samples are reasonably large.

This chapter offers only a brief introduction to a few distribution-free tests. A complete report on such tests could easily fill one or two volumes. As a matter of fact there are at least two excellent texts devoted entirely to this subject, Siegal (1956) and Bradley (1968), as well as many other books which contribute a chapter or two to these special methods (see the Bibliography).

12.2   The Mann-Whitney Test

12.2.1   Testing $\mu_1 = \mu_2$

One of the most versatile of the distribution-free tests is the *Mann-Whitney test*, which can be used as an alternative to the *t*-test for two independent means (Section 11.3.2) when either the equal variances assumption or the normality assumption (or both) are not satisfied. To illustrate this test, let us consider data which has been generated on two independent samples of students who took the same college course from different professors, but took a common final examination. The following data represent final examination scores for the two classes.

Class 1:	63, 72, 58, 74, 62, 63, 69, 54, 59, 66, 75, 65	$\bar{x}_1 = 65.0$
Class 2:	60, 78, 62, 90, 76, 52, 67, 87, 73, 84	$\bar{x}_2 = 72.9$

First, we rank the data as if it were one sample of size 22 assigning the rank "1" to the smallest score, "2" to the second smallest score, and so on. Similar scores are assigned equal ranks according to the procedure described in Section 5.6.1. Replacing each score by its rank gives us

Class 1:	8.5, 14, 3, 16, 6.5, 8.5, 13, 2, 4, 11, 17, 10	$\Sigma = 113.5$
Class 2:	5, 19, 6.5, 22, 18, 1, 12, 21, 15, 20	$\Sigma = 139.5$

We shall let n_1 and n_2 represent the number in each sample, and let R_1 and R_2 represent the sums of the ranks for each sample. For our data, therefore, $n_1 = 12$, $n_2 = 10$, $R_1 = 113.5$, and $R_2 = 139.5$. The theory tells us that if the population means are equal, then the mean values of R_1 and R_2 are

$$\mu_{R_1} = \frac{n_1(n_1 + n_2 + 1)}{2} \quad \text{and} \quad \mu_{R_2} = \frac{n_2(n_1 + n_2 + 1)}{2}$$

For our example, then,

$$\mu_{R_1} = \frac{12(10 + 12 + 1)}{2} = 138 \quad \text{and} \quad \mu_{R_2} = \frac{10(10 + 12 + 1)}{2} = 115$$

Note that the absolute difference between R_1 and μ_{R_1} is equal to the absolute difference between R_2 and μ_{R_2}; i.e., both differ by 24.5. Thus, in practice we do not need to calculate both R_1 and R_2. To test the null hypothesis: $\mu_1 = \mu_2$, we must show whether the difference between R_1

and μ_{R_1} (or R_2 and μ_{R_2}) is small enough to justify accepting the null hypothesis or large enough to reject it. According to statistical theory, if both n_1 and n_2 are sufficiently large (each eight or more) and the null hypothesis is true, then the theoretical sampling distribution for R_i (where $i =$ either one or two) is approximately normal with

$$\text{mean} = \mu_{R_i} \quad \text{and} \quad \text{variance} = \frac{n_1 n_2 (n_1 + n_2 + 1)}{12}$$

In other words,

$$\frac{R_i - \mu_{R_i}}{\sqrt{\dfrac{n_1 n_2 (n_1 + n_2 + 1)}{12}}} \longrightarrow z \qquad (i = 1 \text{ or } 2) \qquad \text{SD16}$$

Therefore, we compute z for R_1 (or R_2)*, and assuming $\alpha = 0.05$, we must reject the null hypothesis if $z > 1.96$ or $z < -1.96$.

In our example, using R_1,

$$z = \frac{113.5 - 138}{\sqrt{\dfrac{(12)(10)(12 + 10 + 1)}{12}}} = \frac{-24.5}{15.2} = -1.61$$

Since the computed z value falls between -1.96 and 1.96, we cannot reject the null hypothesis (see Figure 12.1). That is, the observed difference between $\bar{x}_1$ and $\bar{x}_2$ (7.9) is not large enough to warrant the conclusion that $\mu_1 \neq \mu_2$.

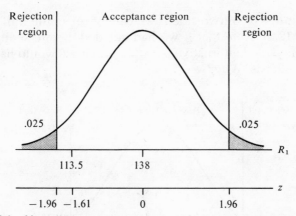

Figure 12.1. Mann-Whitney test; failure to reject the null hypothesis ($\mu_1 = \mu_2$).

The procedure just described has been somewhat oversimplified. Some statisticians suggest a correction for continuity and/or a correction for tied ranks (see Ferguson, 1966, page 359). In addition, if the sample sizes are too small, there is an alternate procedure involving the calculation of a statistic called U. The corresponding exact test, called the *Mann-Whitney U Test*, requires special tables (see Siegal, 1965, page 271).

*Whichever computation appears easier.

12.2.2   Testing $\sigma_1^2 = \sigma_2^2$

A slight variation of the Mann-Whitney test can be applied to two independent samples to test the null hypothesis that $\sigma_1^2 = \sigma_2^2$. The only change in procedure is in the assignment of ranks. The combined sample is ordered as usual. However, we assign the rank of "1" to the *smallest* score, ranks of "2" and "3" to the *largest* and second largest scores, ranks "4" and "5" to the second and third *smallest* scores, ranks "6" and "7" to the third and fourth *largest* scores, and so forth. If $n_1 + n_2$ is an odd number, the middle score is not ranked. In our example, $n_1 + n_2 = 22$; therefore, we assign ranks to all scores as shown below.

Class 1:   4, 5, 8, 12.5, 16.5, 16.5, 20, 21, 19, 18, 14, 11 $\Sigma = 165.5$

Class 2:   1, 9, 12.5, 22, 15, 10, 7, 6, 3, 2 $\Sigma = 87.5$

If the null hypothesis ($\sigma_1^2 = \sigma_2^2$) is true, R_1 and R_2 would tend to be equal. If the variances differ, then the sample from the population with the larger variance will tend to have a smaller rank sum. The two rank sums for our data are $R_1 = 165.5$ and $R_2 = 87.5$. Again, we need to compute only one of these sums to perform the test. Using R_1, for instance, we compute the z value,

$$z = \frac{165.5 - 138}{\sqrt{\dfrac{(12)(10)(12 + 10 + 1)}{12}}} = \frac{27.5}{15.2} = 1.81$$

Therefore, at $\alpha = 0.05$ (two-sided) we cannot reject the null hypothesis (see Figure 12.2). Note that if we had decided, before seeing the data, to test the one-sided alternative $\sigma_1^2 < \sigma_2^2$, then we would have rejected the null hypothesis. Why?

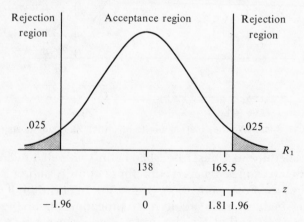

Figure 12.2.   Mann-Whitney test; failure to reject the null hypothesis ($\sigma_1^2 = \sigma_2^2$).

12.3   The Sign Test

The *sign test* can be used to test the null hypothesis that $\mu_1 = \mu_2$ when the samples are *not* independent; such as we experience in the so-called "before and after" studies — where the same sample is measured twice. The sign test offers an alternative to the test suggested in Exercise 15, page 276, for situations where we feel that we cannot assume the equal-variance and/or normality assumptions.

Suppose, for example, that 25 students take a pretest in American history, then attend a six-week course in American history, and finally take another test in the subject. The results might look something like the scores in Table 12.1. Note in the table that an increase in score is assigned a "+" sign, and a decrease a "−" sign. No sign is assigned when the two scores are identical.

Table 12.1.   Determination of Signs for the Pre- and Post-Test Scores in the American History Example

Student	Pre	Post	Sign	Student	Pre	Post	Sign
1	50	68	+	14	50	72	+
2	82	80	−	15	63	54	−
3	38	51	+	16	32	45	+
4	42	66	+	17	41	58	+
5	55	55		18	55	63	+
6	20	38	+	19	45	72	+
7	35	50	+	20	49	49	
8	46	40	−	21	41	38	−
9	70	65	−	22	58	80	+
10	59	56	−	23	68	79	+
11	42	61	+	24	53	62	+
12	40	52	+	25	47	50	+
13	49	49					

If we are hoping to show that there is a significant increase in score after the six-week course, we make the alternate hypothesis one-sided; i.e., $\mu_1 < \mu_2$. If $\mu_1 = \mu_2$, we would expect an equal number of +s and −s, subject only to chance fluctuation. That is, if the null hypothesis is true, we would expect the proportion of +s, denoted by p, to be 0.5. Therefore, we can rephrase our hypotheses as follows:

Null Hypothesis: $p = 0.5$
Alternate Hypothesis: $p > 0.5$

and then proceed according to the method described in Section 11.2.1 (for testing one proportion).

For our data, therefore, to find p we divide the number of +s by the number of +s and −s (i.e., we ignore the cases which involve no change).

Since 22 involve either a positive or negative change, $n = 22$. And since 16 of the 22 are positive changes, $\tilde{p} = 16/22 = 0.73$. Therefore, converting $\tilde{p}$ to z we get

$$z = \frac{0.73 - 0.50}{\sqrt{\dfrac{(0.50)(0.50)}{22}}} = \frac{0.23}{0.107} = 2.15$$

Thus, we can reject the null hypothesis at the 0.05 level of significance (see Figure 12.3).

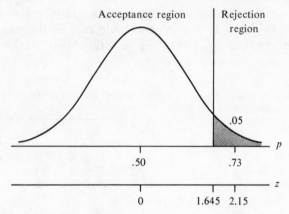

Figure 12.3.   Sign test; rejection the null hypothesis ($p = .5$).

Since it is so easy to use, the sign test is sometimes applied even when the usual classical assumptions are satisfied. If rejection of the null hypothesis is achieved, then there is no need to perform the more involved classical t-test. On the other hand, if rejection is not achieved with the sign test, it is still possible that rejection could occur with the classical test. The reader is reminded again that most distribution-free tests are more conservative than their classical counterparts.

12.4   The Kruskal-Wallis Test

The *Kruskal-Wallis test* is a distribution-free alternative to the classical one-way analysis of variance. That is, it can be used to test the significance among the means of k independent samples. As one might expect, the procedure is an extension of the Mann-Whitney test in that the complete set of data for k samples is ranked according to size. Then within each sample the ranks are summed. If we let R_i represent the sum of the ranks in Sample i, n_i represent the number of observations in Sample i, and $n = n_1 + n_2 + \cdots + n_k$, then the quantity

$$\frac{12}{n(n+1)} \sum_{i=1}^{k} \frac{R_i^2}{n_i} - 3(n+1)$$

is used as the test statistic. If the null hypothesis is true (i.e., if the k population means are equal), and if the samples are large enough (each at least five), then this statistic has a chi-square theoretical sampling distribution with $k - 1$ degrees of freedom. Formally,

$$\frac{12}{n(n+1)} \sum_{i=1}^{k} \frac{R_i^2}{n_i} - 3(n+1) \longrightarrow \chi_{k-1}^2 \qquad \text{SD17}$$

We do *not* have to assume that the populations are normally distributed or that the population variances are equal.

Take, for example, the mercury pollution data that we used to demonstrate the analysis of variance in Section 11.3.3.

Sample 1:   2.2, 3.4, 3.0, 2.6, 3.8
Sample 2:   4.0, 3.2, 3.7, 3.0, 3.6
Sample 3:   2.6, 2.2, 3.0, 2.4, 1.8

If we assign ranks to these 15 observations according to size, with "1" assigned to the smallest observation, the results are

Sample 1:   2.5, 11, 8, 5.5, 14
Sample 2:   15, 10, 13, 8, 12
Sample 3:   5.5, 2.5, 8, 4, 1

Summing the ranks for each sample we get: $R_1 = 41$, $R_2 = 58$, and $R_3 = 21$. Therefore, our computed χ^2 value is

$$\chi^2 = \frac{12}{15(15+1)} \left[\frac{(41)^2}{5} + \frac{(58)^2}{5} + \frac{(21)^2}{5} \right] - 3(15+1) = 6.86$$

If we use the same α that we used in Section 11.3.3 ($\alpha = 0.01$), then we must conclude that the three population means are not significantly different (see Figure 12.4 and compare it to Figure 11.8). Note that this conclusion is just the opposite of our conclusion in Section 11.3.3, which supports the fact that distribution-free tests are more conservative than classical tests. For the Kruskal-Wallis test, if any sample size is smaller than five, the test statistic is not quite chi-square-distributed. In such a case, specially prepared tables are available; for example, see Owen (1962, page 420).

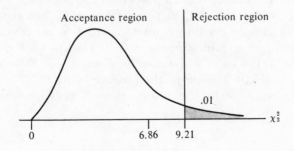

Figure 12.4.   Kruskal-Wallis test; failure to reject the null hypothesis ($\mu_1 = \mu_2 = \mu_3$).

12.5 The Kolmogorov-Smirnov One-Sample Test

An alternative to the classical chi-square, goodness-of-fit test is the *Kolmogorov-Smirnov one-sample test.* This test can be applied to smaller samples, and it generally involves less computation than the chi-square test. However, whereas the chi-square goodness-of-fit test can be applied to both continuous and discrete variables, the Kolmogorov-Smirnov (K-S) test can be applied only to continuous variables.

To begin, we hypothesize that a population has a specific distribution (i.e., that a certain variable has a specific probability distribution). That distribution is arbitrarily divided into k intervals of equal area (probability). A sample of size n is randomly selected from the population and the number of observations which fall in each of the k intervals is recorded. Then, the K-S test is used to compare *cumulative observed and expected relative frequencies* to test the null hypothesis that the observed data has been collected from the specific probability distribution in question. The test statistic is the *maximum absolute difference*, denoted by $D(\text{max})$, between any corresponding pair of observed and expected cumulative relative frequencies.

For example, suppose that a high school teacher gives a standardized test in English grammar to her class of 25 students, with the following results.

56	58	40	77	87
75	61	70	73	71
66	69	67	68	60
72	73	61	64	66
84	72	52	65	67

Suppose that the teacher wants to compare her class to national standards. The test manual reports decile scores for the national high school population, as follows:

Decile:	1	2	3	4	5	6	7	8	9
Score:	45.0	56.8	62.5	66.1	68.7	71.3	74.0	78.5	84.2

Thus, the teacher knows that 10 percent of the population scores below 45.0, 20 percent below 56.8, 30 percent below 62.5, and so on. To see how well the sample fits the population, the teacher computes the observed *cumulative relative frequency* (CRF) below each of the deciles. For example, in the sample she observes one score (40) which is below the first decile (45.0). Therefore, the observed CRF is $1/25 = 0.04$. Below the second decile (56.8) she counts three scores (including the one below 45.0). Thus, the observed CRF below the second decile is 0.12. The rest of the observed CRFs are found the same way. These values are recorded in Table 12.2 along with the expected CRFs which are, of course, 0.10, 0.20, 0.30, etc., since the deciles divide the population distribution into ten equal areas — each equal to 0.10. The *absolute value of the difference* between each pair of observed and expected CRFs, denoted by D, is computed, and also appears in Table 12.2. According to statistical theory, if the sample data have come from a population with the same distribution represented by the expected CRFs, then the *largest D*, called $D(max)$, has a known sampling distribution (see Table K in the appendix, page 333). $D(max)$ for our example is 0.14. Obviously, if the sample data "fit" well, $D(max)$ will be fairly small, and if the fit is poor, $D(max)$ will be significantly large. Table K contains values of $D(max)$ for various values of n (the sample size) and p (the probability that $D(max)_n$ is less than the tabled value). Since $n = 25$ in our example, and assuming that $\alpha = 0.05$, the table shows that $D(max)$ must be at least 0.264 for rejection of the null hypothesis.* Since $D(max) = 0.14$, it must be concluded that the sample distribution does not significantly differ from the national norms.

Table 12.2. Comparison of Observed and Expected CRFs for the English Grammer Example; Showing $D(max) = .14$ (boldface)

	1	2	3	4	5	6	7	8	9	Total
Observed CRF:	.04	.12	.28	.44	.56	.68	.84	.92	.96	1.00
Expected CRF:	.10	.20	.30	.40	.50	.60	.70	.80	.90	1.00
D:	.06	.08	.02	.04	.06	.08	**.14**	.12	.06	.00

A related test, called the *Kolmogorov-Smirnov two-sample test*, can be used to test the hypothesis that two populations have identical distributions (see Roscoe, 1969, page 214).

---

*It may appear that we are testing a one-sided alternate hypothesis; however, since $D(max)$ is determined by taking *absolute values* of differences, we have forced both tails of the "signed differences" distribution into one tail. Therefore, to test at $\alpha = 0.05$, *two-sided*, we look under $p = 0.95$ in Table K.

Exercises
(Use the appropriate distribution-free test)

1. The following data are measurements collected from two independent samples ($n_1 = 16$ and $n_2 = 12$). Test at $\alpha = 0.01$
    (a) the null hypothesis that $\mu_1 = \mu_2$, against the alternate hypothesis that $\mu_1 \neq \mu_2$.
    (b) the null hypothesis that $\sigma_1^2 = \sigma_2^2$, against the alternate hypothesis that $\sigma_1^2 \neq \sigma_2^2$.

Sample 1				Sample 2		
23.6	19.3	22.5	20.5	20.8	24.6	27.1
17.7	20.2	21.9	19.8	21.6	23.1	27.7
21.1	18.1	20.7	19.7	19.0	22.4	25.3
20.6	19.5	19.4	18.8	22.9	26.8	24.5

2. Fifteen fat ladies go on a three-week diet. Their weights before and after the diet are recorded below. Test at $\alpha = 0.01$ the hypothesis that the diet is successful for reducing weight.

Before	After	Before	After
180	172	128	128
165	157	144	131
146	150	156	138
138	138	146	148
142	145	182	160
170	157	155	144
182	151	140	134
164	148		

3. In a sample of 39 high school students who took the College Board Scholastic Aptitude Test (SAT) twice, 23 had higher scores the second time, three got the same score both times, and the rest got lower scores the second time. Determine (at $\alpha = 0.05$) whether or not a person can expect to improve his score on the SAT by taking it a second time.

4. It is hypothesized that a certain variable has a uniform probability distribution (discussed in Chapter 8) which implies that equal intervals contain an equal proportion of the population distribution. The variable is measured on a sample of size 100 and the results are grouped into a frequency distribution.

Class Limits	f_i
20.0–29.9	8
30.0–39.9	10
40.0–49.9	15
50.0–59.9	22
60.0–69.9	20
70.0–79.9	13
80.0–89.9	7
90.0–99.9	5
	100

Test the hypothesis at $\alpha = 0.05$.

5. In a learning experiment three groups of eight children learn to solve a puzzle under different conditions. Group 1 gets positive reinforcement (encouraging comments) throughout the session; Group 2 gets negative reinforcement (scolding, negative criticism); and Group 3 (the control group) is allowed to work alone without outside influence. The number of seconds it takes for each child to complete the puzzle is recorded below.

    Group 1:   52, 61, 48, 75, 83, 78, 66, 69
    Group 2:   62, 84, 78, 76, 92, 106, 90, 85
    Group 3:   58, 60, 55, 61, 66, 59, 67, 63

    Test at $\alpha = 0.01$ the null hypothesis: $\mu_1 = \mu_2 = \mu_3$.

6. Ten names from the list of subscribers to Magazine A are selected at random. The same thing is done for Magazine B. The ages of the people selected are

    A: 18, 26, 22, 35, 32, 21, 27, 28, 30, 25
    B: 28, 40, 34, 38, 45, 32, 36, 31, 37, 42

    Test at $\alpha = 0.03$ the hypothesis that Magazine B appeals to a higher age group than does Magazine A.

7. It is hypothesized that the taproot length of a certain species of plant is normally distributed with mean equal to 50 centimeters and standard deviation five centimeters. Therefore, 20 percent of the population would have taproots less than 45.8 cm long, 40 percent would be less than 48.7 cm, 60 percent would be less than 51.3 cm, and 80 percent would be less than 54.2 cm; (these figures can be verified with reference to Table E and the null hypothesis above). A random sample of 50 taproots are measured, with the following results.

Interval	Frequency
below 45.8	6
45.9–48.6	11
48.7–51.2	18
51.3–54.1	8
54.2 and above	7

    Test the null hypothesis at the 0.02 level of significance.

8. Four types of anesthetic are tested on four randomly chosen groups of patients. The number of seconds necessary after injection to render the patient unconscious is measured. Test at $\alpha = 0.05$ the hypothesis that the four types of anesthetic render the patient unconscious in equal time, given the following sample data (to the nearest tenth of a second).

    Type 1:   3.7, 2.8, 3.8, 3.5, 4.5, 4.2
    Type 2:   5.4, 4.0, 2.7, 6.2, 4.6, 2.5, 5.0, 4.8
    Type 3:   4.3, 3.9, 2.9, 3.1, 3.4
    Type 4:   5.2, 4.7, 3.6, 3.3, 5.1, 3.6

Summary

Distribution-free significance tests offer alternatives to classical significance tests when one or more of the assumptions underlying the classical test cannot be met. These tests are usually easier to understand

and involve fewer computations, although they do waste information and tend to be too conservative in comparison to their classical counterparts.

The *Mann-Whitney test* can be used as an alternative to the *t*-test for testing the difference between two independent means *and* to the *F*-test for testing the difference between two independent variances. The *sign test* replaces the classical *t*-test for testing the difference between two correlated means. The *Kruskal-Wallis test* provides an alternative to the classical analysis of variance *F*-test. And the *Kolmogorov-Smirnov one-sample test* substitutes for the χ^2 goodness-of-fit test for continuous variables. Many other distribution-free tests can be found in Siegal (1956) and in Bradley (1968).

Bibliography

Alder, H. L., and E. B. Roessler. *Introduction to Probability and Statistics*. 4th ed. San Francisco: Freeman. 1968.

Armore, S. J. *Introduction to Statistical Analysis and Inference . . . for Psychology and Education*. New York: Wiley. 1966.

Blakeslee, D. W., and W. G. Chinn. *Introductory Statistics and Probability*. Boston: Houghton Mifflin. 1971.

Bradley, J. V. *Distribution-free Statistical Tests*. Englewood Cliffs, N.J.: Prentice-Hall. 1968.

Chapman, D. G., and R. A. Schaufele. *Elementary Probability Models and Statistical Inference*. Waltham, Mass.: Ginn-Blaisdell. 1970.

Cochran, W. G. *Sampling Techniques*. 3rd ed. New York: Wiley. 1963.

Cornell, F. G. *The Essentials of Educational Statistics*. New York: Wiley. 1956.

Croxton, F. E., D. J. Cowden, and S. Klein. *Applied General Statistics*. 3rd ed. Englewood Cliffs, N.J.: Prentice-Hall. 1967.

Diamond, S. *Information and Error*. New York: Basic Books. 1959.

Edwards, A. L. *Experimental Design in Psychological Research*. rev. ed. New York: Holt, Rinehart, and Winston. 1960.

Ferguson, G. A. *Statistical Analysis in Psychology and Education*. 2nd ed. New York: McGraw-Hill. 1966.

Fischer, F. E. *Basic Concepts of Probability—A Programmed Approach.* Boston: Allyn and Bacon. 1969.

Freund, J. E. *Mathematical Statistics.* Englewood Cliffs, N.J.: Prentice-Hall. 1962.

———. *Modern Elementary Statistics.* 3rd ed. Englewood Cliffs, N.J.: Prentice-Hall. 1967.

———. *Statistics—A First Course.* Englewood Cliffs, N.J.: Prentice-Hall. 1970.

———. and F. J. Williams. *Modern Business Statistics.* Englewood Cliffs, N.J.: Prentice-Hall. 1958.

Fryer, H. C. *Concepts and Methods of Experimental Statistics.* Boston: Allyn and Bacon. 1966.

Garret, H. E. *Statistics in Psychology and Education.* 6th ed. New York: McKay. 1966.

Gemignani, M. C. *Calculus and Statistics.* Reading, Mass.: Addison-Wesley. 1970.

Guenther, W. C. *Concepts of Probability.* New York: McGraw-Hill. 1968.

———. *Concepts of Statistical Inference.* New York: McGraw-Hill. 1965.

Haber, A., and R. P. Runyon. *General Statistics.* Reading, Mass.: Addison-Wesley. 1969.

Hadley, G. *Elementary Statistics.* San Francisco: Holden-Day. 1969.

Hays, W. L., and R. L. Winkler. *Statistics: Probability, Inference, and Decision.* vol. I. New York: Holt, Rinehart, and Winston. 1970.

Hodges, J. L., and E. L. Lehmann. *Basic Concepts of Probability and Statistics.* San Francisco: Holden-Day. 1964.

Hoel, P. G. *Elementary Statistics.* 2nd ed. New York: Wiley. 1966.

Housner, M. *Elementary Probability Theory.* New York: Harper & Row. 1971.

Huff, D. *How to Lie with Statistics.* New York: Norton. 1954.

Hultquist, R. A. *Introduction to Statistics.* New York: Holt, Rinehart, and Winston. 1969.

Huntsberger, D. V. *Elements of Statistical Inference.* 2nd ed. Boston: Allyn and Bacon. 1967.

Langley, R. *Practical Statistics Simply Explained.* rev. ed. New York: Dover. 1970.

Mack, S. F. *Elementary Statistics.* New York: Holt, Rinehart, and Winston. 1960.

McNemar, Q. *Psychological Statistics.* 3rd ed. New York: Wiley. 1962.

Mendenhall, W. *Introduction to Probability and Statistics.* 2nd ed. Belmont, Cal.: Wadsworth. 1967.

Meyer, P. L. *Introductory Probability and Statistical Applications.* Reading, Mass.: Addison-Wesley. 1965.

Mode, E. B. *Elements of Statistics*. 3rd ed. Englewood Cliffs, N.J.: Prentice-Hall. 1961.

Mosimann, J. E. *Elementary Probability for the Biological Sciences*. New York: Appleton-Century-Crofts. 1968.

Mosteller, F., R. E. K. Rourke, and G. B. Thomas. *Probability with Statistical Applications*. 2nd ed. Reading, Mass.: Addison-Wesley. 1970.

O'Toole, A. L. *Elementary Practical Statistics*. New York: Macmillan. 1964.

Owen, D. B. *Handbook of Statistical Tables*. Reading, Mass.: Addison-Wesley. 1962.

Pearson, E. S., and H. O. Hartley. *Biometrika Tables for Statisticians*. vol. I. 3rd ed. Cambridge: Cambridge. 1967.

RAND Corporation. *A Million Random Digits with 100,000 Normal Deviates*. Glencoe, Ill.: Free Press. 1955.

Remington, R. D., and M. A. Schork. *Statistics with Applications to the Biological and Health Sciences*. Englewood Cliffs, N.J.: Prentice-Hall. 1970.

Roscoe, J. T. *Fundamental Research Statistics for the Behavioral Sciences*. New York: Holt, Rinehart, and Winston. 1969.

Schefler, W. C. *Statistics for the Biological Sciences*. Reading, Mass.: Addison-Wesley. 1969.

Schmitt, S. A. *Measuring Uncertainty—An Elementary Introduction to Bayesian Statistics*. Reading, Mass.: Addison-Wesley. 1969.

Siegel, S. *Nonparametric Statistics: For the Behavioral Sciences*. New York: McGraw-Hill. 1956.

Walker, H. M., and J. Lev. *Elementary Statistical Methods*. 3rd ed. New York: Holt, Rinehart, and Winston. 1969.

———. *Statistical Inference*. New York: Henry Holt. 1953.

Walpole, R. E. *Introduction to Statistics*. New York: Macmillan. 1968.

Weinberg, G. H., and J. A. Schumaker. *Statistics—An Intuitive Approach*. 2nd ed. Belmomt, Cal.: Brooks/Cole. 1969.

Wert, J. E., C. O. Neidt, and J. S. Ahmann. *Statistical Methods in Educational and Psychological Research*. New York: Appleton-Century-Crofts. 1954.

Winer, B. J. *Statistical Principles in Experimental Design*. New York: McGraw-Hill. 1962.

Appendix:
Statistical Tables

TABLE A. Squares*

n	0	1	2	3	4	5	6	7	8	9
0	0	1	4	9	16	25	36	49	64	81
1	100	121	144	169	196	225	256	289	324	361
2	400	441	484	529	576	625	676	729	784	841
3	900	961	1024	1089	1156	1225	1296	1369	1444	1521
4	1600	1681	1764	1849	1936	2025	2116	2209	2304	2401
5	2500	2601	2704	2809	2916	3025	3136	3249	3364	3481
6	3600	3721	3844	3969	4096	4225	4356	4489	4624	4761
7	4900	5041	5184	5329	5476	5625	5776	5929	6084	6241
8	6400	6561	6724	6889	7056	7225	7396	7569	7744	7921
9	8100	8281	8464	8649	8836	9025	9216	9409	9604	9801
10	10000	10201	10404	10609	10816	11025	11236	11449	11664	11881
11	12100	12321	12544	12769	12996	13225	13456	13689	13924	14161
12	14400	14641	14884	15129	15376	15625	15876	16129	16384	16641
13	16900	17161	17424	17689	17956	18225	18496	18769	19044	19321
14	19600	19881	20164	20449	20736	21025	21316	21609	21904	22201
15	22500	22801	23104	23409	23716	24025	24336	24649	24964	25281
16	25600	25921	26244	26569	26896	27225	27556	27889	28224	28561
17	28900	29241	29584	29929	30276	30625	30976	31329	31684	32041
18	32400	32761	33124	33489	33856	34225	34596	34969	35344	35721
19	36100	36481	36864	37249	37636	38025	38416	38809	39204	39601
20	40000	40401	40804	41209	41616	42025	42436	42849	43264	43681
21	44100	44521	44944	45369	45796	46225	46656	47089	47524	47961
22	48400	48841	49284	49729	50176	50625	51076	51529	51984	52441
23	52900	53361	53824	54289	54756	55225	55696	56169	56644	57121
24	57600	58081	58564	59049	59536	60025	60516	61009	61504	62001
25	62500	63001	63504	64009	64516	65025	65536	66049	66564	67081
26	67600	68121	68644	69169	69696	70225	70756	71289	71824	72361
27	72900	73441	73984	74529	75076	75625	76176	76729	77284	77841
28	78400	78961	79524	80089	80656	81225	81796	82369	82944	83521
29	84100	84681	85264	85849	86436	87025	87616	88209	88804	89401
30	90000	90601	91204	91809	92416	93025	93636	94249	94864	95481
31	96100	96721	97344	97969	98596	99225	99856	100489	101124	101761
32	102400	103041	103684	104329	104976	105625	106276	106929	107584	108241
33	108900	109561	110224	110889	111556	112225	112896	113569	114244	114921
34	115600	116281	116964	117649	118336	119025	119716	120409	121104	121801
35	122500	123201	123904	124609	125316	126025	126736	127449	128164	128881
36	129600	130321	131044	131769	132496	133225	133956	134689	135424	136161
37	136900	137641	138384	139129	139876	140625	141376	142129	142884	143641
38	144400	145161	145924	146689	147456	148225	148996	149769	150544	151321
39	152100	152881	153664	154449	155236	156025	156816	157609	158404	159201
40	160000	160801	161604	162409	163216	164025	164836	165649	166464	167281
41	168100	168921	169744	170569	171396	172225	173056	173889	174724	175561
42	176400	177241	178084	178929	179776	180625	181476	182329	183184	184041
43	184900	185761	186624	187489	188356	189225	190096	190969	191844	192721
44	193600	194481	195364	196249	197136	198025	198916	199809	200704	201601
45	202500	203401	204304	205209	206116	207025	207936	208849	209764	210681
46	211600	212521	213444	214369	215296	216225	217156	218089	219024	219961
47	220900	221841	222784	223729	224676	225625	226576	227529	228484	229441
48	230400	231361	232324	233289	234256	235225	236196	237169	238144	239121
49	240100	241081	242064	243049	244036	245025	246016	247009	248004	249001

n	0	1	2	3	4	5	6	7	8	9
50	250000	251001	252004	253009	254016	255025	256036	257049	258064	259081
51	260100	261121	262144	263169	264196	265225	266256	267289	268324	269361
52	270400	271441	272484	273529	274576	275625	276676	277729	278784	279841
53	280900	281961	283024	284089	285156	286225	287296	288369	289444	290521
54	291600	292681	293764	294849	295936	297025	298116	299209	300304	301401
55	302500	303601	304704	305809	306916	308025	309136	310249	311364	312481
56	313600	314721	315844	316969	318096	319225	320356	321489	322624	323761
57	324900	326041	327184	328329	329476	330625	331776	332929	334084	335241
58	336400	337561	338724	339889	341056	342225	343396	344569	345744	346921
59	348100	349281	350464	351649	352836	354025	355216	356409	357604	358801
60	360000	361201	362404	363609	364816	366025	367236	368449	369664	370881
61	372100	373321	374544	375769	376996	378225	379456	380689	381924	383161
62	384400	385641	386884	388129	389376	390625	391876	393129	394384	395641
63	396900	398161	399424	400689	401956	403225	404496	405769	407044	408321
64	409600	410881	412164	413449	414736	416025	417316	418609	419904	421201
65	422500	423801	425104	426409	427716	429025	430336	431649	432964	434281
66	435600	436921	438244	439569	440896	442225	443556	444889	446224	447561
67	448900	450241	451584	452929	454276	455625	456976	458329	459684	461041
68	462400	463761	465124	466489	467856	469225	470596	471969	473344	474721
69	476100	477481	478864	480249	481636	483025	484416	485809	487204	488601
70	490000	491401	492804	494209	495616	497025	498436	499849	501264	502681
71	504100	505521	506944	508369	509796	511225	512656	514089	515524	516961
72	518400	519841	521284	522729	524176	525625	527076	528529	529984	531441
73	532900	534361	535824	537289	538756	540225	541696	543169	544644	546121
74	547600	549081	550564	552049	553536	555025	556516	558009	559504	561001
75	562500	564001	565504	567009	568516	570025	571536	573049	574564	576081
76	577600	579121	580644	582169	583696	585225	586756	588289	589824	591361
77	592900	594441	595984	597529	599076	600625	602176	603729	605284	606841
78	608400	609961	611524	613089	614656	616225	617796	619369	620944	622521
79	624100	625681	627264	628849	630436	632025	633616	635209	636804	638401
80	640000	641601	643204	644809	646416	648025	649636	651249	652864	654481
81	656100	657721	659344	660969	662596	664225	665856	667489	669124	670761
82	672400	674041	675684	677329	678976	680625	682276	683929	685584	687241
83	688900	690561	692224	693889	695556	697225	698896	700569	702244	703921
84	705600	707281	708964	710649	712336	714025	715716	717409	719104	720801
85	722500	724201	725904	727609	729316	731025	732736	734449	736164	737881
86	739600	741321	743044	744769	746496	748225	749956	751689	753424	755161
87	756900	758641	760384	762129	763876	765625	767376	769129	770884	772641
88	774400	776161	777924	779689	781456	783225	784996	786769	788544	790321
89	792100	793881	795664	797449	799236	801025	802816	804609	806404	808201
90	810000	811801	813604	815409	817216	819025	820836	822649	824464	826281
91	828100	829921	831744	833569	835396	837225	839056	840889	842724	844561
92	846400	848241	850084	851929	853776	855625	857476	859329	861184	863041
93	864900	866761	868624	870489	872356	874225	876096	877969	879844	881721
94	883600	885481	887364	889249	891136	893025	894916	896809	898704	900601
95	902500	904401	906304	908209	910116	912025	913936	915849	917764	919681
96	921600	923521	925444	927369	929296	931225	933156	935089	937024	938961
97	940900	942841	944784	946729	948676	950625	952576	954529	956484	958441
98	960400	962361	964324	966289	968256	970225	972196	974169	976144	978121
99	980100	982081	984064	986049	988036	990025	992016	994009	996004	998001

TABLE B. Square Roots*

n	A	B		n	A	B
100	100000	316228		150	122474	387298
101	100499	317805		151	122882	388587
102	100995	319374		152	123288	389872
103	101489	320936		153	123693	391152
104	101980	322490		154	124097	392428
105	102470	324037		155	124499	393700
106	102956	325576		156	124900	394968
107	103441	327109		157	125300	396232
108	103923	328634		158	125698	397492
109	104403	330151		159	126095	398748
110	104881	331662		160	126491	400000
111	105357	333167		161	126886	401248
112	105830	334664		162	127279	402492
113	106301	336155		163	127671	403733
114	106771	337639		164	128062	404969
115	107238	339116		165	128452	406202
116	107703	340588		166	128841	407431
117	108167	342053		167	129228	408656
118	108628	343511		168	129615	409878
119	109087	344964		169	130000	411096
120	109545	346410		170	130384	412311
121	110000	347851		171	130767	413521
122	110454	349285		172	131149	414729
123	110905	350714		173	131529	415933
124	111355	352136		174	131909	417133
125	111803	353553		175	132288	418330
126	112250	354965		176	132665	419524
127	112694	356371		177	133041	420714
128	113137	357771		178	113417	421900
129	113578	359166		179	133791	423084
130	114018	360555		180	134164	424264
131	114455	361939		181	134536	425441
132	114891	363318		182	134907	426615
133	115326	364692		183	135277	427785
134	115758	366060		184	135647	428952
135	116190	367423		185	136015	430116
136	116619	368782		186	136382	431277
137	117047	370135		187	136748	432435
138	117473	371484		188	137113	433590
139	117898	372827		189	137477	434741
140	118322	374166		190	137840	435890
141	118743	375500		191	138203	437035
142	119164	376829		192	138564	438178
143	119583	378153		193	138924	439318
144	120000	379473		194	139284	440454
145	120416	380789		195	139642	441588
146	120830	382099		196	140000	442719
147	121244	383406		197	140357	443847
148	121655	384708		198	140712	444972
149	122066	386005		199	141067	446094

*How to Use this Table

Example 1: Find $\sqrt{56.9}$

First *estimate* the answer. Since $7^2 = 49$ and $8^2 = 64$, therefore $\sqrt{56.9}$ must be "7." plus a decimal fraction. Look in the table for n = 569 and note 2 choices in columns A and B: "238537" and "754321." Since we know that the answer must begin with "7."; therefore, $\sqrt{56.9} = 7.54321$.

n	A	B	n	A	B
200	141421	447214	250	158114	500000
201	141774	448330	251	158430	500999
202	142127	449444	252	158745	501996
203	142478	450555	253	159060	502991
204	142829	451664	254	159374	503984
205	143178	452769	255	159687	504975
206	143527	453872	256	160000	505964
207	143875	454973	257	160312	506952
208	144222	456070	258	160624	507937
209	144568	457165	259	160935	508920
210	144914	458258	260	161245	509902
211	145258	459347	261	161555	510882
212	145602	460435	262	161864	511859
213	145945	461519	263	162173	512835
214	146287	462601	264	162481	513809
215	146629	463681	265	162788	514782
216	146969	464758	266	163095	515752
217	147309	465833	267	163401	516720
218	147648	466905	268	163707	517687
219	147986	467974	269	164012	518652
220	148324	469042	270	164317	519615
221	148661	470106	271	164621	520577
222	148997	471169	272	164924	521536
223	149332	472229	273	165227	522494
224	149666	473286	274	165529	523450
225	150000	474342	275	165831	524404
226	150333	475395	276	166132	525357
227	150665	476445	277	166433	526308
228	150997	477493	278	166733	527257
229	151327	478539	279	167033	528205
230	151658	479583	280	167332	529150
231	151987	480625	281	167631	530094
232	152315	481664	282	167929	531037
233	152643	482701	283	168226	531977
234	152971	483735	284	168523	532917
235	153297	484768	285	168819	533854
236	153623	485798	286	169115	534790
237	153948	486826	287	169411	535724
238	154272	487852	288	169706	536656
239	154596	488876	289	170000	537587
240	154919	489898	290	170294	538516
241	155242	490918	291	170587	539444
242	155563	491935	292	170880	540370
243	155885	492950	293	171172	541295
244	156205	493964	294	171464	542218
245	156525	494975	295	171756	543139
246	156844	495984	296	172047	544059
247	157162	496991	297	172337	544977
248	157480	497996	298	172627	545894
249	157797	498999	299	172916	546809

Example 2:   Find $\sqrt{5.69}$

The answer must be between 2 and 3. Thus, $\sqrt{5.69} = 2.38537$. Similarly, $\sqrt{569} = 23.8537$, $\sqrt{0.569} = 0.754321$, $\sqrt{0.0569} = 0.238537$, and so on.

Example 3:   Find $\sqrt{825}$

Since $(20)^2 = 400$ and $(30)^2 = 900$, we know that $\sqrt{825}$ is *in the twenties*. Looking up n = 825 in the table we have a choice between "287228" and "908295." Therefore, $\sqrt{825}$ must be 28.7228.

TABLE B (Cont.)

n	A	B	n	A	B
300	173205	547723	350	187083	591608
301	173494	548635	351	187350	592453
302	173781	549545	352	187617	593296
303	174069	550454	353	187883	594138
304	174356	551362	354	188149	594979
305	174642	552268	355	188414	595819
306	174929	553173	356	188680	596657
307	175214	554076	357	188944	597495
308	175499	554977	358	189209	598331
309	175784	555878	359	189473	599166
310	176068	556776	360	189737	600000
311	176352	557674	361	190000	600833
312	176635	558570	362	190263	601664
313	176918	559464	363	190526	602495
314	177200	560357	364	190788	603324
315	177482	561249	365	191050	604152
316	177764	562139	366	191311	604979
317	178045	563028	367	191572	605805
318	178326	563915	368	191833	606630
319	178606	564801	369	192094	607454
320	178885	565685	370	192354	608276
321	179165	566569	371	192614	609098
322	179444	567450	372	192873	609918
323	179722	568331	373	193132	610737
324	180000	569210	374	193391	611555
325	180278	570088	375	193649	612372
326	180555	570964	376	193907	613188
327	180831	571839	377	194165	614003
328	181108	572713	378	194422	614817
329	181384	573585	379	194679	615630
330	181659	574456	380	194936	616441
331	181934	575326	381	195192	617252
332	182209	576194	382	195448	618061
333	182483	577062	383	195704	618870
334	182757	577927	384	195959	619677
335	183030	578792	385	196214	620484
336	183303	579655	386	196469	621289
337	183576	580517	387	196723	622093
338	183848	581378	388	196977	622896
339	184120	582237	389	197231	623699
340	184391	583095	390	197484	624500
341	184662	583952	391	197737	625300
342	184932	584808	392	197990	626099
343	185203	585662	393	198242	626897
344	185472	586515	394	198494	627694
345	185742	587367	395	198746	628490
346	186011	588218	396	198997	629285
347	186279	589067	397	199249	630079
348	186548	589915	398	199499	630872
349	186815	590762	399	199750	631664

n	A	B	n	A	B
400	200000	632456	450	212132	670820
401	200250	633246	451	212368	671565
402	200499	634035	452	212603	672309
403	200749	634823	453	212838	673053
404	200998	635610	454	213073	673795
405	201246	636396	455	213307	674537
406	201494	637181	456	213542	675278
407	201742	637966	457	213776	676018
408	201990	638749	458	214009	676757
409	202237	639531	459	214243	677495
410	202485	640312	460	214476	678233
411	202731	641093	461	214709	678970
412	202978	641872	462	214942	679706
413	203224	642651	463	215174	680441
414	203470	643428	464	215407	681175
415	203715	644205	465	215639	681909
416	203961	644981	466	215870	682642
417	204206	645755	467	216102	683374
418	204450	646529	468	216333	684105
419	204695	647302	469	216564	684836
420	204939	648074	470	216795	685565
421	205183	648845	471	217025	686294
422	205426	649615	472	217256	687023
423	205670	650385	473	217486	687750
424	205913	651153	474	217715	688477
425	206155	651920	475	217945	689202
426	206398	652687	476	218174	689928
427	206640	653452	477	218403	690652
428	206882	654217	478	218632	691375
429	207123	654981	479	218861	692098
430	207364	655744	480	219089	692820
431	207605	656506	481	219317	693542
432	207846	657267	482	219545	694262
433	208087	658027	483	219773	694982
434	208327	658787	484	220000	695701
435	208567	659545	485	220227	696419
436	208806	660303	486	220454	697137
437	209045	661060	487	220681	697854
438	209284	661816	488	220907	698570
439	209523	662571	489	221133	699285
440	209762	663325	490	221359	700000
441	210000	664078	491	221585	700714
442	210238	664831	492	221811	701427
443	210476	665582	493	222036	702140
444	210713	666333	494	222261	702851
445	210950	667083	495	222486	703562
446	211187	667832	496	222711	704273
447	211424	668581	497	222935	704982
448	211660	669328	498	223159	705691
449	211896	670075	499	223383	706399

TABLE B (Cont.)

n	A	B	n	A	B
500	223607	707107	550	234521	741620
501	223830	707814	551	234734	742294
502	224054	708520	552	234947	742967
503	224277	709225	553	235160	743640
504	224499	709930	554	235372	744312
505	224722	710634	555	235584	744983
506	224944	711337	556	235797	745654
507	225167	712039	557	236008	746324
508	225389	712741	558	236220	746994
509	225610	713442	559	236432	747663
510	225832	714143	560	236643	748331
511	226053	714843	561	236854	748999
512	226274	715542	562	237065	749667
513	226495	716240	563	237276	750333
514	226716	716938	564	237487	750999
515	226936	717635	565	237697	751665
516	227156	718331	566	237908	752330
517	227376	719027	567	238118	752994
518	227596	719722	568	238328	753658
519	227816	720417	569	238537	754321
520	228035	721110	570	238747	754983
521	228254	721803	571	238956	755645
522	228473	722496	572	239165	756307
523	288692	723187	573	239374	756968
524	228910	723878	574	239583	757628
525	229129	724569	575	239792	758288
526	229347	725259	576	240000	758947
527	229565	725948	577	240208	759605
528	229783	726636	578	240416	760263
529	230000	727324	579	240624	760920
530	230217	728011	580	240832	761577
531	230434	728697	581	241039	762234
532	230651	729383	582	241247	762889
533	230868	730068	583	241454	763544
534	231084	730753	584	241661	764199
535	231301	731437	585	241868	764853
536	231517	732120	586	242074	765506
537	231733	732803	587	242281	766159
538	231948	733485	588	242487	766812
539	232164	734166	589	242693	767463
540	232379	734847	590	242899	768115
541	232594	735527	591	243105	768765
542	232809	736206	592	243311	769415
543	233024	736885	593	243516	770065
544	233238	737564	594	243721	770714
545	233452	738241	595	243926	771362
546	233666	738918	596	244131	772010
547	233880	739594	597	244336	772658
548	234094	740270	598	244540	773305
549	234307	740945	599	244745	773951

n	A	B	n	A	B
600	244949	774597	650	254951	806226
601	245153	775242	651	255147	806846
602	245357	775887	652	255343	807465
603	245561	776531	653	255539	808084
604	245764	777174	654	255734	808703
605	245967	777817	655	255930	809321
606	246171	778460	656	256125	809938
607	246374	779102	657	256320	810555
608	246577	779744	658	256515	811172
609	246779	780385	659	256710	811788
610	246982	781025	660	256905	812404
611	247184	781665	661	257099	813019
612	247386	782304	662	257294	813634
613	247588	782943	633	257488	814248
614	247790	783582	664	257682	814862
615	247992	784219	665	257876	815475
616	248193	784857	666	258070	816088
617	248395	785493	667	258263	816701
618	248596	786130	668	258457	817313
619	248797	786766	669	258650	817924
620	248998	787401	670	258844	818535
621	249199	788036	671	259037	819146
622	249399	788670	672	259230	819756
623	249600	789303	673	259422	820366
624	249800	789937	674	259615	820975
625	250000	790569	675	259808	821584
626	250200	791202	676	260000	822192
627	250400	791833	677	260192	822800
628	250599	792465	678	260384	823408
629	250799	793095	679	260576	824015
630	250998	793725	680	260768	824621
631	251197	794355	681	260960	825227
632	251396	794984	682	261151	825833
633	251595	795613	683	261343	826438
634	251794	796241	684	261534	827043
635	251992	796869	685	261725	827647
636	252190	797496	686	261916	828251
637	252389	798123	687	262107	828855
638	252587	798749	688	262298	829458
639	252784	799375	689	262488	830060
640	252982	800000	690	262679	830662
641	253180	800625	691	262869	831264
642	253377	801249	692	263059	831865
643	253574	801873	693	263249	832466
644	253772	802496	694	263439	833067
645	253969	803119	695	263629	833667
646	254165	803741	696	263818	834266
647	254362	804363	697	264008	834865
648	254558	804984	698	264197	835464
649	254755	805605	699	264386	836062

TABLE B (Cont.)

n	A	B	n	A	B
700	264575	836660	750	273861	866025
701	264764	837257	751	274044	866603
702	264953	837854	752	274226	867179
703	265141	838451	753	274408	867756
704	265330	839047	754	274591	868332
705	265518	839643	755	274773	868907
706	265707	840238	756	274955	869483
707	265895	840833	757	275136	870057
708	266083	841427	758	275318	870632
709	266271	842021	759	275500	871206
710	266458	842615	760	275681	871780
711	266646	843208	761	275862	872353
712	266833	843801	762	276043	872926
713	267021	844393	763	276225	873499
714	267208	844985	764	276405	874071
715	267395	845577	765	276586	874643
716	267582	846168	766	276767	875214
717	267769	846759	767	276948	875785
718	267955	847349	768	277128	876356
719	268142	847939	769	277308	876926
720	268328	848528	770	277489	877496
721	268514	849117	771	277669	878066
722	268701	849706	772	277849	878635
723	268887	850294	773	278029	879204
724	269072	850882	774	278209	879773
725	269258	851469	775	278388	880341
726	269444	852056	776	278568	880909
727	269629	852643	777	278747	881476
728	269815	853229	778	278927	882043
729	270000	853815	779	279106	882610
730	270185	854400	780	279285	883176
731	270370	854985	781	279464	883742
732	270555	855570	782	279643	884308
733	270740	856154	783	279821	884873
734	270924	856738	784	280000	885438
735	271109	857321	785	280179	886002
736	271293	857904	786	280357	886566
737	271477	858487	787	280535	887130
738	271662	859069	788	280713	887694
739	271846	859651	789	280891	888257
740	272029	860233	790	281069	888819
741	272213	860814	791	281247	889382
742	272397	861394	792	281425	889944
743	272580	861974	793	281603	890505
744	272764	862554	794	281780	891067
745	272947	863134	795	281957	891628
746	273130	863713	796	282135	892188
747	273313	864292	797	282312	892749
748	273496	864870	798	282489	893308
749	273679	865448	799	282666	893868

n	A	B	n	A	B
800	282843	894427	850	291548	921954
801	283019	894986	851	291719	922497
802	283196	895545	852	291890	923038
803	283373	896103	853	292062	923580
804	283549	896660	854	292233	924121
805	283725	897218	855	292404	924662
806	283901	897775	856	292575	925203
807	284077	898332	857	292746	925743
808	284253	898888	858	292916	926283
809	284429	899444	859	293087	926823
810	284605	900000	860	293258	927362
811	284781	900555	861	293428	927901
812	284956	901110	862	293598	928440
813	285132	901665	863	293769	928978
814	285307	902219	864	293939	929516
815	285482	902774	865	294109	930054
816	285657	903327	866	294279	930591
817	285832	903881	867	294449	931128
818	286007	904434	868	294618	931665
819	286182	904986	869	294788	932202
820	286356	905539	870	294958	932738
821	286531	906091	871	295127	933274
822	286705	906642	872	295296	933809
823	286880	907193	873	295466	934345
824	287054	907744	874	295635	934880
825	287228	908295	875	295804	935414
826	297402	908845	876	295973	935949
827	287576	909395	877	296142	936483
828	287750	909945	878	296311	937017
829	287924	910494	879	296479	937550
830	288097	911043	880	296648	938083
831	288271	911592	881	296816	938616
832	288444	912140	882	296985	939149
833	288617	912688	883	297153	939681
834	288791	913236	884	297321	940213
835	288964	913783	885	297489	940744
836	289137	914330	886	297658	941276
837	289310	914877	887	297825	941807
838	289482	915423	888	297993	942338
839	289655	915969	889	298161	942868
840	289828	916515	890	298329	943398
841	290000	917061	891	298496	943928
842	290172	917606	892	298664	944458
843	290345	918150	893	298831	944987
844	290517	918695	894	298998	945516
845	290689	919239	895	299166	946044
846	290861	919783	896	299333	946573
847	291033	920326	897	299500	947101
848	291204	920869	898	299666	947629
849	291376	921412	899	299833	948156

TABLE B (Cont.)

n	A	B	n	A	B
900	300000	948683	950	308221	974679
901	300167	949210	951	308383	975192
902	300333	949737	952	308545	975705
903	300500	950263	953	308707	976217
904	300666	950789	954	308869	976729
905	300832	951315	955	309031	977241
906	300998	951840	956	309192	977753
907	301164	952365	957	309354	978264
908	301330	952890	958	309516	978775
909	301496	953415	959	309677	979285
910	301662	953939	960	309839	979796
911	301828	954463	961	310000	980306
912	301993	954987	962	310161	980816
913	302159	955510	963	310322	981326
914	302324	956033	964	310483	981835
915	302490	956556	965	310644	982344
916	302655	957079	966	310805	982853
917	302820	957601	967	310966	983362
918	302985	958123	968	311127	983870
919	303150	958645	969	311288	984378
920	303315	959166	970	311448	984886
921	303480	959687	971	311609	985393
922	303645	960208	972	311769	985901
923	303809	960729	973	311929	986408
924	303974	961249	974	312090	986914
925	304138	961769	975	312250	987421
926	304302	962289	976	312410	987927
927	304467	962808	977	312570	988433
928	304631	963328	978	312730	988939
929	304795	963846	979	312890	989444
930	304959	964365	980	313050	989949
931	305123	964883	981	313209	990454
932	305287	965401	982	313369	990959
933	305450	965919	983	313528	991464
934	305614	966437	984	313688	991968
935	305778	966954	985	313847	992472
936	305941	967471	986	314006	992975
937	306105	967988	987	314166	993479
938	306268	968504	988	314325	993982
939	306431	969020	989	314484	994485
940	306594	969536	990	314643	994987
941	306757	970052	991	314802	995490
942	306920	970567	992	314960	995992
943	307083	971082	993	315119	996494
944	307246	971597	994	315278	996995
945	307409	972111	995	315436	997497
946	307571	972625	996	315595	997998
947	307734	973139	997	315753	998499
948	307896	973653	998	315911	998999
949	308058	974166	999	316070	999500

TABLE C. Binomial Coefficients: $\binom{n}{r}$

r \ n	2	3	4	5	6	7	8	9	10	11	12	13
2	1											
3	3	1										
4	6	4	1									
5	10	10	5	1								
6	15	20	15	6	1							
7	21	35	35	21	7	1						
8	28	56	70	56	28	8	1					
9	36	84	126	126	84	36	9	1				
10	45	120	210	252	210	120	45	10	1			
11	55	165	330	462	462	330	165	55	11	1		
12	66	220	495	792	924	792	495	220	66	12	1	
13	78	286	715	1287	1716	1716	1287	715	286	78	13	1
14	91	364	1001	2002	3003	3432	3003	2002	1001	364	91	14
15	105	455	1365	3003	5005	6435	6435	5005	3003	1365	455	105
16	120	560	1820	4368	8008	11440	12870	11440	8008	4368	1820	560
17	136	680	2380	6188	12376	19448	24310	24310	19448	12376	6188	2380
18	153	816	3060	8568	18564	31824	43758	48620	43758	31824	18564	8568
19	171	969	3876	11628	27132	50388	75582	92378	92378	75582	50388	27132
20	190	1140	4845	15504	38760	77520	125970	167960	184756	167960	125970	77520
21	210	1330	5985	20349	54264	116280	203490	293930	352716	352716	293930	203490
22	231	1540	7315	26334	74613	170544	319770	497420	646646	705432	646646	497420
23	253	1771	8855	33649	100947	245157	490314	817190	1144066	1352078	1352078	1144066
24	276	2024	10626	42504	134596	346104	735471	1307504	1961256	2496144	2704156	2496144
25	300	2300	12650	53130	177100	480700	1081575	2042975	3268760	4457400	5200300	5200300

TABLE D.   Values of e^{-m}

m	e^{-m}	m	e^{-m}	m	e^{-m}	m	e^{-m}
.1	.9048	2.1	.1225	4.1	.0166	6.1	.0022
.2	.8187	2.2	.1108	4.2	.0150	6.2	.0020
.3	.7408	2.3	.1003	4.3	.0136	6.3	.0018
.4	.6703	2.4	.0907	4.4	.0123	6.4	.0017
.5	.6065	2.5	.0821	4.5	.0111	6.5	.0015
.6	.5488	2.6	.0743	4.6	.0101	6.6	.0014
.7	.4966	2.7	.0672	4.7	.0091	6.7	.0012
.8	.4493	2.8	.0608	4.8	.0082	6.8	.0011
.9	.4066	2.9	.0550	4.9	.0074	6.9	.0010
1.0	.3679	3.0	.0498	5.0	.0067	7.0	.0009
1.1	.3329	3.1	.0450	5.1	.0061	7.1	.0008
1.2	.3012	3.2	.0408	5.2	.0055	7.2	.0007
1.3	.2725	3.3	.0369	5.3	.0050	7.3	.0007
1.4	.2466	3.4	.0334	5.4	.0045	7.4	.0006
1.5	.2231	3.5	.0302	5.5	.0041	7.5	.0006
1.6	.2019	3.6	.0273	5.6	.0037	7.6	.0005
1.7	.1827	3.7	.0247	5.7	.0033	7.7	.0005
1.8	.1653	3.8	.0224	5.8	.0030	7.8	.0004
1.9	.1496	3.9	.0202	5.9	.0027	7.9	.0004
2.0	.1353	4.0	.0183	6.0	.0025	8.0	.0003

TABLE E.
The Standard Normal Probability Function*

$$P(z < z_p) = p$$

Z	0.00	0.01	0.02	0.03	0.04	0.05	0.06	0.07	0.08	0.09
-3.4	0.0003	0.0003	0.0003	0.0003	0.0003	0.0003	0.0003	0.0003	0.0003	0.0002
-3.3	0.0005	0.0005	0.0005	0.0004	0.0004	0.0004	0.0004	0.0004	0.0004	0.0003
-3.2	0.0007	0.0007	0.0006	0.0006	0.0006	0.0006	0.0006	0.0005	0.0005	0.0005
-3.1	0.0010	0.0009	0.0009	0.0009	0.0008	0.0008	0.0008	0.0008	0.0007	0.0007
-3.0	0.0013	0.0013	0.0013	0.0012	0.0012	0.0011	0.0011	0.0011	0.0010	0.0010
-2.9	0.0019	0.0018	0.0017	0.0017	0.0016	0.0016	0.0015	0.0015	0.0014	0.0014
-2.8	0.0026	0.0025	0.0024	0.0023	0.0023	0.0022	0.0021	0.0021	0.0020	0.0019
-2.7	0.0035	0.0034	0.0033	0.0032	0.0031	0.0030	0.0029	0.0028	0.0027	0.0026
-2.6	0.0047	0.0045	0.0044	0.0043	0.0041	0.0040	0.0039	0.0038	0.0037	0.0036
-2.5	0.0062	0.0060	0.0059	0.0057	0.0055	0.0054	0.0052	0.0051	0.0049	0.0048
-2.4	0.0082	0.0080	0.0078	0.0075	0.0073	0.0071	0.0069	0.0068	0.0066	0.0064
-2.3	0.0107	0.0104	0.0102	0.0099	0.0096	0.0094	0.0091	0.0089	0.0087	0.0084
-2.2	0.0139	0.0136	0.0132	0.0129	0.0125	0.0122	0.0119	0.0116	0.0113	0.0110
-2.1	0.0179	0.0174	0.0170	0.0166	0.0162	0.0158	0.0154	0.0150	0.0146	0.0143
-2.0	0.0228	0.0222	0.0217	0.0212	0.0207	0.0202	0.0197	0.0192	0.0188	0.0183
-1.9	0.0287	0.0281	0.0274	0.0268	0.0262	0.0256	0.0250	0.0244	0.0239	0.0233
-1.8	0.0359	0.0352	0.0344	0.0336	0.0329	0.0322	0.0314	0.0307	0.0301	0.0294
-1.7	0.0446	0.0436	0.0427	0.0418	0.0409	0.0401	0.0392	0.0384	0.0375	0.0367
-1.6	0.0548	0.0537	0.0526	0.0516	0.0505	0.0495	0.0485	0.0475	0.0465	0.0455
-1.5	0.0668	0.0655	0.0643	0.0630	0.0618	0.0606	0.0594	0.0582	0.0571	0.0559
-1.4	0.0808	0.0793	0.0778	0.0764	0.0749	0.0735	0.0722	0.0708	0.0694	0.0681
-1.3	0.0968	0.0951	0.0934	0.0918	0.0901	0.0885	0.0869	0.0853	0.0838	0.0823
-1.2	0.1151	0.1131	0.1112	0.1093	0.1075	0.1056	0.1038	0.1020	0.1003	0.0985
-1.1	0.1357	0.1335	0.1314	0.1292	0.1271	0.1251	0.1230	0.1210	0.1190	0.1170
-1.0	0.1587	0.1562	0.1539	0.1515	0.1492	0.1469	0.1446	0.1423	0.1401	0.1379
-0.9	0.1841	0.1814	0.1788	0.1762	0.1736	0.1711	0.1685	0.1660	0.1635	0.1611
-0.8	0.2119	0.2090	0.2061	0.2033	0.2005	0.1977	0.1949	0.1922	0.1894	0.1867
-0.7	0.2420	0.2389	0.2358	0.2327	0.2296	0.2266	0.2236	0.2206	0.2177	0.2148
-0.6	0.2743	0.2709	0.2676	0.2643	0.2611	0.2578	0.2546	0.2514	0.2483	0.2451
-0.5	0.3085	0.3050	0.3015	0.2981	0.2946	0.2912	0.2877	0.2843	0.2810	0.2776
-0.4	0.3446	0.3409	0.3372	0.3336	0.3300	0.3264	0.3228	0.3192	0.3156	0.3121
-0.3	0.3821	0.3783	0.3745	0.3707	0.3669	0.3632	0.3594	0.3557	0.3520	0.3483
-0.2	0.4207	0.4168	0.4129	0.4090	0.4052	0.4013	0.3974	0.3936	0.3897	0.3859
-0.1	0.4602	0.4562	0.4522	0.4483	0.4443	0.4404	0.4364	0.4325	0.4286	0.4247
-0.0	0.5000	0.4960	0.4920	0.4880	0.4840	0.4801	0.4761	0.4721	0.4681	0.4641
0.0	0.5000	0.5040	0.5080	0.5120	0.5160	0.5199	0.5239	0.5279	0.5319	0.5359
0.1	0.5398	0.5438	0.5478	0.5517	0.5557	0.5596	0.5636	0.5675	0.5714	0.5753
0.2	0.5793	0.5832	0.5871	0.5910	0.5948	0.5987	0.6026	0.6064	0.6103	0.6141
0.3	0.6179	0.6217	0.6255	0.6293	0.6331	0.6368	0.6406	0.6443	0.6480	0.6517
0.4	0.6554	0.6591	0.6628	0.6664	0.6700	0.6736	0.6772	0.6808	0.6844	0.6879
0.5	0.6915	0.6950	0.6985	0.7019	0.7054	0.7088	0.7123	0.7157	0.7190	0.7224
0.6	0.7257	0.7291	0.7324	0.7357	0.7389	0.7422	0.7454	0.7486	0.7517	0.7549
0.7	0.7580	0.7611	0.7642	0.7673	0.7704	0.7734	0.7764	0.7794	0.7823	0.7852
0.8	0.7881	0.7910	0.7939	0.7967	0.7995	0.8023	0.8051	0.8078	0.8106	0.8133
0.9	0.8159	0.8186	0.8212	0.8238	0.8264	0.8289	0.8315	0.8340	0.8365	0.8389
1.0	0.8413	0.8438	0.8461	0.8485	0.8508	0.8531	0.8554	0.8577	0.8599	0.8621
1.1	0.8643	0.8665	0.8686	0.8708	0.8729	0.8749	0.8770	0.8790	0.8810	0.8830
1.2	0.8849	0.8869	0.8888	0.8907	0.8925	0.8944	0.8962	0.8980	0.8997	0.9015
1.3	0.9032	0.9049	0.9066	0.9082	0.9099	0.9115	0.9131	0.9147	0.9162	0.9177
1.4	0.9192	0.9207	0.9222	0.9236	0.9251	0.9265	0.9278	0.9292	0.9306	0.9319
1.5	0.9332	0.9345	0.9357	0.9370	0.9382	0.9394	0.9406	0.9418	0.9429	0.9441
1.6	0.9452	0.9463	0.9474	0.9484	0.9495	0.9505	0.9515	0.9525	0.9535	0.9545
1.7	0.9554	0.9564	0.9573	0.9582	0.9591	0.9599	0.9608	0.9616	0.9625	0.9633
1.8	0.9641	0.9649	0.9656	0.9664	0.9671	0.9678	0.9686	0.9693	0.9699	0.9706
1.9	0.9713	0.9719	0.9726	0.9732	0.9738	0.9744	0.9750	0.9756	0.9761	0.9767
2.0	0.9772	0.9778	0.9783	0.9788	0.9793	0.9798	0.9803	0.9808	0.9812	0.9817
2.1	0.9821	0.9826	0.9830	0.9834	0.9838	0.9842	0.9846	0.9850	0.9854	0.9857
2.2	0.9861	0.9864	0.9868	0.9871	0.9875	0.9878	0.9881	0.9884	0.9887	0.9890
2.3	0.9893	0.9896	0.9898	0.9901	0.9904	0.9906	0.9909	0.9911	0.9913	0.9916
2.4	0.9918	0.9920	0.9922	0.9925	0.9927	0.9929	0.9931	0.9932	0.9934	0.9936
2.5	0.9938	0.9940	0.9941	0.9943	0.9945	0.9946	0.9948	0.9949	0.9951	0.9952
2.6	0.9953	0.9955	0.9956	0.9957	0.9959	0.9960	0.9961	0.9962	0.9963	0.9964
2.7	0.9965	0.9966	0.9967	0.9968	0.9969	0.9970	0.9971	0.9972	0.9973	0.9974
2.8	0.9974	0.9975	0.9976	0.9977	0.9977	0.9978	0.9979	0.9979	0.9980	0.9981
2.9	0.9981	0.9982	0.9982	0.9983	0.9984	0.9984	0.9985	0.9985	0.9986	0.9986
3.0	0.9987	0.9987	0.9987	0.9988	0.9988	0.9989	0.9989	0.9989	0.9990	0.9990
3.1	0.9990	0.9991	0.9991	0.9991	0.9992	0.9992	0.9992	0.9992	0.9993	0.9993
3.2	0.9993	0.9993	0.9994	0.9994	0.9994	0.9994	0.9994	0.9995	0.9995	0.9995
3.3	0.9995	0.9995	0.9995	0.9996	0.9996	0.9996	0.9996	0.9996	0.9996	0.9997
3.4	0.9997	0.9997	0.9997	0.9997	0.9997	0.9997	0.9997	0.9997	0.9997	0.9998

*Reprinted with permission of The Macmillan Company from *Introduction to Statistics* by Ronald E. Walpole. Copyright © by Ronald E. Walpole, 1968.

TABLE F. The Student-t Probability Function (Values of $t_{r;p}$)*

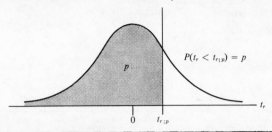

$$P(t_r < t_{r;p}) = p$$

p r	.90	.95	.975	.99	.995
1	3.08	6.31	12.71	31.82	63.66
2	1.89	2.92	4.30	6.96	9.92
3	1.64	2.35	3.18	4.54	5.84
4	1.53	2.13	2.78	3.75	4.60
5	1.48	2.02	2.57	3.36	4.03
6	1.44	1.94	2.45	3.14	3.71
7	1.41	1.89	2.36	3.00	3.50
8	1.40	1.86	2.31	2.90	3.36
9	1.38	1.83	2.26	2.82	3.25
10	1.37	1.81	2.23	2.76	3.17
11	1.36	1.80	2.20	2.72	3.11
12	1.36	1.78	2.18	2.68	3.05
13	1.35	1.77	2.16	2.65	3.01
14	1.35	1.76	2.14	2.62	2.98
15	1.34	1.75	2.13	2.60	2.95
16	1.34	1.75	2.12	2.58	2.92
17	1.33	1.74	2.11	2.57	2.90
18	1.33	1.73	2.10	2.55	2.88
19	1.33	1.73	2.09	2.54	2.86
20	1.33	1.72	2.09	2.53	2.85
21	1.32	1.72	2.08	2.52	2.83
22	1.32	1.72	2.07	2.51	2.82
23	1.32	1.71	2.07	2.50	2.81
24	1.32	1.71	2.06	2.49	2.80
25	1.32	1.71	2.06	2.49	2.79
26	1.32	1.71	2.06	2.48	2.78
27	1.31	1.70	2.05	2.47	2.77
28	1.31	1.70	2.05	2.47	2.76
29	1.31	1.70	2.05	2.46	2.76
30	1.31	1.70	2.04	2.46	2.75
40	1.30	1.68	2.02	2.42	2.70
60	1.30	1.67	2.00	2.39	2.66
80	1.29	1.66	1.99	2.37	2.64
100	1.29	1.66	1.98	2.36	2.63
200	1.29	1.65	1.97	2.35	2.60
500	1.28	1.65	1.96	2.33	2.59
∞	1.282	1.645	1.960	2.326	2.576

*Abridged from Donald B. Owen, *Handbook of Statistical Tables*, 1962, Addison-Wesley, Reading, Mass., pp. 28–30. Courtesy of U.S. Atomic Energy Commission.

TABLE G. The Chi-Square Probability Function (Values of $\chi^2_{r;p}$)*

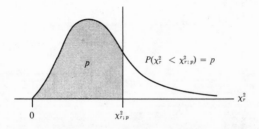

$$P(\chi^2_r < \chi^2_{r;p}) = p$$

r \ p	.005	.01	.025	.05	.95	.975	.99	.995
1	.00	.00	.00	.00	3.84	5.02	6.63	7.88
2	.01	.02	.05	.10	5.99	7.38	9.21	10.60
3	.07	.11	.22	.35	7.81	9.35	11.34	12.84
4	.21	.30	.48	.71	9.49	11.14	13.28	14.86
5	.41	.55	.83	1.15	11.07	12.83	15.09	16.75
6	.68	.87	1.24	1.64	12.59	14.45	16.81	18.55
7	.99	1.24	1.69	2.17	14.07	16.01	18.48	20.28
8	1.34	1.65	2.18	2.73	15.51	17.53	20.09	21.96
9	1.73	2.09	2.70	3.33	16.92	19.02	21.67	23.59
10	2.16	2.56	3.25	3.94	18.31	20.48	23.21	25.19
11	2.60	3.05	3.82	4.57	19.68	21.92	24.73	26.76
12	3.07	3.57	4.40	5.23	21.03	23.34	26.22	28.30
13	3.57	4.11	5.01	5.89	22.36	24.74	27.69	29.82
14	4.07	4.66	5.63	6.57	23.68	26.12	29.14	31.32
15	4.60	5.23	6.26	7.26	25.00	27.49	30.58	32.80
16	5.14	5.81	6.91	7.96	26.30	28.85	32.00	34.27
17	5.70	6.41	7.56	8.67	27.59	30.19	33.41	35.72
18	6.26	7.01	8.23	9.39	28.87	31.53	34.81	37.16
19	6.84	7.63	8.91	10.12	30.14	32.85	36.19	38.58
20	7.43	8.26	9.59	10.85	31.41	34.17	37.57	40.00
21	8.03	8.90	10.28	11.59	32.67	35.48	38.93	41.40
22	8.64	9.54	10.98	12.34	33.92	36.78	40.29	42.80
23	9.26	10.20	11.69	13.09	35.17	38.08	41.64	44.18
24	9.89	10.86	12.40	13.85	36.42	39.36	42.98	45.56
25	10.52	11.52	13.12	14.61	37.65	40.65	44.31	46.93
26	11.16	12.20	13.84	15.38	38.89	41.92	45.64	48.29
27	11.81	12.88	14.57	16.15	40.11	43.19	46.96	49.64
28	12.46	13.56	15.31	16.93	41.34	44.46	48.28	50.99
29	13.12	14.26	16.05	17.71	42.56	45.72	49.59	52.34
30	13.79	14.95	16.79	18.49	43.77	46.98	50.89	53.67
40	20.71	22.16	24.43	26.51	55.76	59.34	63.69	66.77
50	27.99	29.71	32.36	34.76	67.50	71.42	76.15	79.49
60	35.53	37.48	40.48	43.19	79.08	83.30	88.38	91.95
70	43.28	45.44	48.76	51.74	90.53	95.02	100.43	104.22
80	51.17	53.54	57.15	60.39	101.88	106.63	112.33	116.32
90	59.20	61.75	65.65	69.13	113.15	118.14	124.12	128.30
100	67.33	70.06	74.22	77.93	124.34	129.56	135.81	140.17

For $r > 100$: $\chi^2_{r;p} = r\left\{1 - \dfrac{2}{9r} + z_p\sqrt{\dfrac{2}{9r}}\right\}^3$ or $\chi^2_{r;p} = \dfrac{1}{2}\left\{z_p + \sqrt{2r - 1}\right\}^2$, according to the degree of accuracy required.

*Abridged from *Biometrika Tables for Statisticians*, Vol. I, Table 8, pp. 130–131, by permission of the Biometrika Trustees.

TABLE H.
The F Probability Function (Values of $F_{r_1,r_2;p}$)*

For p = 0.95

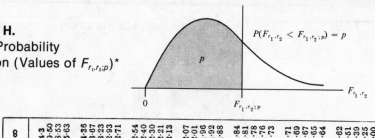

$$P(F_{r_1,r_2} < F_{r_1,r_2;p}) = p$$

r_2 \ r_1	1	2	3	4	5	6	7	8	9	10	12	15	20	24	30	40	60	120	∞
1	161·4	199·5	215·7	224·6	230·2	234·0	236·8	238·9	240·5	241·9	243·9	245·9	248·0	249·1	250·1	251·1	252·2	253·3	254·3
2	18·51	19·00	19·16	19·25	19·30	19·33	19·35	19·37	19·38	19·40	19·41	19·43	19·45	19·45	19·46	19·47	19·48	19·49	19·50
3	10·13	9·55	9·28	9·12	9·01	8·94	8·89	8·85	8·81	8·79	8·74	8·70	8·66	8·64	8·62	8·59	8·57	8·55	8·53
4	7·71	6·94	6·59	6·39	6·26	6·16	6·09	6·04	6·00	5·96	5·91	5·86	5·80	5·77	5·75	5·72	5·69	5·66	5·63
5	6·61	5·79	5·41	5·19	5·05	4·95	4·88	4·82	4·77	4·74	4·68	4·62	4·56	4·53	4·50	4·46	4·43	4·40	4·36
6	5·99	5·14	4·76	4·53	4·39	4·28	4·21	4·15	4·10	4·06	4·00	3·94	3·87	3·84	3·81	3·77	3·74	3·70	3·67
7	5·59	4·74	4·35	4·12	3·97	3·87	3·79	3·73	3·68	3·64	3·57	3·51	3·44	3·41	3·38	3·34	3·30	3·27	3·23
8	5·32	4·46	4·07	3·84	3·69	3·58	3·50	3·44	3·39	3·35	3·28	3·22	3·15	3·12	3·08	3·04	3·01	2·97	2·93
9	5·12	4·26	3·86	3·63	3·48	3·37	3·29	3·23	3·18	3·14	3·07	3·01	2·94	2·90	2·86	2·83	2·79	2·75	2·71
10	4·96	4·10	3·71	3·48	3·33	3·22	3·14	3·07	3·02	2·98	2·91	2·85	2·77	2·74	2·70	2·66	2·62	2·58	2·54
11	4·84	3·98	3·59	3·36	3·20	3·09	3·01	2·95	2·90	2·85	2·79	2·72	2·65	2·61	2·57	2·53	2·49	2·45	2·40
12	4·75	3·89	3·49	3·26	3·11	3·00	2·91	2·85	2·80	2·75	2·69	2·62	2·54	2·51	2·47	2·43	2·38	2·34	2·30
13	4·67	3·81	3·41	3·18	3·03	2·92	2·83	2·77	2·71	2·67	2·60	2·53	2·46	2·42	2·38	2·34	2·30	2·25	2·21
14	4·60	3·74	3·34	3·11	2·96	2·85	2·76	2·70	2·65	2·60	2·53	2·46	2·39	2·35	2·31	2·27	2·22	2·18	2·13
15	4·54	3·68	3·29	3·06	2·90	2·79	2·71	2·64	2·59	2·54	2·48	2·40	2·33	2·29	2·25	2·20	2·16	2·11	2·07
16	4·49	3·63	3·24	3·01	2·85	2·74	2·66	2·59	2·54	2·49	2·42	2·35	2·28	2·24	2·19	2·15	2·11	2·06	2·01
17	4·45	3·59	3·20	2·96	2·81	2·70	2·61	2·55	2·49	2·45	2·38	2·31	2·23	2·19	2·15	2·10	2·06	2·01	1·96
18	4·41	3·55	3·16	2·93	2·77	2·66	2·58	2·51	2·46	2·41	2·34	2·27	2·19	2·15	2·11	2·06	2·02	1·97	1·92
19	4·38	3·52	3·13	2·90	2·74	2·63	2·54	2·48	2·42	2·38	2·31	2·23	2·16	2·11	2·07	2·03	1·98	1·93	1·88
20	4·35	3·49	3·10	2·87	2·71	2·60	2·51	2·45	2·39	2·35	2·28	2·20	2·12	2·08	2·04	1·99	1·95	1·90	1·84
21	4·32	3·47	3·07	2·84	2·68	2·57	2·49	2·42	2·37	2·32	2·25	2·18	2·10	2·05	2·01	1·96	1·92	1·87	1·81
22	4·30	3·44	3·05	2·82	2·66	2·55	2·46	2·40	2·34	2·30	2·23	2·15	2·07	2·03	1·98	1·94	1·89	1·84	1·78
23	4·28	3·42	3·03	2·80	2·64	2·53	2·44	2·37	2·32	2·27	2·20	2·13	2·05	2·01	1·96	1·91	1·86	1·81	1·76
24	4·26	3·40	3·01	2·78	2·62	2·51	2·42	2·36	2·30	2·25	2·18	2·11	2·03	1·98	1·94	1·89	1·84	1·79	1·73
25	4·24	3·39	2·99	2·76	2·60	2·49	2·40	2·34	2·28	2·24	2·16	2·09	2·01	1·96	1·92	1·87	1·82	1·77	1·71
26	4·23	3·37	2·98	2·74	2·59	2·47	2·39	2·32	2·27	2·22	2·15	2·07	1·99	1·95	1·90	1·85	1·80	1·75	1·69
27	4·21	3·35	2·96	2·73	2·57	2·46	2·37	2·31	2·25	2·20	2·13	2·06	1·97	1·93	1·88	1·84	1·79	1·73	1·67
28	4·20	3·34	2·95	2·71	2·56	2·45	2·36	2·29	2·24	2·19	2·12	2·04	1·96	1·91	1·87	1·82	1·77	1·71	1·65
29	4·18	3·33	2·93	2·70	2·55	2·43	2·35	2·28	2·22	2·18	2·10	2·03	1·94	1·90	1·85	1·81	1·75	1·70	1·64
30	4·17	3·32	2·92	2·69	2·53	2·42	2·33	2·27	2·21	2·16	2·09	2·01	1·93	1·89	1·84	1·79	1·74	1·68	1·62
40	4·08	3·23	2·84	2·61	2·45	2·34	2·25	2·18	2·12	2·08	2·00	1·92	1·84	1·79	1·74	1·69	1·64	1·58	1·51
60	4·00	3·15	2·76	2·53	2·37	2·25	2·17	2·10	2·04	1·99	1·92	1·84	1·75	1·70	1·65	1·59	1·53	1·47	1·39
120	3·92	3·07	2·68	2·45	2·29	2·17	2·09	2·02	1·96	1·91	1·83	1·75	1·66	1·61	1·55	1·50	1·43	1·35	1·25
∞	3·84	3·00	2·60	2·37	2·21	2·10	2·01	1·94	1·88	1·83	1·75	1·67	1·57	1·52	1·46	1·39	1·32	1·22	1·00

*Reprinted from *Biometrika Tables for Statisticians*, Vol. I, Table 18, pp. 159–162, by permission of the Biometrika Trustees.

For $p = 0.975$

v_2＼v_1	1	2	3	4	5	6	7	8	9	10	12	15	20	24	30	40	60	120	∞
1	647·8	799·5	864·2	899·6	921·8	937·1	948·2	956·7	963·3	968·6	976·7	984·9	993·1	997·2	1001	1006	1010	1014	1018
2	38·51	39·00	39·17	39·25	39·30	39·33	39·36	39·37	39·39	39·40	39·41	39·43	39·45	39·46	39·46	39·47	39·48	39·49	39·50
3	17·44	16·04	15·44	15·10	14·88	14·73	14·62	14·54	14·47	14·42	14·34	14·25	14·17	14·12	14·08	14·04	13·99	13·95	13·90
4	12·22	10·65	9·98	9·60	9·36	9·20	9·07	8·98	8·90	8·84	8·75	8·66	8·56	8·51	8·46	8·41	8·36	8·31	8·26
5	10·01	8·43	7·76	7·39	7·15	6·98	6·85	6·76	6·68	6·62	6·52	6·43	6·33	6·28	6·23	6·18	6·12	6·07	6·02
6	8·81	7·26	6·60	6·23	5·99	5·82	5·70	5·60	5·52	5·46	5·37	5·27	5·17	5·12	5·07	5·01	4·96	4·90	4·85
7	8·07	6·54	5·89	5·52	5·29	5·12	4·99	4·90	4·82	4·76	4·67	4·57	4·47	4·42	4·36	4·31	4·25	4·20	4·14
8	7·57	6·06	5·42	5·05	4·82	4·65	4·53	4·43	4·36	4·30	4·20	4·10	4·00	3·95	3·89	3·84	3·78	3·73	3·67
9	7·21	5·71	5·08	4·72	4·48	4·32	4·20	4·10	4·03	3·96	3·87	3·77	3·67	3·61	3·56	3·51	3·45	3·39	3·33
10	6·94	5·46	4·83	4·47	4·24	4·07	3·95	3·85	3·78	3·72	3·62	3·52	3·42	3·37	3·31	3·26	3·20	3·14	3·08
11	6·72	5·26	4·63	4·28	4·04	3·88	3·76	3·66	3·59	3·53	3·43	3·33	3·23	3·17	3·12	3·06	3·00	2·94	2·88
12	6·55	5·10	4·47	4·12	3·89	3·73	3·61	3·51	3·44	3·37	3·28	3·18	3·07	3·02	2·96	2·91	2·85	2·79	2·72
13	6·41	4·97	4·35	4·00	3·77	3·60	3·48	3·39	3·31	3·25	3·15	3·05	2·95	2·89	2·84	2·78	2·72	2·66	2·60
14	6·30	4·86	4·24	3·89	3·66	3·50	3·38	3·29	3·21	3·15	3·05	2·95	2·84	2·79	2·73	2·67	2·61	2·55	2·49
15	6·20	4·77	4·15	3·80	3·58	3·41	3·29	3·20	3·12	3·06	2·96	2·86	2·76	2·70	2·64	2·59	2·52	2·46	2·40
16	6·12	4·69	4·08	3·73	3·50	3·34	3·22	3·12	3·05	2·99	2·89	2·79	2·68	2·63	2·57	2·51	2·45	2·38	2·32
17	6·04	4·62	4·01	3·66	3·44	3·28	3·16	3·06	2·98	2·92	2·82	2·72	2·62	2·56	2·50	2·44	2·38	2·32	2·25
18	5·98	4·56	3·95	3·61	3·38	3·22	3·10	3·01	2·93	2·87	2·77	2·67	2·56	2·50	2·44	2·38	2·32	2·26	2·19
19	5·92	4·51	3·90	3·56	3·33	3·17	3·05	2·96	2·88	2·82	2·72	2·62	2·51	2·45	2·39	2·33	2·27	2·20	2·13
20	5·87	4·46	3·86	3·51	3·29	3·13	3·01	2·91	2·84	2·77	2·68	2·57	2·46	2·41	2·35	2·29	2·22	2·16	2·09
21	5·83	4·42	3·82	3·48	3·25	3·09	2·97	2·87	2·80	2·73	2·64	2·53	2·42	2·37	2·31	2·25	2·18	2·11	2·04
22	5·79	4·38	3·78	3·44	3·22	3·05	2·93	2·84	2·76	2·70	2·60	2·50	2·39	2·33	2·27	2·21	2·14	2·08	2·00
23	5·75	4·35	3·75	3·41	3·18	3·02	2·90	2·81	2·73	2·67	2·57	2·47	2·36	2·30	2·24	2·18	2·11	2·04	1·97
24	5·72	4·32	3·72	3·38	3·15	2·99	2·87	2·78	2·70	2·64	2·54	2·44	2·33	2·27	2·21	2·15	2·08	2·01	1·94
25	5·69	4·29	3·69	3·35	3·13	2·97	2·85	2·75	2·68	2·61	2·51	2·41	2·30	2·24	2·18	2·12	2·05	1·98	1·91
26	5·66	4·27	3·67	3·33	3·10	2·94	2·82	2·73	2·65	2·59	2·49	2·39	2·28	2·22	2·16	2·09	2·03	1·95	1·88
27	5·63	4·24	3·65	3·31	3·08	2·92	2·80	2·71	2·63	2·57	2·47	2·36	2·25	2·19	2·13	2·07	2·00	1·93	1·85
28	5·61	4·22	3·63	3·29	3·06	2·90	2·78	2·69	2·61	2·55	2·45	2·34	2·23	2·17	2·11	2·05	1·98	1·91	1·83
29	5·59	4·20	3·61	3·27	3·04	2·88	2·76	2·67	2·59	2·53	2·43	2·32	2·21	2·15	2·09	2·03	1·96	1·89	1·81
30	5·57	4·18	3·59	3·25	3·03	2·87	2·75	2·65	2·57	2·51	2·41	2·31	2·20	2·14	2·07	2·01	1·94	1·87	1·79
40	5·42	4·05	3·46	3·13	2·90	2·74	2·62	2·53	2·45	2·39	2·29	2·18	2·07	2·01	1·94	1·88	1·80	1·72	1·64
60	5·29	3·93	3·34	3·01	2·79	2·63	2·51	2·41	2·33	2·27	2·17	2·06	1·94	1·88	1·82	1·74	1·67	1·58	1·48
120	5·15	3·80	3·23	2·89	2·67	2·52	2·39	2·30	2·22	2·16	2·05	1·94	1·82	1·76	1·69	1·61	1·53	1·43	1·31
∞	5·02	3·69	3·12	2·79	2·57	2·41	2·29	2·19	2·11	2·05	1·94	1·83	1·71	1·64	1·57	1·48	1·39	1·27	1·00

TABLE H (Cont.)

For $p = 0.99$

v_2 \ v_1	1	2	3	4	5	6	7	8	9	10	12	15	20	24	30	40	60	120	∞
1	4052	4999·5	5403	5625	5764	5859	5928	5982	6022	6056	6106	6157	6209	6235	6261	6287	6313	6339	6366
2	98·50	99·00	99·17	99·25	99·30	99·33	99·36	99·37	99·39	99·40	99·42	99·43	99·45	99·46	99·47	99·47	99·48	99·49	99·50
3	34·12	30·82	29·46	28·71	28·24	27·91	27·67	27·49	27·35	27·23	27·05	26·87	26·69	26·60	26·50	26·41	26·32	26·22	26·13
4	21·20	18·00	16·69	15·98	15·52	15·21	14·98	14·80	14·66	14·55	14·37	14·20	14·02	13·93	13·84	13·75	13·65	13·56	13·46
5	16·26	13·27	12·06	11·39	10·97	10·67	10·46	10·29	10·16	10·05	9·89	9·72	9·55	9·47	9·38	9·29	9·20	9·11	9·02
6	13·75	10·92	9·78	9·15	8·75	8·47	8·26	8·10	7·98	7·87	7·72	7·56	7·40	7·31	7·23	7·14	7·06	6·97	6·88
7	12·25	9·55	8·45	7·85	7·46	7·19	6·99	6·84	6·72	6·62	6·47	6·31	6·16	6·07	5·99	5·91	5·82	5·74	5·65
8	11·26	8·65	7·59	7·01	6·63	6·37	6·18	6·03	5·91	5·81	5·67	5·52	5·36	5·28	5·20	5·12	5·03	4·95	4·86
9	10·56	8·02	6·99	6·42	6·06	5·80	5·61	5·47	5·35	5·26	5·11	4·96	4·81	4·73	4·65	4·57	4·48	4·40	4·31
10	10·04	7·56	6·55	5·99	5·64	5·39	5·20	5·06	4·94	4·85	4·71	4·56	4·41	4·33	4·25	4·17	4·08	4·00	3·91
11	9·65	7·21	6·22	5·67	5·32	5·07	4·89	4·74	4·63	4·54	4·40	4·25	4·10	4·02	3·94	3·86	3·78	3·69	3·60
12	9·33	6·93	5·95	5·41	5·06	4·82	4·64	4·50	4·39	4·30	4·16	4·01	3·86	3·78	3·70	3·62	3·54	3·45	3·36
13	9·07	6·70	5·74	5·21	4·86	4·62	4·44	4·30	4·19	4·10	3·96	3·82	3·66	3·59	3·51	3·43	3·34	3·25	3·17
14	8·86	6·51	5·56	5·04	4·69	4·46	4·28	4·14	4·03	3·94	3·80	3·66	3·51	3·43	3·35	3·27	3·18	3·09	3·00
15	8·68	6·36	5·42	4·89	4·56	4·32	4·14	4·00	3·89	3·80	3·67	3·52	3·37	3·29	3·21	3·13	3·05	2·96	2·87
16	8·53	6·23	5·29	4·77	4·44	4·20	4·03	3·89	3·78	3·69	3·55	3·41	3·26	3·18	3·10	3·02	2·93	2·84	2·75
17	8·40	6·11	5·18	4·67	4·34	4·10	3·93	3·79	3·68	3·59	3·46	3·31	3·16	3·08	3·00	2·92	2·83	2·75	2·65
18	8·29	6·01	5·09	4·58	4·25	4·01	3·84	3·71	3·60	3·51	3·37	3·23	3·08	3·00	2·92	2·84	2·75	2·66	2·57
19	8·18	5·93	5·01	4·50	4·17	3·94	3·77	3·63	3·52	3·43	3·30	3·15	3·00	2·92	2·84	2·76	2·67	2·58	2·49
20	8·10	5·85	4·94	4·43	4·10	3·87	3·70	3·56	3·46	3·37	3·23	3·09	2·94	2·86	2·78	2·69	2·61	2·52	2·42
21	8·02	5·78	4·87	4·37	4·04	3·81	3·64	3·51	3·40	3·31	3·17	3·03	2·88	2·80	2·72	2·64	2·55	2·46	2·36
22	7·95	5·72	4·82	4·31	3·99	3·76	3·59	3·45	3·35	3·26	3·12	2·98	2·83	2·75	2·67	2·58	2·50	2·40	2·31
23	7·88	5·66	4·76	4·26	3·94	3·71	3·54	3·41	3·30	3·21	3·07	2·93	2·78	2·70	2·62	2·54	2·45	2·35	2·26
24	7·82	5·61	4·72	4·22	3·90	3·67	3·50	3·36	3·26	3·17	3·03	2·89	2·74	2·66	2·58	2·49	2·40	2·31	2·21
25	7·77	5·57	4·68	4·18	3·85	3·63	3·46	3·32	3·22	3·13	2·99	2·85	2·70	2·62	2·54	2·45	2·36	2·27	2·17
26	7·72	5·53	4·64	4·14	3·82	3·59	3·42	3·29	3·18	3·09	2·96	2·81	2·66	2·58	2·50	2·42	2·33	2·23	2·13
27	7·68	5·49	4·60	4·11	3·78	3·56	3·39	3·26	3·15	3·06	2·93	2·78	2·63	2·55	2·47	2·38	2·29	2·20	2·10
28	7·64	5·45	4·57	4·07	3·75	3·53	3·36	3·23	3·12	3·03	2·90	2·75	2·60	2·52	2·44	2·35	2·26	2·17	2·06
29	7·60	5·42	4·54	4·04	3·73	3·50	3·33	3·20	3·09	3·00	2·87	2·73	2·57	2·49	2·41	2·33	2·23	2·14	2·03
30	7·56	5·39	4·51	4·02	3·70	3·47	3·30	3·17	3·07	2·98	2·84	2·70	2·55	2·47	2·39	2·30	2·21	2·11	2·01
40	7·31	5·18	4·31	3·83	3·51	3·29	3·12	2·99	2·89	2·80	2·66	2·52	2·37	2·29	2·20	2·11	2·02	1·92	1·80
60	7·08	4·98	4·13	3·65	3·34	3·12	2·95	2·82	2·72	2·63	2·50	2·35	2·20	2·12	2·03	1·94	1·84	1·73	1·60
120	6·85	4·79	3·95	3·48	3·17	2·96	2·79	2·66	2·56	2·47	2·34	2·19	2·03	1·95	1·86	1·76	1·66	1·53	1·38
∞	6·63	4·61	3·78	3·32	3·02	2·80	2·64	2·51	2·41	2·32	2·18	2·04	1·88	1·79	1·70	1·59	1·47	1·32	1·00

For $p = 0.995$

$r_2 \backslash r_1$	1	2	3	4	5	6	7	8	9	10	12	15	20	24	30	40	60	120	∞
1	16211	20000	21615	22500	23056	23437	23715	23925	24091	24224	24426	24630	24836	24940	25044	25148	25253	25359	25465
2	198·5	199·0	199·2	199·2	199·3	199·3	199·4	199·4	199·4	199·4	199·4	199·4	199·4	199·5	199·5	199·5	199·5	199·5	199·5
3	55·55	49·80	47·47	46·19	45·39	44·84	44·43	44·13	43·88	43·69	43·39	43·08	42·78	42·62	42·47	42·31	42·15	41·99	41·83
4	31·33	26·28	24·26	23·15	22·46	21·97	21·62	21·35	21·14	20·97	20·70	20·44	20·17	20·03	19·89	19·75	19·61	19·47	19·32
5	22·78	18·31	16·53	15·56	14·94	14·51	14·20	13·96	13·77	13·62	13·38	13·15	12·90	12·78	12·66	12·53	12·40	12·27	12·14
6	18·63	14·54	12·92	12·03	11·46	11·07	10·79	10·57	10·39	10·25	10·03	9·81	9·59	9·47	9·36	9·24	9·12	9·00	8·88
7	16·24	12·40	10·88	10·05	9·52	9·16	8·89	8·68	8·51	8·38	8·18	7·97	7·75	7·65	7·53	7·42	7·31	7·19	7·08
8	14·69	11·04	9·60	8·81	8·30	7·95	7·69	7·50	7·34	7·21	7·01	6·81	6·61	6·50	6·40	6·29	6·18	6·06	5·95
9	13·61	10·11	8·72	7·96	7·47	7·13	6·88	6·69	6·54	6·42	6·23	6·03	5·83	5·73	5·62	5·52	5·41	5·30	5·19
10	12·83	9·43	8·08	7·34	6·87	6·54	6·30	6·12	5·97	5·85	5·66	5·47	5·27	5·17	5·07	4·97	4·86	4·75	4·64
11	12·23	8·91	7·60	6·88	6·42	6·10	5·86	5·68	5·54	5·42	5·24	5·05	4·86	4·76	4·65	4·55	4·44	4·34	4·23
12	11·75	8·51	7·23	6·52	6·07	5·76	5·52	5·35	5·20	5·09	4·91	4·72	4·53	4·43	4·33	4·23	4·12	4·01	3·90
13	11·37	8·19	6·93	6·23	5·79	5·48	5·25	5·08	4·94	4·82	4·64	4·46	4·27	4·17	4·07	3·97	3·87	3·76	3·65
14	11·06	7·92	6·68	6·00	5·56	5·26	5·03	4·86	4·72	4·60	4·43	4·25	4·06	3·96	3·86	3·76	3·66	3·55	3·44
15	10·80	7·70	6·48	5·80	5·37	5·07	4·85	4·67	4·54	4·42	4·25	4·07	3·88	3·79	3·69	3·58	3·48	3·37	3·26
16	10·58	7·51	6·30	5·64	5·21	4·91	4·69	4·52	4·38	4·27	4·10	3·92	3·73	3·64	3·54	3·44	3·33	3·22	3·11
17	10·38	7·35	6·16	5·50	5·07	4·78	4·56	4·39	4·25	4·14	3·97	3·79	3·61	3·51	3·41	3·31	3·21	3·10	2·98
18	10·22	7·21	6·03	5·37	4·96	4·66	4·44	4·28	4·14	4·03	3·86	3·68	3·50	3·40	3·30	3·20	3·10	2·99	2·87
19	10·07	7·09	5·92	5·27	4·85	4·56	4·34	4·18	4·04	3·93	3·76	3·59	3·40	3·31	3·21	3·11	3·00	2·89	2·78
20	9·94	6·99	5·82	5·17	4·76	4·47	4·26	4·09	3·96	3·85	3·68	3·50	3·32	3·22	3·12	3·02	2·92	2·81	2·69
21	9·83	6·89	5·73	5·09	4·68	4·39	4·18	4·01	3·88	3·77	3·60	3·43	3·24	3·15	3·05	2·95	2·84	2·73	2·61
22	9·73	6·81	5·65	5·02	4·61	4·32	4·11	3·94	3·81	3·70	3·54	3·36	3·18	3·08	2·98	2·88	2·77	2·66	2·55
23	9·63	6·73	5·58	4·95	4·54	4·26	4·05	3·88	3·75	3·64	3·47	3·30	3·12	3·02	2·92	2·82	2·71	2·60	2·48
24	9·55	6·66	5·52	4·89	4·49	4·20	3·99	3·83	3·69	3·59	3·42	3·25	3·06	2·97	2·87	2·77	2·66	2·55	2·43
25	9·48	6·60	5·46	4·84	4·43	4·15	3·94	3·78	3·64	3·54	3·37	3·20	3·01	2·92	2·82	2·72	2·61	2·50	2·38
26	9·41	6·54	5·41	4·79	4·38	4·10	3·89	3·73	3·60	3·49	3·33	3·15	2·97	2·87	2·77	2·67	2·56	2·45	2·33
27	9·34	6·49	5·36	4·74	4·34	4·06	3·85	3·69	3·56	3·45	3·28	3·11	2·93	2·83	2·73	2·63	2·52	2·41	2·29
28	9·28	6·44	5·32	4·70	4·30	4·02	3·81	3·65	3·52	3·41	3·25	3·07	2·89	2·79	2·69	2·59	2·48	2·37	2·25
29	9·23	6·40	5·28	4·66	4·26	3·98	3·77	3·61	3·48	3·38	3·21	3·04	2·86	2·76	2·66	2·56	2·45	2·33	2·21
30	9·18	6·35	5·24	4·62	4·23	3·95	3·74	3·58	3·45	3·34	3·18	3·01	2·82	2·73	2·63	2·52	2·42	2·30	2·18
40	8·83	6·07	4·98	4·37	3·99	3·71	3·51	3·35	3·22	3·12	2·95	2·78	2·60	2·50	2·40	2·30	2·18	2·06	1·93
60	8·49	5·79	4·73	4·14	3·76	3·49	3·29	3·13	3·01	2·90	2·74	2·57	2·39	2·29	2·19	2·08	1·96	1·83	1·69
120	8·18	5·54	4·50	3·92	3·55	3·28	3·09	2·93	2·81	2·71	2·54	2·37	2·19	2·09	1·98	1·87	1·75	1·61	1·43
∞	7·88	5·30	4·28	3·72	3·35	3·09	2·90	2·74	2·62	2·52	2·36	2·19	2·00	1·90	1·79	1·67	1·53	1·36	1·00

TABLE I. Random Numbers*

	1-5	6-10	11-15	16-20	21-25	26-30	31-35	36-40	41-45	46-50
1	85517	81525	33438	23564	46044	02504	89381	82450	81551	61720
2	89930	20229	59952	98576	75591	75432	91309	19748	52871	92347
3	55169	18607	24222	06817	05215	44311	34831	42787	41341	88196
4	91267	73443	35256	68078	24655	68274	56594	14685	62681	66259
5	26938	15002	34410	91512	36622	06868	32011	71056	68257	54203
6	22689	97094	40368	94294	28268	35656	41149	10282	63703	84245
7	76349	05925	04463	71004	37569	28204	35397	81719	31963	30337
8	91192	58195	86891	42544	16020	63359	05973	97272	75316	49045
9	83473	58889	01823	50312	93468	22102	48249	47723	70903	84422
10	64124	35774	28222	60999	23686	43597	94396	84390	29584	69433
11	76521	84991	49426	12111	06763	49042	15964	86074	97284	46743
12	44738	69418	79383	25793	26643	70818	17346	46320	19027	96850
13	52757	99418	19992	25236	46923	74979	24780	96067	77691	07740
14	58757	12052	81068	83780	12124	57615	05700	97929	15013	82826
15	23930	66624	90167	92422	46866	50364	36212	18533	85896	29736
16	98930	42707	96469	56992	61578	24259	23505	92807	60976	01133
17	20099	82492	54019	64512	61255	72641	17134	03152	15071	95974
18	14847	91085	49572	72542	83096	96816	63087	76302	33439	87944
19	42969	08484	94248	92601	27134	62723	77449	84087	37010	31988
20	01630	09011	95842	71735	80459	09329	02608	33484	49952	32252
21	88477	85677	84732	88537	70313	61552	78566	96429	35883	99179
22	78653	51095	93236	91342	21206	51564	81596	13707	72882	71635
23	81104	65536	97753	65119	73229	11085	24283	38749	18885	59673
24	36900	97906	60076	94963	70769	96144	24295	05472	02235	19819
25	48144	45773	77461	78269	15134	75572	06809	74920	59392	50729
26	84382	34002	56599	54073	29088	16908	83077	52552	37304	22277
27	13808	26121	80436	93807	73649	22384	12732	39419	91461	24227
28	13427	20686	31527	00699	27039	51823	15872	70613	85831	66863
29	83418	51853	41660	53304	64050	59379	35426	28991	67243	47260
30	07789	90809	84452	09516	45101	48390	76618	55521	69530	93097
31	27393	41840	70033	62658	32479	90055	92876	44990	36137	65595
32	36220	50640	09003	15513	83034	21067	28461	75005	91305	12509
33	48519	76444	98117	36097	80447	05232	97887	39084	32567	36152
34	33692	39550	85325	49677	83497	80718	32630	50671	18337	52420
35	81686	75404	04069	27478	22303	33660	92863	44580	11315	96501
36	89449	86119	40064	52417	06620	65364	29568	17286	25913	55171
37	93994	48605	27009	07304	89374	24101	43196	72820	23229	35324
38	72415	63235	99631	53549	52361	54395	64668	75371	00546	05864
39	99578	12458	46210	63562	28765	32812	71947	46329	60305	09825
40	13623	92018	06663	39756	94489	38162	11583	85859	53896	08242
41	66236	72173	10799	57605	70045	65998	63907	82489	49935	36400
42	78414	24018	81353	24447	20397	64671	91753	56744	57706	19954
43	74925	99155	42821	24403	62415	00500	10563	56204	94736	64341
44	12575	90337	31619	58703	95469	01316	96833	84505	84945	66743
45	90587	86517	56636	18289	67298	53393	61389	03635	39002	15648
46	56660	26132	81159	63498	74431	58536	25630	87276	37735	78409
47	46326	12403	83023	99726	46829	04834	56749	44509	59318	17776
48	15275	14227	27956	64859	49954	86585	43391	04922	32085	90106
49	94650	50140	63633	95947	27329	40567	29214	37700	80557	45112
50	67991	40573	90349	75550	26299	49928	17646	93583	36920	00226

*Reprinted by permission from Rand Corporation, *A Million Random Digits With 100,000 Normal Deviates* (Glencoe, Ill.: Free Press of Glencoe, 1955), pp. 266–267.

TABLE I (Cont.)

331

	1-5	6-10	11-15	16-20	21-25	26-30	31-35	36-40	41-45	46-50
1	71697	84395	91705	58188	67452	80847	71128	46973	15992	97747
2	35739	01715	66192	27218	74026	19270	24706	08000	69662	18064
3	79403	11945	10260	53954	59918	73014	09431	33324	55821	32309
4	43287	62243	35804	35245	84321	72384	00122	64516	27241	95803
5	27301	52127	04924	45355	69884	63401	27852	68143	26367	15500
6	73808	13547	94767	16877	99037	23335	04648	25835	16787	40873
7	96818	50168	99701	04633	62496	93835	09270	37256	77615	10454
8	16060	57724	82092	50495	41834	14154	21618	00999	78680	73308
9	27046	25046	87192	15077	00268	94098	65690	39876	62144	87435
10	48505	20660	90682	59018	90236	03236	86001	05408	36975	68606
11	59033	86705	65910	58500	27531	79960	35790	25009	95852	89419
12	38037	61410	60515	44512	14600	67952	32878	27261	30453	16630
13	31996	49725	10172	77184	27277	71306	42951	59626	99940	59098
14	12668	91233	88787	59535	49642	84125	05679	42127	15690	00370
15	19160	29346	93245	84815	11543	07769	48415	91665	42586	80304
16	21637	27788	51842	38511	03525	60016	34857	90686	39202	91632
17	51113	23525	64349	56773	17907	37489	44219	87051	21017	50955
18	96047	45107	37319	31059	40345	65414	82007	45383	21791	26460
19	52531	72695	12560	09520	21023	41753	12336	38114	44918	82150
20	52891	45252	02577	80275	11178	68593	78207	35104	21405	46166
21	81988	90968	54114	92531	98125	90247	42142	22189	94485	14363
22	76282	11043	02218	35497	83924	01429	64932	66931	29280	52849
23	28702	51106	89848	02546	52539	17681	00262	35208	16332	65716
24	74963	59244	74394	30052	37935	68531	41029	60917	79866	26512
25	25123	86332	59738	55438	56053	58803	12775	72656	18529	25090
26	01027	79775	37666	16816	71697	99021	24676	25916	58558	76849
27	76378	25003	59725	66505	41450	58874	35674	75313	98516	52018
28	82302	38174	73793	13159	21180	38425	81352	11464	01761	32357
29	86443	56206	90218	98785	92430	33004	54651	11512	66613	48970
30	30224	37414	02926	45499	91692	04396	04984	16874	75919	17504
31	26539	47073	29754	00982	44265	62551	10824	86463	19247	65690
32	03643	53529	11430	95645	72943	24475	76795	71817	75517	63344
33	68498	69345	46554	09903	19054	57996	65323	27796	93981	74804
34	25031	35246	85213	06681	53858	28752	02218	02726	52187	44077
35	51664	98609	48290	20914	02946	59144	34025	07166	62568	65912
36	59570	09001	40484	77932	97256	38756	23982	09885	00896	50389
37	01925	11981	42809	09625	70639	02505	00711	91376	70697	01669
38	73148	44748	61150	43328	48033	77415	83811	55755	42436	84540
39	98851	61614	07571	37791	80094	62994	88640	79680	82716	23483
40	10168	38203	18288	86384	36804	01865	83627	57148	16850	81053
41	31050	73000	52752	49807	27295	57224	50371	97555	20876	46263
42	46155	91045	03033	28469	40065	09597	05488	65163	18308	06694
43	38560	02861	16097	25428	38168	20369	63582	54261	11156	17843
44	90666	06547	05117	13076	85568	10835	23817	07933	28951	03939
45	64122	27975	46798	92347	52031	98236	26392	37653	59724	49577
46	70100	58026	34226	36441	62150	18683	69024	36681	29199	84694
47	96245	22660	22274	16722	30621	30035	25347	78369	28181	97784
48	89627	99474	65841	87477	03964	01170	65620	39097	97428	96616
49	56832	55686	65531	59171	01300	52802	91762	40164	82533	47894
50	50240	04386	14679	12478	24005	83447	64196	84605	33379	41740

TABLE J. Values of $w_r = \frac{1}{2} \log_e \frac{1+r^*}{1-r}$

r	w_r	r	w_r	r	w_r	r	w_r
.00	.000	.25	.255	.50	.549	.75	.973
.01	.010	.26	.266	.51	.563	.76	.996
.02	.020	.27	.277	.52	.576	.77	1.020
.03	.030	.28	.288	.53	.590	.78	1.045
.04	.040	.29	.299	.54	.604	.79	1.071
.05	.050	.30	.310	.55	.618	.80	1.099
.06	.060	.31	.321	.56	.633	.81	1.127
.07	.070	.32	.332	.57	.648	.82	1.157
.08	.080	.33	.343	.58	.662	.83	1.188
.09	.090	.34	.354	.59	.678	.84	1.221
.10	.100	.35	.365	.60	.693	.85	1.256
.11	.110	.36	.377	.61	.709	.86	1.293
.12	.121	.37	.388	.62	.725	.87	1.333
.13	.131	.38	.400	.63	.741	.88	1.376
.14	.141	.39	.412	.64	.758	.89	1.422
.15	.151	.40	.424	.65	.775	.90	1.472
.16	.161	.41	.436	.66	.793	.91	1.528
.17	.172	.42	.448	.67	.811	.92	1.589
.18	.182	.43	.460	.68	.829	.93	1.658
.19	.192	.44	.472	.69	.848	.94	1.738
.20	.203	.45	.485	.70	.867	.95	1.832
.21	.213	.46	.497	.71	.887	.96	1.946
.22	.224	.47	.510	.72	.908	.97	2.092
.23	.234	.48	.523	.73	.929	.98	2.298
.24	.245	.49	.536	.74	.950	.99	2.647

*Reprinted by permission from Paul G. Hoel, *Elementary Statistics*, 2nd ed. (New York: Wiley, 1966), Table VI, p. 331.

TABLE K. The Probability Function for the Kolmogorov-Smirnov One-Sample Statistic (Values of $D(\max)_{n;p}$)*

$$P(D(\max)_n < D(\max)_{n;p}) = p$$

n \ p	.80	.90	.95	.98	.99
1	.900	.950	.975	.990	.995
2	.684	.776	.842	.900	.929
3	.565	.636	.708	.785	.829
4	.493	.565	.624	.689	.734
5	.447	.509	.563	.627	.669
6	.410	.468	.519	.577	.617
7	.381	.436	.483	.538	.576
8	.358	.410	.454	.507	.542
9	.339	.387	.430	.480	.513
10	.323	.369	.409	.457	.489
11	.308	.352	.391	.437	.468
12	.296	.338	.375	.419	.449
13	.285	.325	.361	.404	.432
14	.275	.314	.349	.390	.418
15	.266	.304	.338	.377	.404
16	.258	.295	.327	.366	.392
17	.250	.286	.318	.355	.381
18	.244	.279	.309	.346	.371
19	.237	.271	.301	.337	.361
20	.232	.265	.294	.329	.352
21	.226	.259	.287	.321	.344
22	.221	.253	.281	.314	.337
23	.216	.247	.275	.307	.330
24	.212	.242	.269	.301	.323
25	.208	.238	.264	.295	.317
26	.204	.233	.259	.290	.311
27	.200	.229	.254	.284	.305
28	.197	.225	.250	.279	.300
29	.193	.221	.246	.275	.295
30	.190	.218	.242	.270	.290
35	.177	.202	.224	.251	.269
40	.165	.189	.210	.235	.252
45	.156	.179	.198	.222	.238
50	.148	.170	.188	.211	.226
60	.136	.155	.172	.193	.207
70	.126	.144	.160	.179	.192
80	.118	.135	.150	.167	.179
90	.111	.127	.141	.158	.169
100	.106	.121	.134	.150	.161

*Adapted from Donald B. Owen, *Handbook of Statistical Tables*, 1962, Addison-Wesley, Reading, Mass., pp. 424–425. Courtesy of U.S. Atomic Energy Commission.

Answers
to Exercises

1. (a) {1, 2, 3, 4, 6, 7, 9}
    (b) {2, 3, 4, 5, 8}
    (c) {2, 4}
    (d) {8}
    (e) {1, 2, 3, 4, 5, 7, 8, 9}
    (f) {2, 4, 5}
    (g) {1, 3, 7, 9}
    (h) ϕ
    (i) {(2,5),(2,8),(3,5),(3,8),(4,5),(4,8),(5,5),(5,8)}
    (j) {(5,2),(5,3),(5,4),(5,5),(8,2),(8,3),(8,4),(8,5)}
2. No; no; yes
3. (a), (b), (c), (f)
4. {2}, {✳},{Δ}, {m,2},{m,✳}, {m,Δ},{2,✳},{✳,Δ}, {m,2,✳}, {m,2,Δ},{m,✳,Δ}, {2,✳,Δ}
5. mutually exclusive (disjoint)
6. (a) mutually exclusive

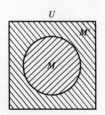

    (b) mutually exclusive
    (c) not mutually exclusive

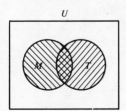

(d) mutually exclusive

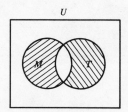

(e) mutually exclusive

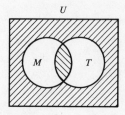

7. all are true
8. (a) {0,1}   (b) U   (c) {2,3,4,5, . . .}
   (d) G   (e) ϕ
9. 32; 2^n
11.

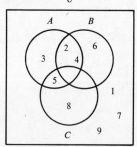

Pages 12–13

1. (a) $y_1 + y_2 + y_3 + y_4 + y_5$
   (b) $x_3^2 + x_4^2 + x_5^2 + x_6^2$
   (c) $x_1y_1 + x_2y_2 + x_3y_3 + x_4y_4 + x_5y_5 + x_6y_6 + x_7y_7$
   (d) $(2x_1 + 5) + (2x_2 + 5) + (2x_3 + 5)$
   (e) $x_2^2f_2 + x_3^2f_3 + x_4^2f_4$

2. (a) $\sum_{i=1}^{17} y_i^2$ (b) $\sum_{i=1}^{100} 2z_i$

   (c) $\sum_{i=2}^{5} x_iy_i^2$ (d) $\sum_{i=1}^{3} (x_i^2 + 5)^3$

3. (a) 15   (b) 3
    (c) 85   (d) 10
4. (a) −9   (b) 23
    (c) 42   (d) 6
5. 607

Page 24

1. mean = 10.4, median = 9.0, mode = 8.0
2. mean = median = mode = 10
3. mean = 3.00, median = 2.95, mode = 3.20
4. (a) mean = $14,563, median = $8,500, mode = $8,500
    (b) The mean is greatly affected by a few extreme values; the median and mode are not.
    (c) Probably the median and mode, since they represent best the salary of the "typical" or "average" person in the sample. However, having all three measures of central location gives us the most complete picture.
    (d) mean = $14,563; the median and mode cannot be calculated from the information given.
5. 0.280
6. 10.05 seconds
7. (a) 30   (b) 0.40
8. (a) 30

$$(b) \quad 109; \frac{\text{total distance}}{\text{total time}} = \frac{4}{2/100 + 2/120} = \frac{2}{1/100 + 1/120},$$

which is the harmonic mean of 100 and 120.

Page 32

1. range = 20, m.d. = 4.15, $s_x^2 = 33.03$, $s_x = 5.75$
2. (a) 5.20   (b) 5.20
3. $s_x^2 = .18$, $s_x = .43$
4. $\bar{x} = 11.77$, $s_x^2 = 0.75$, $s_x = 0.87$
5. (a) No
    (b) $s_x^2 = 0$ when all measurements are equal.
    (c) Yes
6. $\sum_{i=1}^{3} x_i^2 = x_1^2 + x_2^2 + x_3^2$,

$$\left(\sum_{i=1}^{3} x_i\right)^2 = (x_1 + x_2 + x_3)^2 = x_1^2 + x_2^2 + x_3^2 + 2(x_1 x_2 + x_1 x_3 + x_2 x_3)$$

7. Not enough information given.
8. 7.39
9. Store #3 (The means are all about equal, but the variances tell us that steak prices vary more widely in store #3)

Page 37

1. (a) $u_i = (1)x_i + (-50)$ (b) $u_i = 2x_i + 5$
   (c) $u_i = \frac{1}{10} x_i + \left(-\frac{1}{5}\right)$
2. (a) $\bar{u} = 105$, $s_u^2 = 64$, $s_u = 8$ (b) $\bar{u} = 25$, $s_u^2 = 16$, $s_u = 4$
   (c) $\bar{u} = 5$, $s_u^2 = 0.16$, $s_u = 0.4$ (d) $\bar{u} = 0$, $s_u^2 = 1$, $s_u = 1$
3. 5.20
4. $\bar{x} = 425.50$, $s_x^2 = 6.29$, $s_x = 2.51$
5. $\bar{y} = 26$, $s_y^2 = 0.36$, $s_y = 0.6$
6. median = 26.5, mode = 28, range = 4
7. $\bar{u} = 0$, $s_u^2 = 1$, $s_u = 1$
8. mean = 64.4, variance = 81, standard deviation = 9

Pages 46–47

1. 8, 9, 10, . . . , 14
2. 7.4, 7.5, 7.6, . . . , 14.6
3. Errors are:
   (a) Not all classes are of equal length.
   (b) The second and third classes overlap.
   (c) The fifth and sixth classes leave a gap.

4.

Class Limits	Class Marks
16.00–22.99	19.495
23.00–29.99	26.495
30.00–36.99	33.495
37.00–43.99	40.495
44.00–50.99	47.495
51.00–57.99	54.495

   Class Length = 7.00

5. (a) The approximate center is in the 19–24 class; the extremes are (at the outside) 1 and 42; the dispersion is not uniform – in fact measurements occur most frequently in the 7–12 and 31–36 classes. The distribution could be roughly described as "bimodal."

   (b)

Class Boundaries	Class Marks
0.5– 6.5	3.5
6.5–12.5	9.5
12.5–18.5	15.5
18.5–24.5	21.5
24.5–30.5	27.5
30.5–36.5	33.5
36.5–42.5	39.5

   Class Length = 6

6.

Class Boundaries	Class Limits	Class Mark (x_i)	Frequency (f_i)	Percentage
5.5–9.5	6–9	7.5	9	10.0
9.5–13.5	10–13	11.5	15	16.7
13.5–17.5	14–17	15.5	24	26.7
17.5–21.5	18–21	19.5	22	24.4
21.5–25.5	22–25	23.5	10	11.1
25.5–29.5	26–29	27.5	6	6.7
29.5–33.5	30–33	31.5	2	2.2
33.5–37.5	34–37	35.5	0	0.0
37.5–41.5	38–41	39.5	1	1.1
41.5–45.5	42–45	43.5	1	1.1
			90	100.0

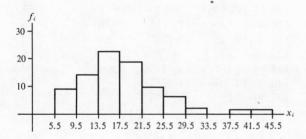

7.

Class Limits	Frequencies	Class Limits	Frequencies
72.0–73.9	2	72.0–75.4	4
74.0–75.9	3	75.5–78.9	7
76.0–77.9	2	79.0–82.4	15
78.0–79.9	6	82.5–85.9	20
80.0–81.9	11	86.0–89.4	10
82.0–83.9	13	89.5–92.9	5
84.0–85.9	9	93.0–96.4	1
86.0–87.9	6		
88.0–89.9	4		62
90.0–91.9	4		
92.0–93.9	1		
94.0–95.9	1		
	62		

8.

Class Boundaries	Class Limits	Class Mark (x_i)	Frequency (f_i)	Percentage
0.4245–0.4275	0.425–0.427	0.426	6	11.1
0.4275–0.4305	0.428–0.430	0.429	6	11.1
0.4305–0.4335	0.431–0.433	0.432	8	14.8
0.4335–0.4365	0.434–0.436	0.435	15	27.8
0.4365–0.4395	0.437–0.439	0.438	11	20.4
0.4395–0.4425	0.440–0.442	0.441	6	11.1
0.4425–0.4455	0.443–0.445	0.444	2	3.7
			54	100.0

Class Length = 0.003

9.

Class Boundaries	Class Limits	Class Mark (x_i)	Frequency (f_i)	Percentage
87.495– 94.995	87.50– 94.99	91.245	3	4.7
94.995–102.495	95.00–102.49	98.745	17	26.6
102.495–109.995	102.50–109.99	106.245	16	25.0
109.995–117.495	110.00–117.49	113.745	10	15.6
117.495–124.995	117.50–124.99	121.245	9	14.1
124.995–132.495	125.00–132.49	128.745	4	6.3
132.495–139.995	132.50–139.99	136.245	2	3.1
139.995–147.495	140.00–147.49	143.745	1	1.6
147.495–154.995	147.50–154.99	151.245	2	3.1
			64	100.1

Class Length = 7.50

10. (a)

#5 Data	Cumulative Frequency	Percentage
less than 1	0	0.0
less than 7	7	4.6
less than 13	59	38.8
less than 19	77	50.7
less than 25	80	52.6
less than 31	95	62.5
less than 37	143	94.1
less than 43	152	100.0

#8 Data	Cumulative Frequency	Percentage
less than 0.425	0	0.0
less than 0.428	6	11.1
less than 0.431	12	22.2
less than 0.434	20	37.0
less than 0.437	35	64.8
less than 0.440	46	85.2
less than 0.443	52	96.3
less than 0.446	54	100.0

#5 Data	Cumulative Frequency	Percentage
more than 0	152	100.0
more than 6	145	95.4
more than 12	93	61.2
more than 18	75	49.3
more than 24	72	47.4
more than 30	57	37.5
more than 36	9	5.9
more than 42	0	0.0

#8 Data	Cumulative Frequency	Percentage
more than 0.424	54	100.0
more than 0.427	48	88.9
more than 0.430	42	77.8
more than 0.433	34	63.0
more than 0.436	19	35.2
more than 0.439	8	14.8
more than 0.442	2	3.7
more than 0.445	0	0.0

10. (b)

Frequency histogram (Exercise 5)

Frequency histogram (Exercise 8)

10. (c)

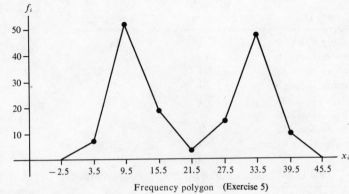

Frequency polygon   (Exercise 5)

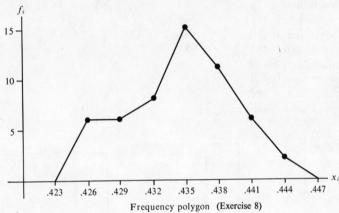

Frequency polygon   (Exercise 8)

11.

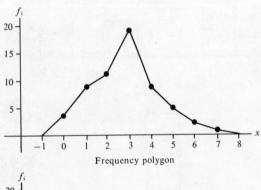

Frequency polygon

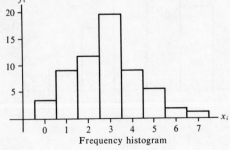

Frequency histogram

Page 53

1. (a) $\bar{x} = 2.82$, median = 3, mode = 3
   (b) range = 7, $s_x^2 = 2.29$, $s_x = 1.51$

2.

x	f	
6.6	1	
6.7	2	
6.8	5	$\bar{x} = 7.1$
6.9	6	median = 7.1
7.0	10	modes = 7.0 and 7.1
7.1	10	range = 1.0
7.2	8	$s_x^2 = 0.053$
7.3	4	$s_x = 0.23$
7.4	5	
7.5	4	
7.6	1	
	56	

3. (a) 0.52   (b) 0.52
4. (a) mean deviation = 1.15
   (b) the standard deviation

Pages 60–61

1. $\hat{\bar{x}} = 11.59$, $\hat{s}_x^2 = 24.45$, $\hat{s}_x = 4.94$ (same for both definitions, except for possible rounding error)
2. (same as answers to Exercise 1)
3. $\hat{\bar{x}} = 17.78$, $\hat{s}_x^2 = 49.39$, $\hat{s}_x = 7.03$
4. $\hat{\bar{x}} = 41.43$, median est. = 42.22, mode est. = 44.5, $\hat{s}_x^2 = 153.49$, $\hat{s}_x = 12.39$
5. (a) For $l = 2.0$:
   $\hat{\bar{x}} = 83.24$, median est. = 83.03, mode est. = 82.95, $\hat{s}_x^2 = 22.54$, $\hat{s}_x = 4.75$
   For $l = 3.5$:
   $\hat{\bar{x}} = 83.18$, median est. = 83.33, mode est. = 84.2, $\hat{s}_x^2 = 22.65$, $\hat{s}_x = 4.76$
   (b) $\hat{\bar{x}} = 0.4345$, median est. = 0.4349, mode est. = 0.4350, $\hat{s}_x^2 = 0.0000234$, $\hat{s}_x = 0.0048$
   (c) $\hat{\bar{x}} = 111.17$, median est. = 108.59, mode est. = 98.75, $\hat{s}_x^2 = 189.67$, $\hat{s}_x = 13\,77$
6. Since the assumption in Definition 4.4 is that "in each class all measurements equal the class mark," it follows that each class mark (x_i) is in essence the *mean of the measurements* in *subset* (class) i, f_i *is the number of measurements* in subset $i (f_1 + f_2 + f_3 + \cdots + f_k = n)$, and therefore, from Theorem 2.1 *the overall mean $\bar{x}$ of the n measurements ... is*

$$\bar{x} = \frac{\sum_{i=1}^{k} f_i x_i}{n}, \quad \text{or equivalently} \quad \frac{\sum_{i=1}^{k} x_i f_i}{n}$$

which is the formula given in Definition 4.4.

Pages 62–63

1. median estimate, 50th percentile, 5th decile, 2nd quartile
2. $P_6 = 7.50$, $P_{82} = 23.77$, third decile $(P_{30}) = 14.24$, first quartile $(P_{25}) = 13.05$

3. Since $P_{10} = 56.17$, $P_{30} = 65.82$, $P_{70} = 77.71$, and $P_{90} = 86.91$,
    therefore, the cut-off scores for each grade are

    A: 87 and above
    B: 78 to 86
    C: 66 to 77
    D: 57 to 65
    E: 56 and below

4. No. The histogram area between P_{30} and P_{50} equals the histogram area between P_{50} and P_{70}, but the corresponding distances along the x axis are not necessarily equal.

Page 65

1. -2.46 (left)
    2.16 (right)
    -1.72 (left)
    0.81 (right)
    0.09 (right)
    -0.05 (left)
    0

2. (a) $K = -0.36$ (b) $K = -0.42$ (c) $K = 0.42$

3. The only information we have in 2(c) is the interval-grouped frequency distribution. In 2(a) and 2(b) the data were grouped without loss of information.

Page 70

1. (a) linear; $y = (1)x + 1$ (b) not linear
    (c) linear; $y = (-1)x + 2.5$ (d) linear; $y = 8x + 0$
    (e) linear; $y = (0)x + 5$ (f) not linear

2.

(a)

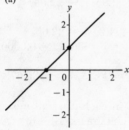

(b)

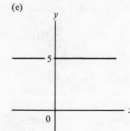

(c)

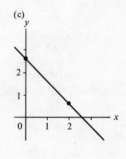

(d)

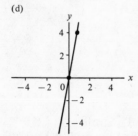

(e)

(f)

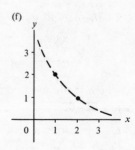

3. (a) slope = 1, intercept = 1   (b) ———
  (c) slope = −1, intercept = 2.5   (d) slope = 8, intercept = 0
  (e) slope = 0, intercept = 5   (f) ———

Pages 74–75

1. (a) $h = 0.152w + 43.25$
   (b)

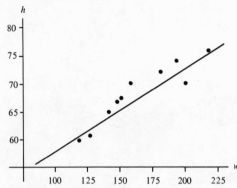

   (c) 69.85 inches
2. $x = 1.053y + 1.34$
3. (a) $c = 6.219t - 36.60$
   (b)

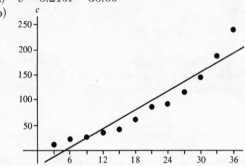

(c) 87.78
(a) 27.8   (b) can't tell from the y-on-x regression equation.
(a) $y = -0.80x - 0.40$; $x = -0.67y - 0.50$
(b)

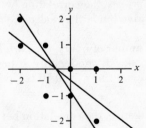

6. (a) $y = 8.10x + 58.88$, where y = score, x = hours
   (b) 139.88
   (c) One of the dangers of predicting outside the range of the given values of x.

Pages 81–82

1. (a)

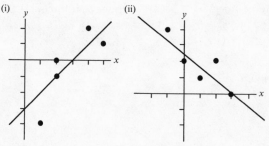

(b) i. $y = x - 3$ ii. $y = -0.8x + 2.6$
(c) see (a)
(d) i. 0.85   ii. −0.85
2. (i) $Q_R = 4.0$, $Q_M = 14.8$ (ii) $Q_R = 2.4$, $Q_M = 8.8$
3. (a) 0.98   (b) 0.96   (c) 0.87

Page 84

1. 0.09, 0.27, −0.29, −0.52, 0.67, −0.95
2. −0.50; $r_{xy} = -0.50$ has 6.25 times the strength of $r_{wz} = 0.20$
3. Either +0.50 or −0.50; 25 percent
4. No; No; Yes; Yes

Pages 89–90

1. 0.95
2. 0.71; Students with high high school averages (x) tend to get higher college freshman grade-point averages (y) than students with low high school averages. However, only 50.41 percent of the variation in y can be attributed to its linear relationship with x.
3. −0.79
4. (a) Judges A and C   (c) Judges B and C
5. −0.57
6. $r_{xz \cdot y} = 0.70$ for cities with equal whiskey consumption, the correlation between number of priests and population is positive and moderately strong (accounting for 49 percent of the variance).

   $r_{yz \cdot x} = 0.50$; for cities with equal number of priests, the correlation between whiskey consumption and population is positive but fairly weak (accounting for only 25 percent of the variance).
7. $r_{12 \cdot 3} = 0.70$; for those who study the same number of hours, the correlation between freshman year index and intelligence is moderately strong (49 percent of the variance accounted for), compared to $r_{12} = 0.50$ (25 percent accounted for).

   $r_{13 \cdot 2} = 0.66$; for those who are of equal intelligence, the correlation between freshman year index and number of hours studied is moderately strong (44 percent of the variance accounted for), compared to $r_{13} = 0.40$ (16 percent accounted for).

$r_{23.1} = -0.63$; for those who are of approximately equal standing in the freshman class, the correlation between intelligence and hours studied is moderately strong (40 percent of the variance accounted for), compared to $r_{23} = -0.30$ (9 percent accounted for).

Page 98

1. (a) $P(B)$ (b) $P(B \text{ and } A)$, or $P(A \text{ and } B)$
    (c) $P(\text{not-}A)$ (d) $P(B \text{ or } A)$, or $P(A \text{ or } B)$
    (e) $P(A|B)$
2. (a) the probability that the person is not a male.
    (b) the probability that the person is a male and a republican.
    (c) the probability that the person is either a male or a republican (or both). (the probability that at least one of the events "person is a male" and "person is a republican" occurs.)
    (d) the probability that the person is a male, given that the person is a republican.
    (e) the probability that the person is a republican, given that the person is a male.
3. (a) the probability that events A, B, and C all occur
    (b) the probability that at least one of the events A, B, and C occurs

Page 100

1. 0.46
2. 352/365, or 0.96
3. batting average
4. Answers will vary, but the three relative frequencies should be approximately in the ratio 1:2:1 for A, B, and C, respectively. To improve on the reliability of these estimates, flip the coins some more, say another 100 times, and recalculate the relative frequencies based on the 200 flips.

Pages 105–106

1. 2^m
2. (a) yes   (b) yes   (c) yes   (d) no   (e) yes
3. $S = \{h1, h2, h3, h4, h5, h6, t1, t2, t3, t4, t5, t6\}$
4. $S = \{hhh, hht, hth, thh, htt, tht, tth, ttt\}$
5. (a) $A = \{htt, tht, tth\}$; $P(A) = 3/8$
    (b) $B = \{hhh, hht, hth, thh\}$; $P(B) = 1/2$
    (c) $C = \{thh, tht, tth, ttt\}$; $P(C) = 1/2$
6. (a) $W = \{(5,5), (4,6), (6,4)\}$; $P(W) = 1/12$
    (b) $X = \{(2,1), (2,2), (2,3), (2,4), (2,5), (2,6), (1,2), (3,2), (4,2), (5,2), (6,2)\}$; $P(X) = 11/36$
    (c) $Y = \{(5,1), (5,2), (5,3), (5,4), (5,5), (5,6)\}$; $P(Y) = 1/6$
    (d) $Z = \{(1,1), (1,2), (2,1), (2,2), (1,3), (3,1), (2,3), (3,2), (1,4), (4,1)\}$; $P(Z) = 5/18$
7. (a) $S = \{(1,2), (1,3), (2,1), (2,3), (3,1), (3,2)\}$
    (b) $S = \{(1,1), (1,2), (1,3), (2,1), (2,2), (2,3), (3,1), (3,2), (3,3)\}$

8. (a) $K = \{(1,1), (1,3), (3,1), (3,3)\}; P(K) = 4/9$
        $L = \{(1,1), (1,2), (1,3), (2,1), (2,2), (3,1)\}; P(L) = 2/3$
        $G = \{(1,1), (2,2), (3,3)\}; P(G) = 1/3$
    (b) $K = \{(1,3), (3,1)\}; P(K) = 1/3$
        $L = \{(1,2), (1,3), (2,1), (3,1)\}; P(L) = 2/3$
        $G = \phi; P(G) = 0$
9. (a) $P(A) = 4/11, P(B) = 3/5, P(C) = 3/5$
    (b) 3 to 5, 1 to 1, 17 to 83
    (c) 11/26

Pages 111–113

1. (a) 40,320   (b) 210   (c) 10
    (d) 495   (e) 2450   (f) 1
    (g) 120   (h) 1

2. (a) $\dbinom{n}{n} = \dfrac{n!}{(n-n)!n!} = \dfrac{n!}{0!n!} = \dfrac{n!}{n!} = 1$

    (b) $\dbinom{n}{1} = \dfrac{n!}{(n-1)!1!} = \dfrac{n!}{(n-1)!} = \dfrac{(n)(n-1)!}{(n-1)!} = n$

    (c) $\dbinom{n}{n-1} = \dfrac{n!}{[n-(n-1)]!(n-1)!} = \dfrac{n!}{(n-n+1)!(n-1)!}$

        $= \dfrac{n!}{1!(n-1)!} = n$

    (d) $\dbinom{n}{0} = \dfrac{n!}{(n-0)!0!} = \dfrac{n!}{n!0!} = \dfrac{n!}{n!} = 1$

    (e) $_nP_{n-1} = \dfrac{n!}{[n-(n-1)]!} = \dfrac{n!}{(n-n+1)!} = \dfrac{n!}{1!} = n!$

    (f) $_nP_0 = \dfrac{n!}{(n-0)!} = \dfrac{n!}{n!} = 1$

    (g) $_nP_1 = \dfrac{n!}{(n-1)!} = n$

    (h) $\dbinom{n}{n-r} = \dfrac{n!}{[n-(n-r)]!(n-r)!} = \dfrac{n!}{(n-n+r)!(n-r)!}$

        $= \dfrac{n!}{r!(n-r)!} = \dbinom{n}{r}$

3. (a) 720   (b) 144   (c) 24
4. 190
5. (a) 67,600,000   (b) 67,600   (c) 0.001
6. 648
7. (a) 216   (b) 125   (c) 91   (d) 27   (e) 15

8. (a) 4096   (b) 375   (c) 375/4096
    (d) 1500   (e) 375/1024   (f) 675/2048
9. (a) 1680   (b) 180   (c) 3/28
    (d) 720   (e) 3/7   (f) 3/7
10. (a) 4896   i. 720   ii. 336   iii. 1680
    (b) 816   i. 120   ii. 56   iii. 280
11. (a) 1/4896   (b) 1/816   (c) 35/102   (d) 35/306
12. (a) 220   (b) 28/55   (c) 424/495
13. 1/230

14. $_nP_r = \dfrac{n!}{(n-r)!}$

$$= \frac{n(n-1)(n-2) \ldots (n-r+1)(n-r)(n-r-1)(n-r-2) \ldots (1)}{(n-r)(n-r-1)(n-r-2) \ldots (1)}$$

$$\therefore {}_nP_r = n(n-1)(n-2) \ldots (n-r+1)$$

Pages 115–117

1. (a) X' (b) $X \cap Y$

(c) $X \cap Y'$ (d) $X \cup Y$

(e) $X' \cup Y$ (f) $X' \cup Y'$

(g) $X' \cap Y$

2.　i.

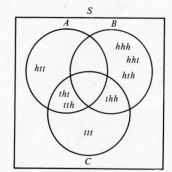

ii. (a) 0 　　(b) 1/4 　　(c) 1/8
　　(d) 7/8 　　(e) 7/8 　　(f) 5/8
　　(g) 0 　　(h) 0 　　(i) 1/4
　　(j) 1/3 　　(k) 1/4 　　(l) 1/4
　　(m) 3/8 　　(n) 7/8

3. (a) 0 　　(b) 1/3 　　(c) 5/36
　 (d) 7/12 　　(e) 4/33 　　(f) 1/36

4. $S = \{r_1r_2, r_1w_1, r_1w_2, r_1w_3, r_2r_1, r_2w_1, r_2w_2, r_2w_3, w_1r_1, w_1r_2, w_1w_2, w_1w_3, w_2r_1, w_2r_2,$
$w_2w_1, w_2w_3, w_3r_1, w_3r_2, w_3w_1, w_3w_2\}$

$R = \{r_1r_2, r_1w_1, r_1w_2, r_1w_3, r_2r_1, r_2w_1, r_2w_2, r_2w_3, w_1r_1, w_1r_2, w_2r_1, w_2r_2, w_3r_1, w_3r_2\}$

$W = \{r_1w_1, r_1w_2, r_1w_2, r_2w_1, r_2w_2, r_2w_3, w_1r_1, w_1r_2, w_2r_1, w_2r_2, w_3r_1, w_3r_2\}$

(a) 3/5 　　(b) 7/10 　　(c) 1
(d) 6/7 　　(e) 3/10 　　(f) 3/4

5. $S = \{r_1r_1, r_1r_2, r_1w_1, r_1w_2, r_1w_3, r_2r_1, r_2r_2, r_2w_1, r_2w_2, r_2w_3, w_1r_1, w_1r_2, w_1w_1, w_1w_2,$
$w_1w_3, w_2r_1, w_2r_2, w_2w_1, w_2w_2, w_2w_3, w_3r_1, w_3r_2, w_3w_1, w_3w_2, w_3w_3\}$

$R = \{r_1r_1, r_1r_2, r_1w_1, r_1w_2, r_1w_3, r_2r_1, r_2r_2, r_2w_1, r_2w_2, r_2w_3, w_1r_1, w_1r_2, w_2r_1, w_2r_2,$
$w_3r_1, w_3r_2\}$

$W = \{r_1w_1, r_1w_2, r_1w_3, r_2w_1, r_2w_2, r_2w_3, w_1r_1, w_1r_2, w_2r_1, w_2r_2, w_3r_1, w_3r_2\}$

(a) 12/25 　　(b) 16/25 　　(c) 1
(d) 3/4 　　(e) 9/25 　　(f) 9/13

6. (a) 699/1369 　　(b) 338/1369 　　(c) 161/1369
　 (d) 161/338 　　(e) 161/699 　　(f) 748/1369
　 (g) 364/1369 　　(h) 818/1369

Pages 122–123

1. (a) 0.7 　(b) 0.06 　(c) 0.15 　(d) 0.64 　(e) 0.8
2. (a) 0.50 　(b) 0.10 　(c) 0.80 　(d) 0.25
　 (e) 0.75 　(f) 0.20 　(g) 0.60 　(h) 0.20
　 (i) 0.90 　(j) 0.125

3. (a)

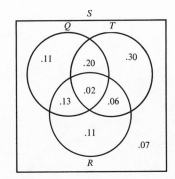

(b) i. $0.46 + 0.58 + 0.32 - 0.22 - 0.15 - 0.08 + 0.02 = 0.93$
    ii. $0.11 + 0.20 + 0.30 + 0.02 + 0.13 + 0.06 + 0.11 = 0.93$

4. (a) $\dfrac{(5)(3)(4)(3)}{(8)(7)(6)(5)} = \dfrac{3}{28}$

   (b) $(5/8)(3/7)(4/6)(3/5) = 3/28$

   (c) $375/4096$

5. (a) $429/41650$ (b) $429/41650$ (c) 4

   (d) $A \cup B \cup C \cup D$ (e) $858/20825$

6. (a) $3/256$ (b) $3/256$ (c) 4

   (d) $A \cup B \cup C \cup D$ (e) $3/64$

7. (a) $A \cup B$ and $A' \cap B'$ are complements of each other.

   (b) $A \cup B \cup C$ and $A' \cap B' \cap C'$ are complements of each other.

   (c) $E_1 \cup E_2 \cup \ldots \cup E_k$ and $E_1' \cap E_2' \cap \ldots \cap E_k'$ are complements of each other. Since $E_1 \cup E_2 \cup \ldots \cup E_k$ represents the event that *at least one of the k events occurs*, it follows that its complement is the event that *none of the k events occur*, which is precisely the meaning of $E_1' \cap E_2' \cap \ldots \cap E_k'$.

8. $1023/1024$

9. By Theorem 6.3: $P(E \cap F) = P(E) \cdot P(F|E)$ and $P(F \cap E) = P(F) \cdot P(E|F)$; since $P(E \cap F) = P(F \cap E)$, therefore $P(E) \cdot P(F|E) = P(F) \cdot P(E|F)$.

Pages 128–129

1. (a) II   (b) III   (c) III   (d) I
   (e) I   (f) I, II   (g) II, I, I   (h) II, II, I
   (i) III, I

2. (a) 0   (b) 0   (c) 0.77   (d) 0.97
   (e) 0   (f) 0.03   (g) 0.38   (h) 0.20

3. (a) 0.30   (b) 0.50   (c) 0.96   (d) 0.20
   (e) 0.05   (f) 0.98   (g) 0.27   (h) 0.80

4. They are not mutually exclusive because they can both occur, and they are not independent because the probability of B depends upon the outcome of the first draw.

Page 134

1. (a) 0.30   (b) 0.60   (c) 0.10   (d) 0.50   (e) 0.40
2. (a) 0.06   (b) 0.25
3. (a) 77/144   (b) 32/77
4. 10/11
5. 0.40

Pages 135–136

1. (a) 0.55   (b) 0.18   (c) 0.07
2. (a) 0.95   (b) 0.35   (c) 0.583
3. (a) 0.224   (b) 0.488   (c) 0.776
4. 2041/2401
5. (a) 0.995   (b) 0.0046   (c) 0.0004   (d) 0.9996
6. (a) $(0.80)^9(0.20)$ (b) $(10)(0.80)^9(0.20)$ (c) $1 - (0.80)^{10}$
7. (a) 1.00   (b) 0.00   (c) 0.75   (d) 0.33
8. 0.76
9. (a) 27/256   (b) 27/64
10. (a) 3/23   (b) 57/77

Pages 142–143

1.

y	$g(y)$	z	$h(z)$
-3	1/8	0	1/2
-1	3/8	1	1/2
1	3/8		
3	1/8		

In each case, the values of the function are never negative, and the sum of the values equals one.

2. (b) 5/14   (c) 6/7   (d) 5/14   (e) $f(x) = \dfrac{x+1}{14}$, $x = 1, 2, 3, 4$

3.

x:	2	3	4	5	6	7	8	9	10	11	12
$f(x)$:	1/36	2/36	3/36	4/36	5/36	6/36	5/36	4/36	3/36	2/36	1/36

or $f(x) = \dfrac{6 - |7 - x|}{36}$, $x = 2, 3, 4, \ldots, 12$

4.

y	$f(y)$
0	25/36
1	10/36
2	1/36

5. (a)

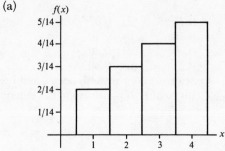

(b)

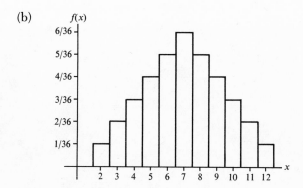

(c)

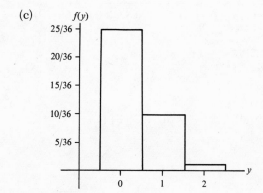

6.

z	$g(z)$
0	1/18
1	2/18
2	5/18
3	10/18

7.

x	$f(x)$
0	5/15
1	8/15
2	2/15

8.

x	$f(x)$
0	9/25
1	12/25
2	4/25

9.

x	$f(x)$
0	0.84
1	0.10
5	0.05
10	0.01

10.

y	$f(y)$
-2	0.84
-1	0.10
3	0.05
8	0.01

11.

y	$f(y)$
-4	3/6
2	2/6
5	1/6

Pages 148–149

1. $\mu_x = 3.3$, $\sigma_x^2 = 8.21$, $\sigma_x = 2.87$
2. (a) $\mu_x = 7$, $\sigma_x^2 = 5.83$, $\sigma_x = 2.41$ (b) $\mu_y = 0.33$, $\sigma_y^2 = 0.28$, $\sigma_y = 0.53$
   (c) $\mu_x = 0.8$, $\sigma_x^2 = 0.43$, $\sigma_x = 0.66$ (d) $\mu_x = 0.8$, $\sigma_x^2 = 0.48$, $\sigma_x = 0.69$
3. (a) $-\$.50$
   (b) In the long run A would lose 50 cents per game on the average.
   (c) $\$.50$ (d) 0
4. (a) 3.06   (b) 1.27
5. $\$.45$
6. yes

7. $\mu_w = 15$, $\sigma_w{}^2 = 6.25$, $\sigma_w = 2.5$

8. $\mu_y = 90$, $\sigma_y{}^2 = 225$, $\sigma_y = 15$

9. (a)

x	$f(x)$
1	1/10
2	2/10
3	3/10
4	4/10

(b)

y	$g(y)$
2	1/10
5	2/10
8	3/10
11	4/10

(c) $\mu_x = 3$, $\mu_y = 8$, $\sigma_x = 1$, $\sigma_y = 3$

10. -2.86, -0.71, 0, 1.43

11. (a) 2   (b) 0.89   (c) 1.12

12.

x:	2	3	4	5	6	7	8	9	10	11	12
x':	-2.07	-1.66	-1.24	-0.83	-0.41	0	0.41	0.83	1.24	1.66	2.07

Pages 157–159

1. (a) $a^5 + 5a^4b + 10a^3b^2 + 10a^2b^3 + 5ab^4 + b^5$

(b) $16x^4 - 32x^3 + 24x^2 - 8x + 1$

(c) $\dfrac{y^3}{8} + \dfrac{15y^2}{4} + \dfrac{75y}{2} + 125$

2. (a) 220   (b) 10   (c) 720   (d) $\binom{8}{5}\left(\dfrac{2}{5}\right)^3\left(\dfrac{3}{5}\right)^5$

(e) $\binom{265}{149}(0.25)^{116}(0.75)^{149}$

3. 35

4. (a) 720   (b) 15

5. 10

6. 56

7. 495

8. (a) 1215/4096   (b) 1/9   (c) 37/256

(d) $1 - (5/7)^{200} - 200(2/7)(5/7)^{199}$

9. (a) 75/216   (b) 125/216   (c) 16/216

10. (a) $63/256 = 0.2461$ (b) $319/512 = 0.6230$

11. (a) $1789/65536 = 0.0273$ (b) $135/2048 = 0.0659$

12. $\binom{6}{3}\left(\dfrac{91}{216}\right)^3\left(\dfrac{125}{216}\right)^3$

13. $2000/16807 = 0.1190$

14. Results are exactly the same as before.

15. (a) Ex. 10: $b(x) = \binom{10}{x}\left(\dfrac{1}{2}\right)^x\left(\dfrac{1}{2}\right)^{10-x}$, $x = 0, 1, 2, \ldots, 10$

Ex. 11: $b(x) = \binom{8}{x}\left(\dfrac{1}{4}\right)^x\left(\dfrac{3}{4}\right)^{8-x}$, $x = 0, 1, 2, \ldots, 8$

Ex. 12: $b(x) = \binom{6}{x}\left(\dfrac{91}{216}\right)^x\left(\dfrac{125}{216}\right)^{6-x}$, $x = 0, 1, 2, \ldots, 6$

Ex. 13: $b(x) = \binom{4}{x}\left(\dfrac{2}{7}\right)^x\left(\dfrac{5}{7}\right)^{4-x}$, $x = 0, 1, 2, 3, 4$

(b) Ex. 10: $\mu_x = 5$, $\sigma_x{}^2 = 2.5$, $\sigma_x = 1.58$

Ex. 11: $\mu_x = 2$, $\sigma_x{}^2 = 1.5$, $\sigma_x = 1.22$

Ex. 12: $\mu_x = 2.53$, $\sigma_x{}^2 = 1.46$, $\sigma_x = 1.21$

Ex. 13: $\mu_x = 1.14$, $\sigma_x{}^2 = 0.82$, $\sigma_x = 0.91$

16. $\mu_y = 28/9$, $\sigma_y{}^2 = 196/81$, $\sigma_y = 14/9$

17. (a) 277,200   (b) 15,120

Pages 162–163

1. $h(x) = \dfrac{\dbinom{5}{x}\dbinom{7}{3-x}}{\dbinom{12}{3}}$, $x = 0, 1, 2, 3$;

x	$h(x)$
0	7/44
1	21/44
2	14/44
3	2/44

2. (a) 99/2303   (b) 1/46060   (c) 16261/46060

3. (a) $h(y) = \dfrac{\dbinom{13}{y}\dbinom{39}{13-y}}{\dbinom{52}{13}}$, $y = 0, 1, 2, \ldots , 13$

   (b) $\dfrac{\dbinom{13}{2}\dbinom{39}{11}}{\dbinom{52}{13}}$

   (c) $\displaystyle\sum_{y=5}^{13} \dfrac{\dbinom{13}{y}\dbinom{39}{13-y}}{\dbinom{52}{13}}$

   (d) $\dfrac{\dbinom{13}{2}\dbinom{39}{2}}{\dbinom{52}{4}} \cdot \dfrac{11}{48}$

4. (a) $91/323 = 0.2817$ (b) $59/969 = 0.0609$ (c) 0.2646, 0.0837

5. (a) $\dfrac{\dbinom{300}{4}\dbinom{200}{1}}{\dbinom{500}{5}}$ (b) 162/625

6. (a) $h(x) = \dfrac{\dbinom{Q}{x}\dbinom{R-Q}{k-x}}{\dbinom{R}{k}}$, $x = 0, 1, 2, \ldots , Q$

   (b) $b(x) = \dbinom{k}{x}\left(\dfrac{Q}{R}\right)^x\left(1 - \dfrac{Q}{R}\right)^{k-x}$, $x = 0, 1, 2, \ldots , k$

7. (a) Ex. 1: $\mu_x = 1.25$, $\sigma_x{}^2 = 0.59$, $\sigma_x = 0.77$
   (b) Ex. 2: $\mu_x = 0.4$, $\sigma_x{}^2 = 0.34$, $\sigma_x = 0.58$
   (c) Ex. 3: $\mu_x = 3.25$, $\sigma_x{}^2 = 1.83$, $\sigma_x = 1.35$

8. $\mu_x = 3.75$, $\sigma_x{}^2 = 1.80$, $\sigma_x = 1.34$

Pages 166–167

1. (a) 0.1722   (b) 0.959   (c) 0.1312
2. 0.6906
3. $\mu_x = 2.56$, $\sigma_x{}^2 = 2.56$, $\sigma_x = 1.6$
4. 0.5291
5. 0.1792
6. (a) 0.144   (b) 0.018   (c) 0.382
7. 0.3345

Page 172

1. (a)

x \ y	0	1	2	$f(x)$
0	10/36	10/36	5/36	25/36
1	0	6/36	4/36	10/36
2	0	0	1/36	1/36
$g(y)$	10/36	16/36	10/36	1

   (b) No

2. (a)

x \ y	0	1	2	3	$f(x)$
0	0	1/32	4/32	9/32	14/32
1	1/32	2/32	5/32	10/32	18/32
$g(y)$	1/32	3/32	9/32	19/32	1

   (b) 9/16; 9/32; 9/32; 11/16; 1/3
3. (a) $\mu_x = 1.6$; $\mu_y = 1.8$; $\sigma_x = 0.49$; $\sigma_y = 1.66$; $\rho_{xy} = 0.20$
   (b) 0.60; 0.30; 0.27; 0.90
   (c) No
5. Since $\Sigma\ xyj(x,y) = 1.0$, $\mu_x = 2.0$, and $\mu_y = 0.5$, therefore $\rho_{xy} = 0$. However, x and y are not independent (for example: $f(0) \cdot g(3) \neq j(0,3)$).

Pages 181–182

1. (a) 1/4   (b) 1/2   (c) 0   (d) 5
   (e) 6.2   (f) 0.573
2. (a)

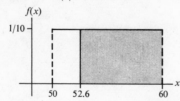

   (b) 0.26   (c) 0.44   (d) 0.316

3. (a)

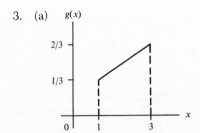

(b) $g(x)$ is never negative and the area under the graph $= 1.0$.
(c) 5/12   (d) 39/48
4. $\mu_w = 17$, $\sigma_w^2 = 16$
5. (a) $-3, -0.2, 1.2$ (b) $\mu_{x'} = 0$, $\sigma_{x'}^2 = 1$
6. (a) 5.5   (b) 5.83   (c) 2.16
7. (a) $\mu_x = 55$, $\sigma_x = 2.89$
    (b) 1.04
    (c) μ_x and the median are equal for all uniform probability functions.

Pages 186–187

1. (a) 0.1033   (b) 0.7549   (c) 0.9969   (d) 0.0537
    (e) 0.2818   (f) 0.4859
2. (a) 1.93   (b) -0.36 (c) -1.74 (d) 0.71
    (e) 0.92   (f) 1.96   (g) 2.33   (h) 2.58
3. (a) 0.79   (b) -1.11 (c) -1.96 (d) -2.33
    (e) -1.65
4. (a) 1.40; 1.40   (b) 0.80; -0.80 (c) 0.2119, 0.9192, 0.7073
    (d) 0.3848   (e) 80.8
5. (a) 0.6826   (b) 0.9544   (c) 0.9974
6. (a) 0   (b) 0.2420   (c) 0.3674   (d) 0.3674
    (e) 48   (f) They are equal.
7. 4.31
8. (a) 0.1112   (b) 0.2743   (c) 0   (d) 0.3830
9. 33.72%
10. 23.64%
11. 3.67%
12. (a) 0.0475   (b) 0.0918

Pages 192–193

1. (a) $93/128 = 0.7266$ (b) 0.7258
2. (a) 0.2510   (b) 0.6255
3. 0.0287
4. (a) 0.6101   (b) 0.7333
5. 0.0392
6. 0.1056
7. (a) 0.1841 (no correction needed)   (b) 0.1423
8. 81
9. 0.0951
10. 0.7967 (no correction needed)
11. 0.6628

Pages 201–202

1. 0.8034
2. (a) 1.282   (b) 1.645   (c) 1.960   (d) 2.326
   (e) 2.576
4. 0.95
5. (a) −2.02, −2.11, 1.97, 2.44   (b) 10.85, 26.22, 15.51, 42.14
   (c) 3.06, 4.67, 0.25, 3.43
6. (a) 0.005, −3.05, 3.05   (b) 0.01, −2.68, 2.68
   (c) 0.025, −2.18, 2.18   (d) 0.05, −1.78, 1.78
7. (a) mean = 0, variance = 1.25, standard deviation = 1.12
   (b) mean = 8, variance = 16, standard deviation = 4
   (c) mean = 1.5, variance = 3.15, standard deviation = 1.77
8. $\chi^2_{1;\,1-p} = (z_{1-p/2})^2$
9. $F_{1,\,r_2;\,1-p} = (t_{r_2;\,1-p/2})^2$

Page 212

1. 120
2. 1/4845
3. (a) 1,2,3,4   1,2,5,6   2,3,4,5
         1,2,3,5   1,3,4,5   2,3,4,6
         1,2,3,6   1,3,4,6   2,3,5,6
         1,2,4,5   1,3,5,6   2,4,5,6
         1,2,4,6   1,4,5,6   3,4,5,6
   (b) 1/15
   (c) 2/3
   (d) yes
   (e) 2/5
   (f) yes
4. No; there is no way to guarantee that each of the 10 numbers is equally likely to be selected.
5. 3196, 2958, 1902, 1501, 1507, 3343, 3701, 3588
6. Assign a second four-digit number to each piston; e.g. let 0000 and 5000 represent the same piston, 0001 and 5001, 0002 and 5002, etc.
7. 6, 7, 0, 2, 3, 3, 7, 5, 9, 4, which simulate the outcomes $h, t, h, h, t, t, t, t, t, h$.
8. Assuming that: $x = 0$ is represented by 000–124
                  $x = 1$ is represented by 125–499
                  $x = 2$ is represented by 500–874
                  $x = 3$ is represented by 875–999,
   the simulated outcomes are 2, 1, 2, 1, 1, 2, 3, 1, 1, 1
9. Assuming that: 00–26 represent red
                  27–64 represent blue
                  65–99 represent green,
   the simulated outcomes include four red, eight blue, and eight green marbles.

Page 216

1. (a) No; it is not true that every possible sample of size 52 has equal probability.
   (b) No; the sample chooses itself.
   (c) Yes   (d) Yes   (e) Yes
2. (a) Values of $\bar{x}$ for the 20 possible samples:

108	133	138	144
112	134	139	159
114	135	140	161
116	136	141	163
131	137	142	167

   (b) Values of $\bar{x}$ for the 12 possible samples:

131	135	138	141
133	136	139	142
134	137	140	144

   (c) The population mean = 137.5. In (a) the possible sample means vary from 108 to 167, but in (b) no sample mean differs from the population mean by more than 6.5.
3. Designate: 000–159 to represent the A group
                160–399 to represent the B group
                400–819 to represent the C group
                820–959 to represent the D group
                960–999 to represent the E group,
   then use the random number table.
4. (a) Bad, because the sample selects itself.
   (b) Bad, because not only does the sample select itself, but the political preference of those who subscribe to the magazine may differ greatly from the preferences of the general public.

Page 218

1. (a) 10
   (b)

x:	0	1	2	3	4	5	6	7	8	9
$f(x)$:	1/10	1/10	1/10	1/10	1/10	1/10	1/10	1/10	1/10	1/10

   (c) i. $\mu_x = 4.5$, $\sigma_x^2 = 8.25$, $\sigma_x = 2.87$
       ii. $\mu_x = 4.5$, $\sigma_x^2 = 8.25$, $\sigma_x = 2.87$
2. (a) 500
   (b) same as (b) in Exercise 1
   (c) same as (c) in Exercise 1
3. (a) 6.5   (b) 5.2
4. $\mu_x = 0.69$, $\sigma_x^2 = 4.59$, $\sigma_x = 2.14$

Pages 224–225

1. (a) 10
   (b)

$\bar{x}$:	2.5	4.0	5.5	7.0	8.5	10.0	11.5
$f(\bar{x})$:	1/10	1/10	2/10	2/10	2/10	1/10	1/10

   (c) $\mu_{\bar{x}} = 7$, $\sigma_{\bar{x}}^2 = 3.02$
   (d) $\mu_{\bar{x}} = 7$, $\sigma_{\bar{x}}^2 = 3.02$
   (e) $\sigma_{\bar{x}} = 1.74$
2. (a) 15; the samples (1,1), (4,4), (7,7), (10,10), and (13,13) are added. Note, however, that each of these five is only half as likely to occur as each of the original ten listed in Exercise 1(b). Why?
   (b)

$\bar{x}$:	1.0	2.5	4.0	5.5	7.0	8.5	10.0	11.5	13.0
$f(\bar{x})$:	1/25	2/25	3/25	4/25	5/25	4/25	3/25	2/25	1/25

   (c) $\mu_{\bar{x}} = 7$, $\sigma_{\bar{x}}^2 = 4.02$
   (d) $\mu_{\bar{x}} = 7$, $\sigma_{\bar{x}}^2 = 4.02$
   (e) $\sigma_{\bar{x}} = 2.00$
3. $\mu_{\bar{x}} = 7$, $\sigma_{\bar{x}}^2 = 1.38$, $\sigma_{\bar{x}} = 1.17$
4. $\mu_{\bar{x}} = 7$, $\sigma_{\bar{x}}^2 = 2.05$, $\sigma_{\bar{x}} = 1.43$
5. $\mu_{\bar{x}} = 7$, $\sigma_{\bar{x}}^2 = 2.06$, $\sigma_{\bar{x}} = 1.44$
6. $\mu_{\bar{x}} = 7$, $\sigma_{\bar{x}}^2 = 2.06$, $\sigma_{\bar{x}} = 1.44$
7. $\mu_{\bar{x}} = 50$, $\sigma_{\bar{x}}^2 = 2.25$, $\sigma_{\bar{x}} = 1.50$
8. $\mu_{\bar{x}} = 50$, $\sigma_{\bar{x}}^2 = 2.56$, $\sigma_{\bar{x}} = 1.60$
9. 0.8944
10. (a) 0   (b) 0.1151   (c) 0.5328
11. It is increased by a factor of four.
12. (a) med. f(med.); the variance = 5.4, compared to 3.0 for $x_{\bar{x}}^2$.

4	3/10
7	4/10
10	3/10

    (b) Because the population is not normally distributed.
13. (a) 0.4483   (b) 0.2643   (c) 0.1056
14. (a) $\mu_{\bar{x}} = 12$, $\sigma_{\bar{x}} = 0.05$ (b) production is stopped   (c) 0.1336
15. Stratified sampling.

Pages 238–240

1. $46.02 < \mu_x < 47.58$
2. (a) 24.56   (b) $22.95 < \mu_x < 26.17$
3. (a) $123.24 < \mu_x < 125.96$ (b) $54.02 < \sigma_x^2 < 86.40$
4. (a) $5.03 < \mu_x < 5.49$ (b) $5.00 < \mu_x < 5.52$
5. $2.76 < \mu_x < 3.90$; $1.88 < \sigma_x^2 < 4.85$; $1.37 < \sigma_x < 2.20$

6. $11.28 < \mu_x < 12.26$
    $0.42 < \sigma_x^2 < 1.66$
    $0.65 < \sigma_x < 1.29$
7. The error is at most 1.28
8. At least 75.
9. At least 27.
10. (a) $14.6 < \mu_x < 17.0$; $2.36 < \sigma_x^2 < 11.21$; $1.54 < \sigma_x < 3.35$
11. 93% confidence.
12. At least 68.
13. (a) an increase   (b) an increase   (c) a decrease
14. $0.499 < p < 0.621$
15. $0.228 < p < 0.418$
16. (a) 0.20   (b) $0.156 < p < 0.244$
17. $0.076 < p < 0.124$
18. $1.38 < \sigma_x < 2.20$
19. $0.63 < \rho_{xy} < 0.93$
20. $0.56 < \rho_{xy} < 0.76$

Pages 254–255

1. (a) none   (b) Type I
    (c) He can avoid a Type II error if the null hypothesis is in fact true and he
         accepts it, or if the null hypothesis is false and he reserves judgment.
3. (a) 0.2938   (b) 0.4282; 0.5665; 0.1700
    (c)

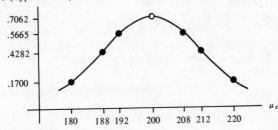

P (Type II error)

    (d) Because the Type I error probability is higher here.
4. (a) N.H.: $\mu_x = 50$; Alt. H.: $\mu_x \neq 50$
    (b) Accept N.H. if $-1.96 \le z \le 1.96$
         Reject N.H. if $z < -1.96$ or $z > 1.96$
    (c) Accept N.H. if $40.20 \le x \le 59.80$
         Reject N.H. if $x < 40.20$ or $x > 59.80$
    (d) 0.05
    (e) 0.4840
    (f) Accept N.H. (or possibly reserve judgment)
5. (a) N.H. $\mu_x = 50$; Alt. H.: $\mu_x > 50$
    (b) Accept N.H. if $z \le 1.645$
         Reject N.H. if $z > 1.645$
    (c) Accept N.H. if $x \le 58.23$
         Reject N.H. if $x > 58.23$
    (d) 0.05   (e) 0.3594   (f) Reject N.H.

6. (a) N.H.: $\mu_x = 20$; Alt. H.: $\mu_x \neq 20$
   (b) Accept N.H. if $-2.33 \leq z \leq 2.33$
       Reject N.H. if $z < -2.33$ or $z > 2.33$
   (c) Accept N.H. if $10.47 \leq x \leq 29.53$
       Reject N.H. if $x < 10.47$ or $x > 29.53$
   (d) 0.02   (e) 0.8944   (f) Accept N.H. (or reserve judgment)
7. Reject the null hypothesis that $\mu_x = 540$
8. (a) For Ex. 5; N.H.: $p = 1/2$
                Alt. H.: $p > 1/2$
   (b) For Ex. 6; N.H.: $p = 1/6$.
                Alt. H.: $p \neq 1/6$
   (c) For Ex. 7; N.H.: $p = 0.60$
                Alt. H.: $p < 0.60$

Pages 265–266

1. Reject N.H. ($z = -2.73$, compared to $z_{0.05} = -1.645$)
2. Reject the claim ($z = 2.38$, compared to $z_{0.975} = 1.96$)
3. Reject the claim; i.e., the product is overrated ($z = -2.78$, compared to $z_{0.01} = -2.33$)
4. Reject N.H.; i.e., this year's freshmen are better prepared ($z = -4.23$, compared to $z_{0.01} = -2.33$)
5. Cannot reject N.H.; i.e., not quite strong enough evidence that drug is effective ($z = 1.62$, compared to $z_{0.95} = 1.645$)
6. Cannot reject N.H.; i.e., no significant difference in quality ($z = 1.30$, compared to $z_{0.995} = 2.58$)
7. Cannot reject N.H.; i.e., no significant difference in opinion ($z = 1.56$, compared to $z_{0.975} = 1.96$)
8. Cannot reject N.H.; i.e., cannot support expert's belief ($z = -0.80$, compared to $z_{0.03} = -1.88$)
9. Cannot reject N.H.; two types are not significantly different in effectiveness ($z = -0.86$, compared to $z_{0.025} = -1.96$)
10. Reject N.H.; ($\chi^2 = 29.16$, compared to $\chi^2_{2;0.95} = 5.99$)
11. Cannot reject N.H.; i.e., the true proportions are not different ($\chi^2 = 3.33$, compared to $\chi^2_{3;0.95} = 7.81$)
12. Cannot reject N.H.; i.e., no difference in effectiveness among fertilizers ($\chi^2 = 4.67$, compared to $\chi^2_{2;0.95} = 5.99$)
13. $\chi^2 = 2.44$; $z^2 = (1.56)^2 = 2.43$
14. $\chi^2 = 29.16$

Pages 275–277

1. Cannot reject N.H. ($z = -2.50$, compared to $z_{0.005} = -2.58$)
2. 0.0037
3. Reject N.H. ($z = 5.61$, compared to $z_{0.99} = 2.33$)
4. (a) Reject N.H. ($z = 1.67$, compared to $z_{0.95} = 1.645$)
   (b) Accept N.H. or reserve judgment ($t = 1.67$, compared to $t_{24;0.95} = 1.71$)
5. Reject N.H. ($t = 2.74$, compared to $t_{19;0.99} = 2.54$)

6. Reject N.H.; i.e., State U. freshmen more intelligent than average ($z = 2.86$, compared to $z_{0.99} = 2.33$)
7. Reject N.H.; ($t = 2.38$, compared to $t_{113;0.025} = 1.98$)
8. Cannot reject N.H.; i.e., cannot support the claim ($t = -2.19$, compared to $t_{18;0.01} = -2.55$)
9. Reject N.H. ($t = -3.09$, compared to $t_{38;0.01} = -2.43$)
10. Cannot reject N.H. ($z = 1.16$, compared to $z_{0.975} = 1.96$)
11. Reject N.H. ($t = 2.86$, compared to $t_{43;0.995} = 2.69$)
12. Cannot reject N.H. ($t = -2.20$, compared to $t_{61;0.01} = -2.39$)
13. Assume $\sigma_1{}^2 = \sigma_2{}^2$
14. Reject N.H. ($z = -2.37$, compared to $z_{0.01} = -2.33$)
15. Cannot reject N.H. ($t = 1.72$, compared to $t_{7;0.95} = 1.89$)
16. Reject N.H. ($t = -4.64$, compared to $t_{9;0.005} = -3.25$)
17. Cannot reject N.H. ($F = 1.50$, compared to $F_{2;9,0.95} = 4.26$)
18. Reject N.H. ($F = 8.90$, compared to $F_{3,16;0.99} = 5.29$)
19. (a) Cannot reject N.H. ($t = 1.50$, compared to $t_{8;0.975} = 2.31$)
    (b) Cannot reject N.H. ($F = 2.25$, compared to $F_{1,8;0.95} = 5.32$)

Page 281

1. (a) $\chi^2_{24;0.95} = 36.42$
   (b) $\chi^2 = 38.55$
   (c) Reject N.H.; i.e., conclude that $\sigma_x{}^2 > 8$
2. Cannot reject N.H. ($\chi^2 = 121.35$, compared to $\chi^2_{99;0.975} = 128.42$)
3. Reject N.H. ($F = 2.70$, compared to $F_{29,29;0.99} = 2.42$)
4. (a) Cannot reject N.H. ($t = -1.56$, compared to $t_{24;0.025} = -2.06$)
   (b) Cannot reject N.H. ($\chi^2 = 38.88$, compared to $\chi^2_{24;0.975} = 39.36$)
   (c) Select a larger sample.
5. (a) Accept N.H.; it is reasonable to assume $\sigma_1{}^2 = \sigma_2{}^2$ ($F = 2.11$, compared to $F_{24,24;0.975} = 2.27$)
   (b) Reject N.H. ($t = -2.26$, compared to $t_{48;0.025} = -2.01$)
6. (a) Ex. 10: Accept N.H. ($F = 1.22$, compared to $F_{19,24;0.975} = 2.35$)
   (b) Ex. 11: Reject N.H. ($F = 0.23$, compared to $F_{27,34;0.025} = 0.47$)

Pages 284–285

1. (a) Cannot reject N.H. ($t = 1.37$, compared to $t_{28;0.975} = 2.05$)
   (b) Reject N.H. ($t = -2.77$, compared to $t_{8;0.025} = -2.31$)
   (c) Reject N.H. ($t = 2.13$, compared to $t_{198;0.975} = 1.97$)
2. Cannot reject N.H. ($z = -1.32$, compared to $z_{0.025} = -1.96$)
3. (a) Reject N.H.; i.e., ρ is greater than 0.50 ($z = 2.05$, compared to $z_{0.95} = 1.645$)
   (b) No; even if ρ is greater than 0.50, there is no valid cause and effect implication.
4. Reject N.H.; i.e., the two variables are correlated ($t = 4.07$, compared to $t_{98;0.995} = 2.63$)
5. Cannot reject N.H. ($z = 1.94$, compared to $z_{0.99} = 2.33$)
6. Reject N.H. ($z = 1.72$, compared to $z_{0.95} = 1.645$)
7. Reject N.H. ($z = 2.69$, compared to $z_{0.995} = 2.58$)

Pages 290–292

1. Cannot reject N.H.; i.e., evidence does indicate that the two variables are dependent ($\chi^2 = 3.04$, compared to $\chi^2_{4;0.95} = 9.49$)
2. (a) 6   (b) $\chi^2_{6;0.99} = 16.81$ (c) $\chi^2 = 42.00$
   (d) Reject N.H.; i.e., strong evidence that the two variables are not independent.
3. The relationship tested need not be linear.
4. Cannot reject N.H. ($\chi^2 = 3.33$, compared to $\chi^2_{3;0.95} = 7.81$)
5. χ^2 (corrected) $= 2.05$, compared to χ^2 (uncorrected) $= 2.44$
6. Cannot reject N.H.; i.e., fits well enough ($\chi^2 = 4.17$, compared to $\chi^2_{4;0.95} = 9.49$)
7. Since the digits in the random number table are in fact *random*, and therefore uniformly distributed, the probability is 0.99 that your computed χ^2 is less than $\chi^2_{9;0.99}$, which equals 21.67.
8. Reject N.H.; ı.e., not a good fit ($\chi^2 = 11.70$, compared to $\chi^2_{5;0.95} = 11.07$)
9. Cannot reject N.H.; i.e., there is not sufficient reason to doubt the theory ($\chi^2 = 3.97$, compared to $\chi^2_{3;0.95} = 7.81$)

Pages 302–303

1. (a) Reject N.H. ($z = -3.34$, using R_1, compared to $z_{0.005} = -2.58$)
   (b) Cannot reject N.H. ($z = 0.84$, using R_1, compared to $z_{0.995} = 2.58$)
2. Cannot reject N.H. ($z = -1.74$, compared to $z_{0.01} = -2.33$)
3. Reject N.H. ($z = 1.67$, compared to $z_{0.95} = 1.645$)
4. Reject N.H. ($D(\max) = 0.135$, compared to $D(\max)_{100;0.95} = 0.134$)
5. Reject N.H. ($\chi^2 = 11.28$, compared to $\chi^2_{2;0.99} = 9.21$)
6. Reject N.H. ($z = 3.17$, using R_1, compared to $z_{0.97} = 1.88$)
7. Cannot reject N.H. ($D(\max) = 0.10$, compared to $D(\max)_{50;0.98} = 0.211$)
8. Cannot reject N.H. ($\chi^2 = 3.31$, compared to $\chi^2_{3;0.95} = 7.81$)

Index

74 75   10 9 8 7 6 5 4 3